A.S.S.R.s, A.O.s and N.O.s are administrative units containing ethnic groups and, in theory, as they develop, they may be promoted from an A.S.S.R. to an S.S.R., from an A.O. to an A.S.S.R. and from an N.O. to an A.O. Oblasts are usually designed to include well co-ordinated economic regions. Krays are administrative units for frontier and thinly populated areas. As the economy of the U.S.S.R. changes, so the composition and boundaries of Oblasts and Krays constantly alter.

U.S.S.R.

S.S.R. — R.S.F.S.R.

A.S.S.R. — A.O. — Oblast

A.S.S.R. — Kray — Oblast

A.O. — N.O. — Oblast — Area — N.O.

Wrangel I.
(To Magadan Oblast)

Chukchi N.O.

Severnaya Zemlya
(To Taymyr N.O.)

New Siberian Islands
(To Yakut A.S.S.R.)

Koryak N.O.

Kamchatka
(To Khabarovsk Kray)

Taymyr N.O.

Evenki N.O.

R E P U B L I C

Khabarovsk Kray

Kuril Islands
(To Sakhalin Oblast)

Sakhalin

S O C I A L I S T

F E D E R A T E D

Tomsk

Krasnoyarsk Kray

Irkutsk

Amur

Jewish A.O.

Maritime Kray

Kemerovo

Ust-Orda Buryat-Mongol N.O.

Buryat-Mongol A.S.S.R.

Chita

Aga Buryat-Mongol N.O.

ovosibirsk

Khakass A.O.

Altay Kray

Gorno-Altay A.O.

Tuva A.S.S.R.

East Kazakh.

Bayan-Ulegey

Ubsa-Nor

Khöbsogol

Bulgan

Selenge

Tub

Ulan-Bator

Hentey

Sukhe-Bator

Choibalsan

REPUBLIC

Dzabkhan

Ara-Hangay

MONGOLIAN PEOPLE'S

Hobdo

Ara-Hangay

Obor-Hangay

Dund-Gobi

Dornot-Gobi

Bayan Hongor

Gobi-Altay

Umuni-Gobi

IRGIZ S.S.R.
OBLASTS
Frunze
Osh
Tyan'-Shan'

KAZAKH S.S.R.

Tashkent

Osh

Andizhan

Fergana

Samarkand

Leninabad

Osh

Dyushambe

Gorno

0 50 Miles

FERGANA VALLEY

Scale 1 : 25 m. 1 inch to 400 miles approx.

Copyright Oxford University Press

OXFORD
REGIONAL ECONOMIC ATLAS

The U.S.S.R.
and Eastern Europe

PREPARED BY
THE ECONOMIST INTELLIGENCE UNIT
AND THE
CARTOGRAPHIC DEPARTMENT OF
THE CLARENDON PRESS

OXFORD UNIVERSITY PRESS

Oxford University Press, Ely House, London W.1

GLASGOW NEW YORK TORONTO MELBOURNE WELLINGTON

CAPE TOWN SALISBURY IBADAN NAIROBI LUSAKA ADDIS ABABA

BOMBAY CALCUTTA MADRAS KARACHI LAHORE DACCA

KUALA LUMPUR SINGAPORE HONG KONG TOKYO

© *Oxford University Press*

Reprinted 1960, 1963, 1969

**See Appendix for geographical changes (p. 134) and Back Endpaper for
Population of major towns (Post 1959).**

DRAWN AND PHOTOGRAPHED BY THE
CARTOGRAPHIC DEPARTMENT OF THE CLARENDON PRESS
AND PRINTED IN GREAT BRITAIN BY
MESSRS. COOK, HAMMOND, AND KELL, LONDON

REGIONAL ECONOMIC ATLASES

The series of *Oxford Regional Economic Atlases* are planned to cover the following areas:

The U.S.S.R. and Eastern Europe
The Middle East and North Africa
Africa
Canada and the U.S.A.
Europe
Latin America
India, China and Japan
S.E. Asia, Australia and New Zealand

The atlases are in series with *The Oxford Economic Atlas of the World*.

This volume covers the following countries:

The Union of Soviet Socialist Republics
East Germany
Poland
Czechoslovakia
Hungary
Romania
Bulgaria
Albania
Mongolian People's Republic
Yugoslavia

The Eastern European countries (except Yugoslavia) are included with the U.S.S.R. since their economies are becoming more and more closely integrated. Yugoslavia is not now in close economic relations with the other countries, but is included as being a geographical part of Eastern Europe. It is intended that these countries of E. Europe should also be covered by the *Oxford Regional Economic Atlas of Europe*. Although there might have been some case for including the other Communist countries of Asia obvious practical considerations of scale made it necessary to omit from this volume all but the Mongolian People's Republic.

This atlas is arranged in three main parts:

1. GENERAL REFERENCE MAPS (with layer colouring) providing comparatively large-scale cover of the more developed parts of the region (pp. 4–19).

2. TOPIC MAPS of the whole region (pp. 20–102). These deal with Physical Geography (geology, soils, climate, etc.), Agriculture, Minerals, Industries and Human Geography (transport, population, ethnography, etc.), and form the major part of the work. All these topics are presented on a common base map of the whole region (scale 1/25M). This facilitates the comparison of different topics. Most of these topic maps are preceded by supplementary texts providing an historical outline and notes on current production, etc.

3. GAZETTEER (pp. 110–134). This gives references in terms of latitude and longitude for some 5,500 place names. It also serves as an index, showing for any particular town both the general reference map and the topic maps on which it occurs.
    In addition, an index map showing the administrative sub-divisions of the area appears on the front endpaper.

For further details of the arrangement of the atlas see pp. VIII—IX.

ACKNOWLEDGEMENTS

Geographical Advisers

Mr. Theodore Shabad, New York.

Professor C. D. Harris, University of Chicago.

Dr. E. W. Russell, formerly Reader in Soil Science,
Department of Agriculture, Oxford University.

We also acknowledge help from the following individuals and organizations

Mr. M. Aurousseau, Secretary, Permanent Committee for Geographical Names, London
British Electricity Authority
British Iron and Steel Federation
Colonial Geological Survey
Dr. T. K. Derry, London
Mr. D. J. Footman, St. Antony's College, Oxford
Forestry Commission
Mr. C. F. W. R. Gullick, Lecturer in Economic Geography, Oxford University
Professor Holland Hunter, Associate Professor of Economics, Haverford College, Haverford, Pa., U.S.A.
Mrs. M. Houldsworth, Oxford
Ministry of Fuel and Power
Mr. E. J. Parsons, Map Room, Bodleian Library, Oxford
Dr. G. Etzel Pearcy, Geographic Attaché, Foreign Service, U.S.A.
Petroleum Information Bureau
Petroleum Press Service
Capt. J. Rousset, St. Antony's College, Oxford
Dr. K. S. Sandford, Reader in Geology, Department of Geology and Mineralogy, Oxford University
Mr. F. Seton, Nuffield College, Oxford
Mr. J. Simmons, Taylor Institution, Oxford
Professor Dr. E. Thiel, Munich University, Germany
Lt.-Col. G. Wheeler, Central Asian Research Centre, London
Mr. P. J. Wiles, New College, Oxford

BOOKS OF REFERENCE

ATLASES

Atlas Republiký Československé. Prague, 1936
Atlas des deutschen Lebensraumes in Mitteleuropa. Leipzig, 1937
Atlas F.N.R.J. Belgrade, 1952
Atlas Istorii S.S.S.R. I–III. Moscow, 1949
Atlas Mira. Moscow, 1954
Atlas Polski I–IV. Warsaw, 1954
Atlas Ziem Odzyskannych. Warsaw, 1947

Bol'shoy Sovetskiy Atlas Mira, Vols I–II. Moscow 1937–1939
Geograficheskiy Atlas dlya Uchiteley Sredney Shkoly. Moscow, 1954
Shepherd, W. R. *Historical Atlas.* New York, 1929
Weltatlas: Die Staaten der Erde und Ihre Wirtschaft. Leipzig, 1952
William-Olsson, W. *Economic Map of Europe.* Stockholm, 1953

BOOKS ON SPECIAL SUBJECTS

Alexandrowicz, C. *Comecon: The Soviet Retort to the Marshall Plan.* London, 1950
Allen, W. E. D. *The Ukraine, a History.* Cambridge, 1940
Anon. *Die Bevölkerungsbilanz der Sowjetischen Besatzungszone.* Bonn, 1951
Armstrong, T. *The Northern Sea Route.* Cambridge, 1952
Auty, P. *Building a New Yugoslavia.* London, 1954
Barbag, S. *Gospodarcza Geografia Polski.* Warsaw, 1953
Baykov, A. *The Development of the Soviet Economic System.* Oxford, 1947
Bettelheim, C. *L'économie Sovietique.* Paris, 1950
Bergson, A. *The Soviet National Income and Product in 1937.* New York, 1953
Bergson, A. (Ed.). *Soviet Economic Growth: Conditions and Perspectives.* New York, 1954
Bergson, A., and Heymann, H. *Soviet National Income and Product 1940–1948.* New York, 1954
Chardonnet, J. *Geographie économique de l'Europe danubienne et de la Pologne* (2 Vols.). Paris, 1948–1949
Condoide, M. V. *The Soviet Financial System: Its Development and Relations with the Western World.* Ohio, 1951
Dobb, M. *Soviet Economic Development since 1917.* London, 1948
Frumkin, G. *Population Changes in Europe since 1939.* Geneva, 1951
Galenson, W. *Labour Productivity in Soviet and American Industry.* New York, 1955
Garbutt, P. E. *The Russian Railways.* London, 1949
Harris, S. E., and others. *Appraisals of Russian Economic Statistics.* Harvard, 1947.
Hassmann, H. *Oil in the Soviet Union.* Princeton, 1953
History of the Communist Party of the Soviet Union (Bolskeviks). Moscow, 1943
Hodgman, D. R. *Soviet Industrial Production 1928–1951.* Harvard, 1954
Hoeffding, O. *Soviet National Income and Product in 1928.* New York, 1954
Institut National de la Statistique et des Études Économiques. *L'Asia Sovietique. L'Extreme Nord Sovietique.* }Paris, 1949 *La Pologne.* Paris, 1953 }
Jasny, N. *The Socialized Agriculture of the U.S.S.R.* Oxford, 1949

Kolarz, W. *Russia and her Colonies.* London, 1952
The Peoples of the Soviet Far East. London, 1954
Kramer, M. *Die Landwirtschaft in der Sowjetischen Besatzungszone.* Bonn, 1951
Lorimer, F. *The Population of the Soviet Union: History and Prospects.* Geneva, 1946
Mandel W. *The Soviet Far East.* New York, 1944
Martovych, O. R. *National Problems in the U.S.S.R.* Edinburgh, 1953
McKitterick, T. E. M. *Russian Economic Policy in Eastern Europe.* London, 1948
Mid-European Studies Centre. *The Hungarian Oil Industry.* New York, 1954
Population changes in Poland, 1939–1950. New York, 1955
Mirchuk, I. (Ed.). *Ukraine and its people.* Munich, 1949
Nettl, J. P. *The Eastern Zone and Soviet Policy in Germany, 1945–50.* Oxford, 1951
Prokopovicz, S. N. *Histoire économique de l'U.R.S.S.* Paris, 1952
Riazov, N. N., and Titelbaum, N. P. *Statistika Sovetskoy Torgovli (Statistics of Soviet Trade).* Moscow, 1951
Rothstein, A. *Man and Plan in the Soviet Economy.* London, 1948
Rupp, F. *Die Reparationsleistungen der Sowjetischen Besatzungszone.* Bonn, 1951
Schwartz, H. *Russia's Soviet Economy.* London, 1951
Seidel, W. *Die Eisenbahn in der Sowjetzone.* *Der Kraftverkehr in der Sowjetzone.* }Bonn, 1952
Skimkin, D. B. *Minerals—A Key to Soviet Power.* Harvard, 1953.
Smithsonian Institute World Weather Records, 1931–1940. Washington, 1947
Taylor, J. *The Economic Development of Poland.* New York, 1952
Thiel, E. *Sowjet-Fernost.* Munich, 1953
Tubert, General. *L'Ouzbekistan, Républic Sovietique.* Paris, 1951
Voznesensky, N. A. *The Economy of the U.S.S.R. during World War II.* Washington, 1948
Warriner, D. *Yugoslavia Rebuilds.* London, 1946
World Railways, 1954–1955. London, 1955
Zavalain, T. *How Strong is Russia?* London, 1951

GENERAL REFERENCE, OR DESCRIPTIVE, BOOKS

Balzak, S. S., Vasyutin, V. F., and Feigin, Ya.G. *Economic Geography of the U.S.S.R.* (Translated from the Russian. Edited by C. D. Harris.) New York, 1949

Baranskiy, N. N. *Ekonomicheskaya Geografiya S.S.S.R.* (Economic Geography of the U.S.S.R.). Moscow, 1949
(German edition: Die ökonomische Geographie der U.d.S.S.R., Berlin 1954)

Berg, L. S. *Natural Regions of the Soviet Union* (Translated from the Russian). New York, 1950

Bol'shaya Sovetskaya Entsiklopedia. 50 Vols. 2nd edition Moscow, 1949

Borisov, A. A. *Klimaty SSSR.* (*Climates of the U.S.S.R.*). Moscow, 1948

Columbia-Lippincott Gazetteer of the World. Columbia, 1952

Dickinson, R. E. *Germany.* London, 1953

Gregory, J. S. and Shave, D. W. *The U.S.S.R. A Geographical Survey.* London, 1947

Jorre, G. *L'U.R.S.S. La Terre et les Hommes.* Paris, 1946

Kendrew, W. G. *The Climates of the Continents.* Oxford, 1953

Konstantinov, F. T. *Narodnaya Respublika Bulgaria.* Moscow, 1952

Leimbach, W. *Die Söwjetunion: Natur, Volk und Wirtschaft.* Stuttgart, 1950

Mikhailov, N. N. *Soviet Russia. The Land and its People.* New York, 1948

Murzayev, E. M. *Mongol'skaya Narodnaya Respublika: Fizikogeograficheskoye Opisanie.* Moscow, 1948

Osteuropa-Handbuch. Band: Jugoslawien. Ed. Markert, W. Köln, 1954

Rosenberg, H. *Deutschland.* Frankfurt, 1953.

Shabad, T. *Geography of the U.S.S.R.: A Regional Survey.* New York, 1951

Simmons, E. J. (Ed.). *U.S.S.R.: A Concise Handbook.* New York, 1947

Wanklyn, H. *Czechoslovakia.* London, 1954

Wanklyn, H. *The Eastern Marchlands of Europe.* London, 1941

Woytinskiy, W. S. and Woytinskiy, E. S. *World Population and Production, Trends and Outlook.* New York, 1953

PERIODICALS

British Iron and Steel Federation. *Statistical Yearbooks; Part II Overseas Countries*

University of Birmingham, Faculty of Commerce and Social Sciences. *Bulletins of Soviet Economic Development* (irregular)

Colonial Geological Survey. *Statistical Summary of the Mineral Industry* (annual)

Commonwealth Economic Committee. *Annual reviews: Fruit; Grain Crops; Industrial Fibres; Meat; Plantation Crops; Vegetable Oils and Oilseeds*

Ministry of Foreign Trade. *Czechoslovak Economic Bulletin.* Prague (monthly)

Economic Commission for Europe: *Economic Bulletin for Europe* (quarterly); *Economic Survey of Europe* (annual); *Quarterly Bulletins: Coal Statistics for Europe; Steel Statistics for Europe.*

Gosfinizdat. *Finantsy S.S.S.R. (Finance of the U.S.S.R.).* Moscow (monthly)

Food and Agriculture Organization. *Commodity Series* (irregular); *Yearbook of Food and Agriculture Statistics; Yearbook of Forest Products Statistics; Unasylva* (bi-monthly). Rome

Hungarian Working People's Party. *Information Bulletin.* Budapest (monthly)

Communist Party of the Soviet Union. *Kommunist* (*formerly Bolshevik*). Moscow (bi-monthly)

La Documentation Française. *Notes et Études Documentaires.* Paris (irregular)

Deutsche Gesellschaft für Osteuropakunde. *Osteuropa.* Stuttgart (bi-monthly)

Gosplan. *Planovoye Khozyaistvo.* (*Planned Economy.*) Moscow (bi-monthly)

Chamber for Foreign Trade. *Polish Foreign Trade.* Warsaw (bi-monthly)

Ministry of Agriculture. *Sotsialisticheskoye Selskoye Khozyaistvo* (*Socialist Rural Economy*). Moscow (monthly)

University of Glasgow, Department for the Study of the Social and Economic Institutions of the U.S.S.R. *Soviet Studies* (quarterly).

Svensk Trävaru–Tidning (*Swedish Timber and Wood Pulp Journal*) Stockholm (bi-monthly, in Swedish and English)

United Nations Yearbook

Academy of Sciences. *Voprosy Ekonomiki* (*Economic Questions*). Moscow (monthly)

Wirtschaft des Ostblocks. Bonn

SOURCES AND METHOD

The use of Soviet and East European statistical material involves special difficulties which do not arise with most other countries. While actual output figures are published for some products, the more usual practice is to give the achievement for each year as a percentage of an earlier year; consequently any error that may occur in arriving at a figure for a base-year is liable to increase as time goes on. For some items not even percentage figures are published.

Primary sources used in preparing this volume consist chiefly of official reports on the progress of the economic plans, and articles in the Soviet and East European Press and periodicals. Much use has also been made of a large number of secondary sources, of which the principal ones are listed on pp. IV–VI. It has frequently been necessary to reconcile conflicting estimates in the light of probabilities, and occasionally to use estimates which are not fully supported by evidence; where the element of doubt about the figures used is particularly great, this is indicated in the text or tables.

The Economist Intelligence Unit claims only that the figures given are as accurate as careful analysis of the sources permits. While acknowledging its debt to the individuals and published works listed, the Unit is wholly responsible for all the economic material in the Atlas.

The information shown on the maps is in metric tons (2205 pounds).

Agricultural production is given in metric tons per hectare: 1 hectare = 2·471 acres.

The figures used in compiling the maps are for the year 1954, or 1953 where later figures are not available.

TRANSLITERATION

Place names have been transliterated into Roman form in accordance with the system agreed between the Permanent Committee on Geographical Names and the United States Board of Geographic Names.

PROJECTION

The projection used on the base map of the whole region is the *Conical Orthomorphic*, with standard parallels at 46½° N and 64½° N. The origin of the projection is on 56° N and to this parallel a scale reduction of 1·25 per cent has been applied, making the scale true along the two standard parallels. Over the whole populated area of U.S.S.R. and Eastern Europe the projection involves no greater " scale error " or distortion than 1·25 per cent—a negligible amount in relation to the large area covered.

TOPICS

CONTENTS

(For ease of reference, maps marked with an asterisk are shown in the Atlas on a common base map of U.S.S.R. and Eastern Europe at the scale of 1:25 million)

† See also back endpapers for population of major towns (post 1959)

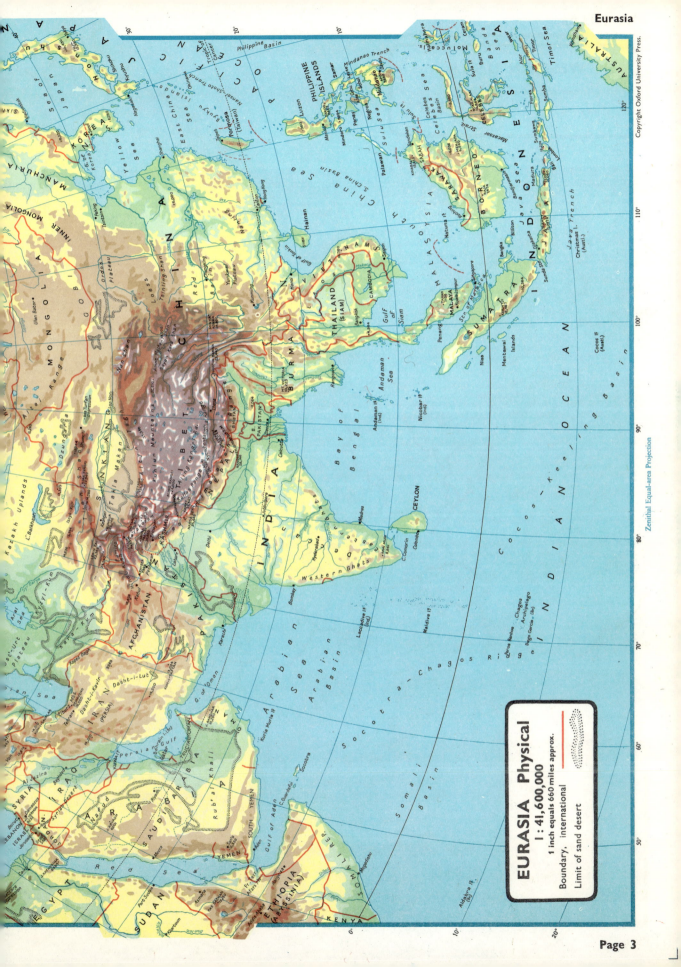

Zenithal Equal-area Projection

EURASIA Physical
1 : 41,600,000
1 inch equals 660 miles approx.
Boundary, international
Limit of sand desert

WESTERN U.S.S.R.

1 inch to 252 miles approx.   1:16,000,000

Miles

Kilometres

Zenithal Equidistant Projection   Origin 50° N 35° E

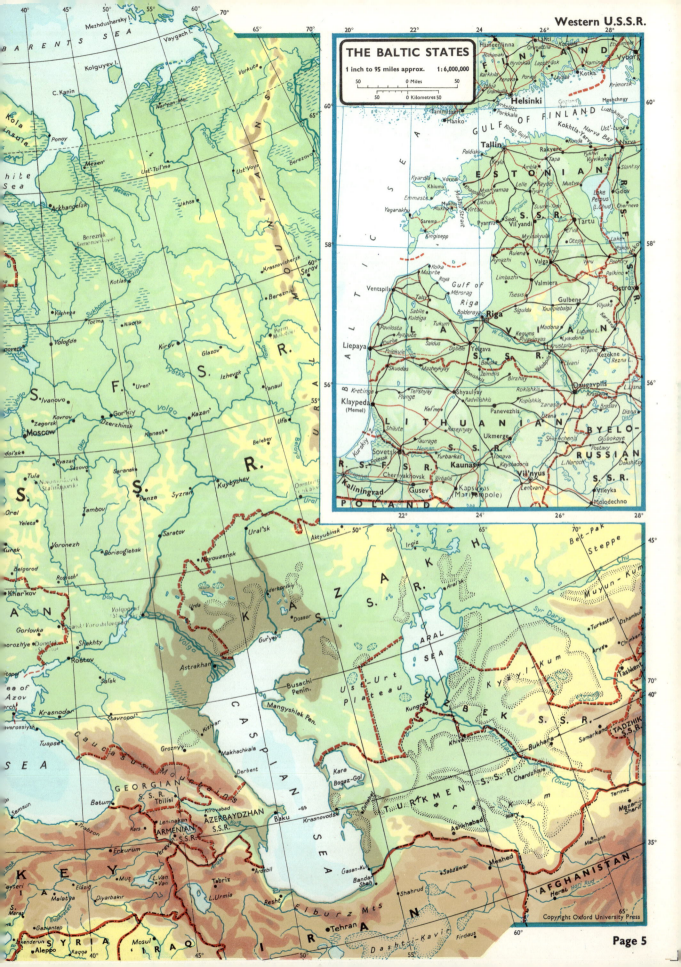

EAST GERMANY
and POLAND
1 inch to 63 miles approx. 1:4,000,000

For Altitude Layers and Legend see p 10

Administrative Districts 1963

EAST GERMANY	CZECHOSLOVAKIA		
1 Karl-Marx-Sdt.	7 Banská Bystrica	13 Jihlava	19 Ostrava
2 Dresden	8 Bratislava	14 Karlovy Vary	20 Pardubice
3 Frankfurt	9 Brno	15 Košice	21 Plzeň
4 Halle	10 České Budějovice	16 Liberec	22 Praha
5 Leipzig	11 Gottwaldov	17 Nitra	23 Prešov
6 Magdeburg	12 Hradec Králové	18 Olomouc	24 Usti
	(See Endpaper)		25 Zilina

Conical Orthomorphic Projection.
Origin 47°N. Standard parallel 55°N.

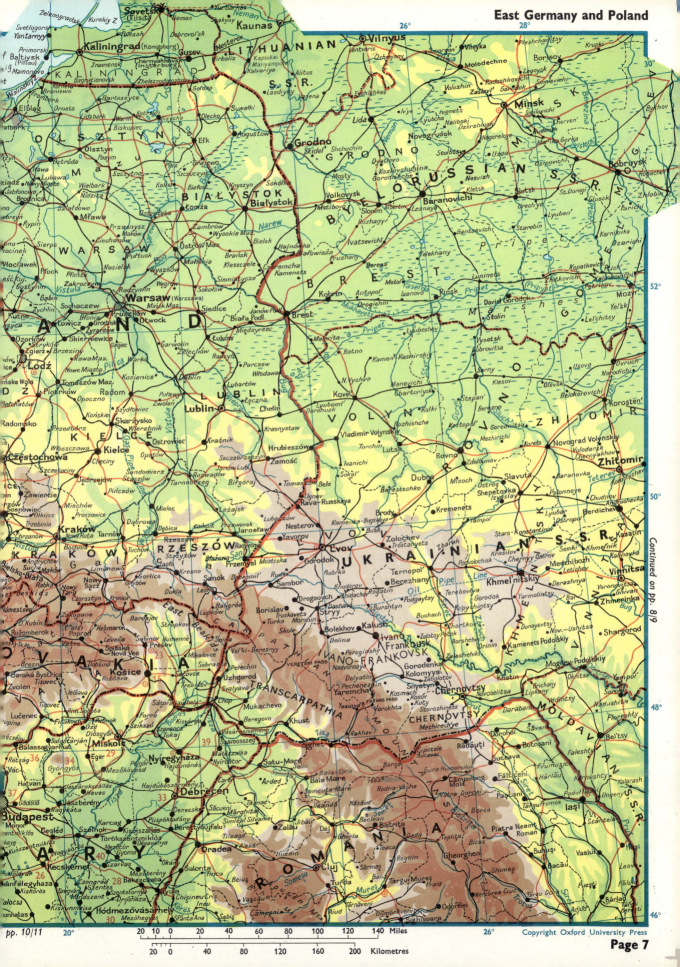

Continued on pp. 8/9

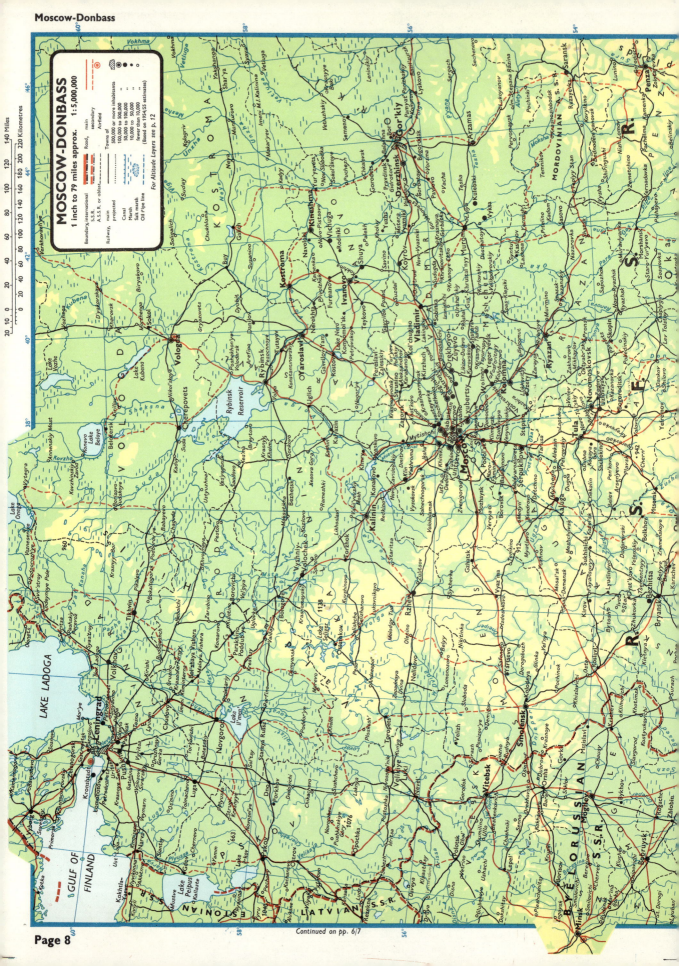

MOSCOW-DONBASS

1 inch to 79 miles approx.   1:5,000,000

Continued on pp. 6/7

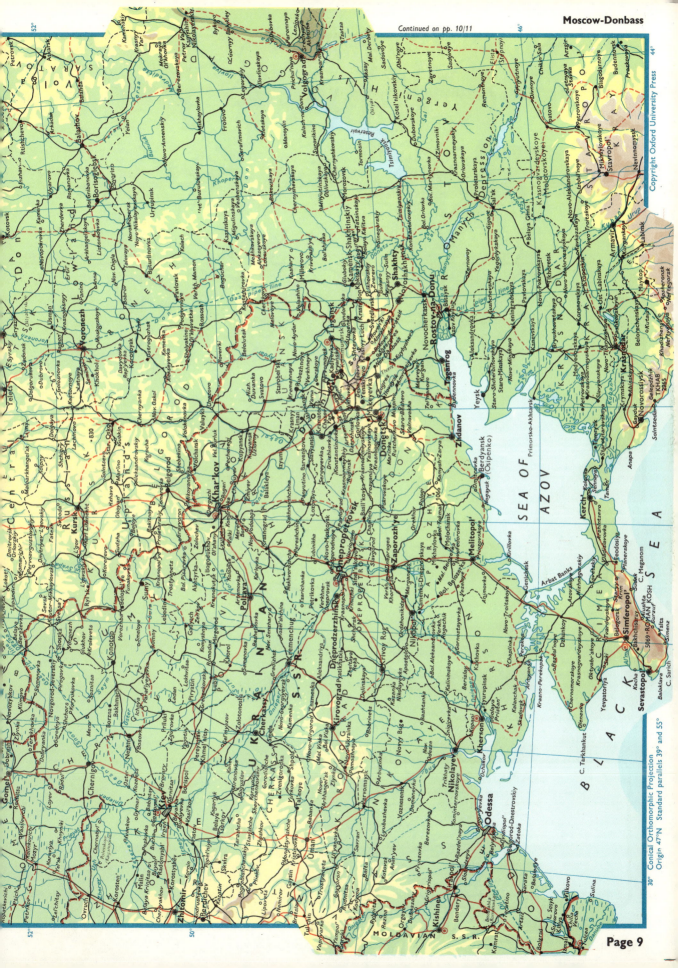

Continued on pp. 10/11

30° Conical Orthomorphic Projection
Origin 47°N Standard parallels 39° and 55°

SEA OF AZOV

BLACK SEA

Administrative Districts
HUNGARY

26 Bács-Kiskun	35 Komárom
27 Baranya	36 Nógrád
28 Békés	37 Pest
29 Borsod-Abaúj	38 Somogy
30 Csongrád	39 Szabolcs-Szatmar
31 Fejér	40 Szolnok
32 Györ-Sopron	41 Tolna
33 Hajdu-Bihar	42 Vas
34 Heves	43 Veszprém
	44 Zala

DANUBIA

1 inch to 63 miles approx.   1:4,000,000

Boundary	international	Canal
	S.S.R.	Marsh
	A.S.S.R. oblast or district	Pipe line
Railway	broad gauge	Towns of
	narrow gauge	500,000 or more inhabitants
	projected	100,000 to 500,000
Road	Autobahn	50,000 to 100,000
	main	10,000 to 50,000
	secondary	fewer than 10,000
Airfield		

(Based on censuses and estimates 1946-1955)

Feet
12,000
9000
6000
3000
1500
1000
600
300
Sea Level
Land Depr.
600
6000

Conical Orthomorphic Projection.
Origin 47°N. Standard parallel 39°N.

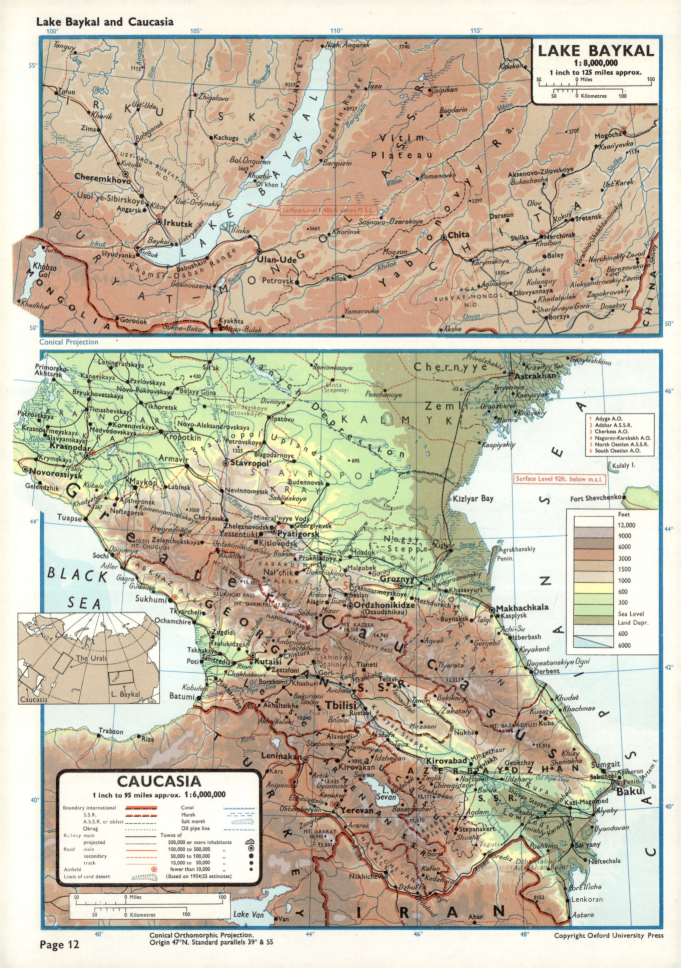

LAKE BAYKAL
1 : 8,000,000
1 inch to 125 miles approx.

Conical Projection

Surface Level 1,486 ft above M.S.L.

1 Adyge A.O.
2 Adzhar A.S.S.R.
3 Cherkess A.O.
4 Nagorno-Karabakh A.O.
5 North Osetian A.S.S.R.
6 South Osetian A.O.

Surface Level 92ft. below m.s.l.

Feet
12,000
9000
6000
3000
1500
1000
600
300
Sea Level
Land Depr.
600
6000

The Urals
Caucasia
L. Baykal

CAUCASIA
1 inch to 95 miles approx. 1:6,000,000

Boundary international
S.S.R.
A.S.S.R. or oblast
Okrug
Railway main
projected
Road main
secondary
track
Airfield
Limit of sand desert

Canal
Marsh
Salt marsh
Oil pipe line

Towns of
500,000 or more inhabitants
100,000 to 500,000 ,,
50,000 to 100,000 ,,
10,000 to 50,000 ,,
fewer than 10,000 ,,
(Based on 1954/55 estimates)

50   0 Miles   100
50   0 Kilometres   100

Conical Orthomorphic Projection.
Origin 47°N. Standard parallels 39° & 55

Copyright Oxford University Press

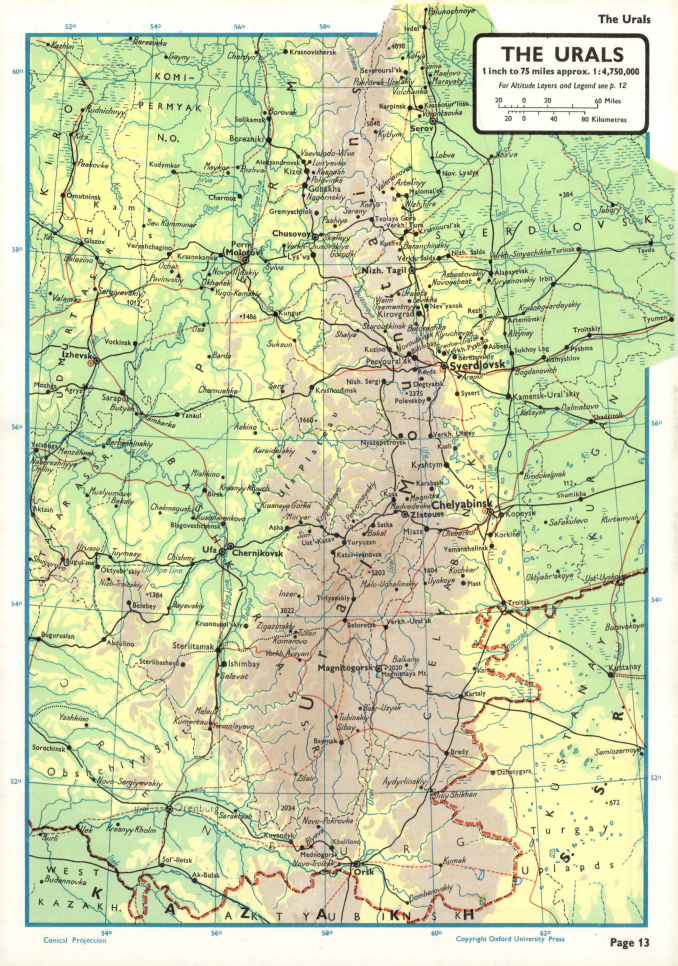

THE URALS
1 inch to 75 miles approx. 1: 4,750,000

For Altitude Layers and Legend see p. 12

20   0   20   60 Miles
20   0   40   80 Kilometres

Palunochnoye
Ivdel'
4898
Katya
Sama
Maslovo
Marsyaty
Volchanka
Karpinsk
Krasnotur'insk
Severoural'sk
Pokrovsk-Ural'skiy
Vorontsovka
KOMI-
Kazhim
Berezovka
Gayny
Cherdyn'
Krasnovishersk
5048
Serov
Kytlym
PERMYAK
Rudnichnyy
N.O.
Borovsk
Solikamsk
Lobva
Nov. Lyalya
Kirs
Peskovka
Bereźniki
Sos'va
384
Omutninsk
Kudymkar
Maykor
Pozhva
Aleksandrovsk
Vsevolodo-Vil'va
Lun'yevka
Kizel
Polovinka
Valer'yanovsk
Malomal'sk
Artel'nyy
Tabory
Chermoz
Gubakha
Nagornskiy
Kos'va
Is
Nizh. Tura
Glazov
Gremyachinsk
Pashiya
Sarany
Teplaya Gora
Verkh. Tura
Krasnoural'sk
Balezino
Vereshchagino
Krasnokamsk
Chusovoy
Skal'nyy
Kushva
Baranchinskiy
SVERDLOVSK
Ochan
Novo-Il'inskiy
Molotov
Verkh.-Chusovskiye
Gorodki
Nizh. Salda
Verkh.-Sinyachikha
Turinsk
Tavda
Perm
Lys'va
Verkh. Salda
Sev. Kommunar
Pavlovskiy
Okhansk
Yugo-Kamskiy
Nizh. Tagil
Asbestovskiy
Alapayevsk
Zyryanovskiy
Irbit
Valamaz
Sergiyevskiy
1012
Visim
Uralets
Levikha
Novoasbest
Izhevsk
Votkinsk
Kungur
1486
Isementnyy
Nev'yansk
Rezh
Krasnogvardeyskiy
Artemovskiy
Tyumen
Mozhga
Agryz
Osa
Kirovgrad
Staroutkinsk
Belorechka
Klyuchevsk
Sredne-Uralsk
Izumrud
Altynay
Troitskiy
Barda
Suksun
Shalya
Kuzino
Novouткinsk
Bilimbay
Verkh. Pyshma
Asbest
Sukhoy Log
Pyshma
Sarapul
Butysh
Chernushka
Sars
Krasnoufimsk
Pervoural'sk
Berezovskiy
Kamyshlov
Yanaul
Kambarka
Askino
1660
Nizh. Sergi
2375
Polevskoy
Degtyarsk
Aramil
Revda
Sverdlovsk
Sysert'
Bogdanovich
Kamensk-Ural'skiy
Yelabuga
Menzelinsk
Derbeshinskiy
Verkh. Ufaley
Kasli
Kataysk
Dalmatovo
Shadrinsk
Naberezhnyye
Chelny
Muslyumovo
Mishkino
Nyazepetrovsk
Krasnyy Klyuch
Bakaly
Birsk
Karabash
312
Brodokalmak
Aktash
Chekmagush
Kustharenkovo
Krasnaya Gorka
Kusa
Magnitka
Kyshtym
Shumikha
Blagoveshchensk
Min'yar
Kopachevo
Pervomayskiy
Nedvedevka
Zlatoust
Chelyabinsk
Kopeysk
Safakulevo
Kurtamysh
Urussu
Tuymazy
Chishmy
Asha
Sim
Satka
Bakal
Miass
Cheberkul'
Korkino
Oktyabr'skoye
Ust'-Uyskoye
Shugurovo
Bugul'ma
Oktyabr'skiy
Oil Pipe Line
Ufa
Chernikovsk
Ust'-Katav
Yuryuzan
Katav-Ivanovsk
Yemanzhelinsk
Kochkar'
Plast
Nizh-Troitskiy
1384
5203
1604
Uyskoye
Belebey
Rayevskiy
Inzer
Tirlyanskiy
Malo-Ubhalinsky
Buguruslan
Abdulino
3022
Beloretsk
Verkh.-Ural'sk
Troitsk
Borovskoye
Sterlitamak
Krasnousol'skiy
Zigazinskiy
Tukan
Komarovo
Verkh. Avzyan
Balkany
2020
Varna
Kustanay
Sterlibashevo
Ishimbay
Salavat
Magnitogorsk
Magnitnaya Mt.
Kartaly
Sorochinsk
Meleuz
Kumertau
Yermolayevo
Tubinskiy
Sibay
Bakr-Uzyak
Baymak
Bredy
Semiozernoye
Yashkino
Novo-Sergiyevskiy
Zilair
Aydyrlinskiy
Dzhetygara
872
Orenburg
2034
Siniy Shikhan
Turgay
Burli
Illek
Krasnyy Kholm
Saraktash
Novo-Pokrovka
Uplands
Kuvandyk
Blyava
Khalilovo
Budennovka
Sol'-Iletsk
Mednogorsk
Novo-Troitsk
Kumak
Ak-Bulak
Orsk
Dombarovskiy
WEST
KAZAKH.
A   K   T   Y   A   U   B   I   N   S   K   H

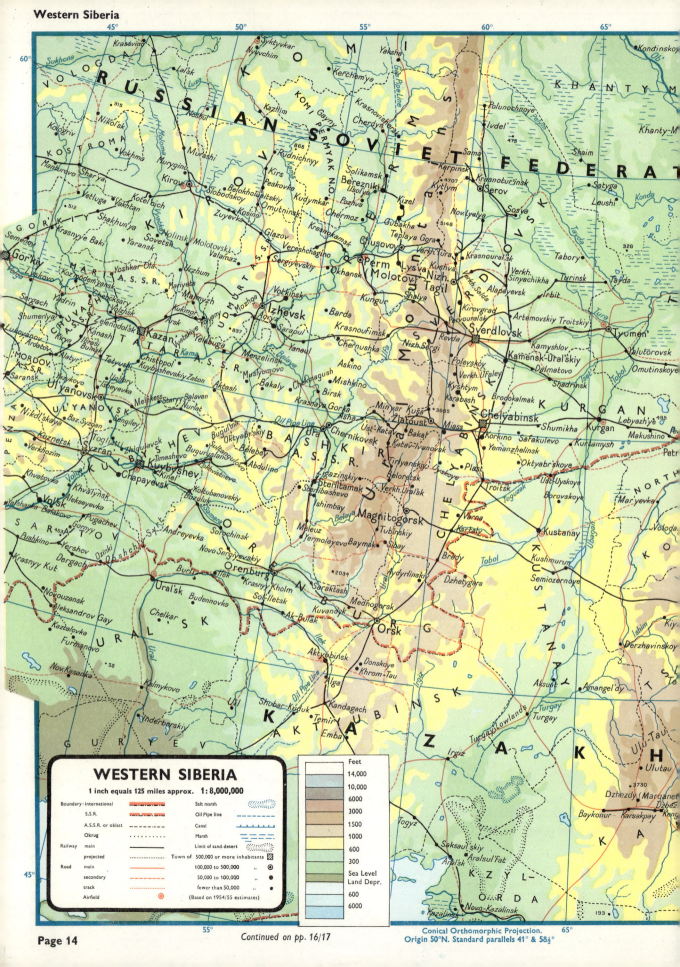

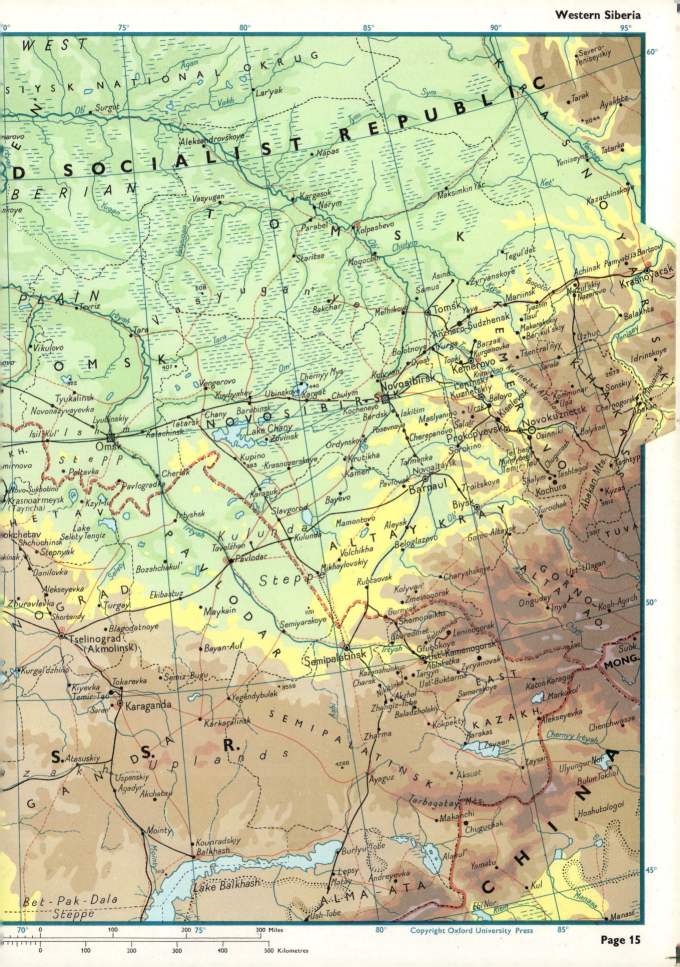

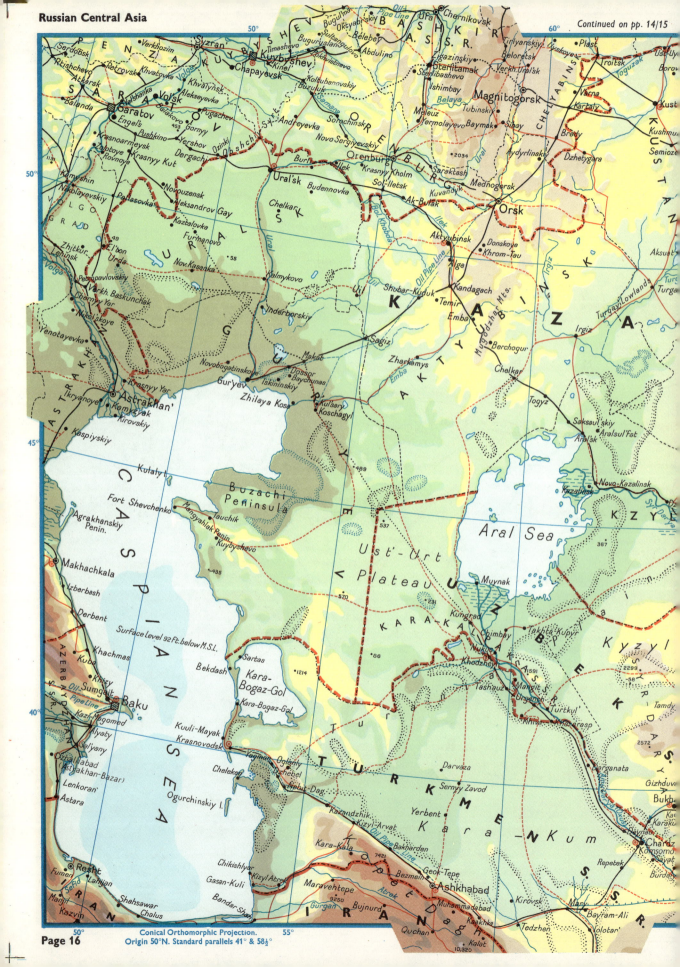

Conical Orthomorphic Projection.
Origin 50°N. Standard parallels 41° & 58½°

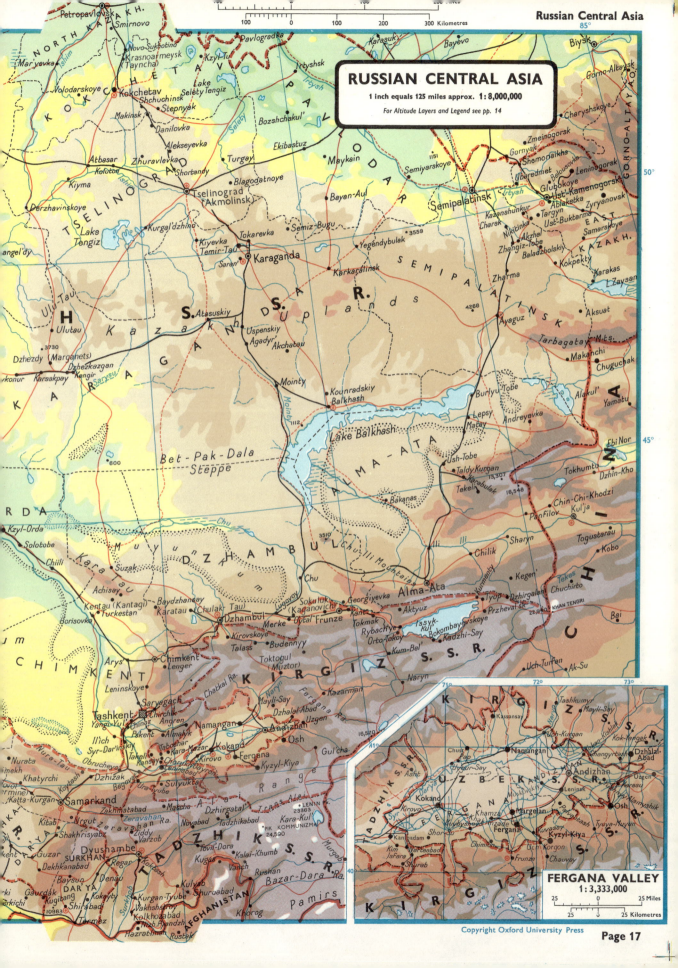

RUSSIAN CENTRAL ASIA

1 inch equals 125 miles approx. 1: 8,000,000

For Altitude Layers and Legend see pp. 14

85°

50°

45°

FERGANA VALLEY

1: 3,333,000

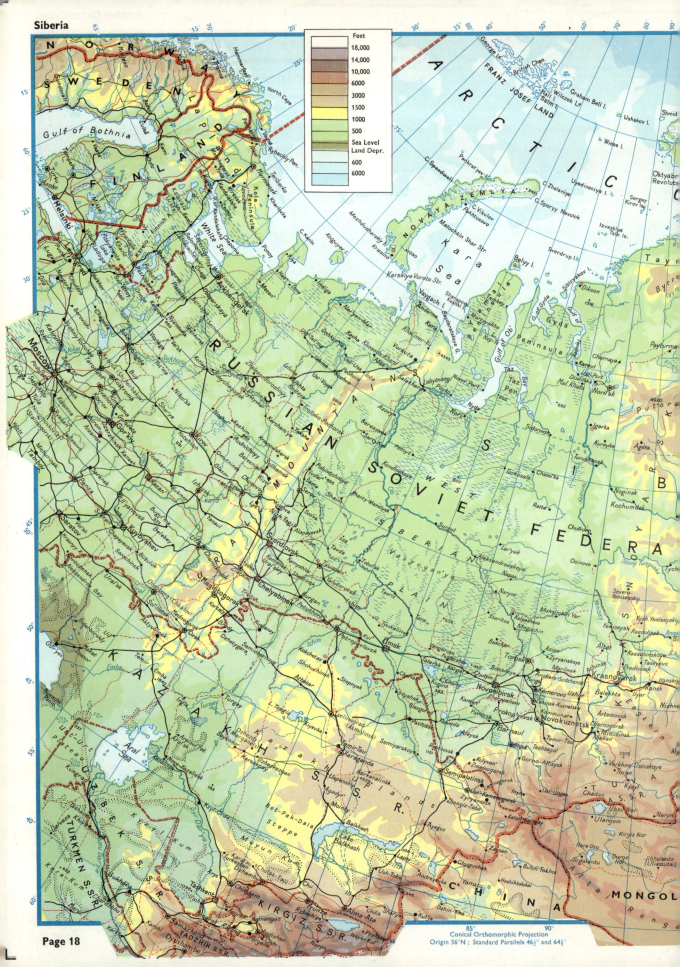

Siberia

Conical Orthomorphic Projection
Origin 56°N ; Standard Parallels 46½° and 64½°

SIBERIA

1 inch to 300 miles approx. 1:19,000,000

Boundary international		Airfield
province or state		Canal
Railway main		Marsh
minor		Limit of sand desert
projected		Town of 500,000 or more inhabitants
Road main		100,000 to 500,000 ,,
secondary		50,000 to 100,000 ,,
track		fewer than 50,000 ,,
		(Based on 1954/55 estimates)

Copyright Oxford University Press

Page 19

GEOLOGY

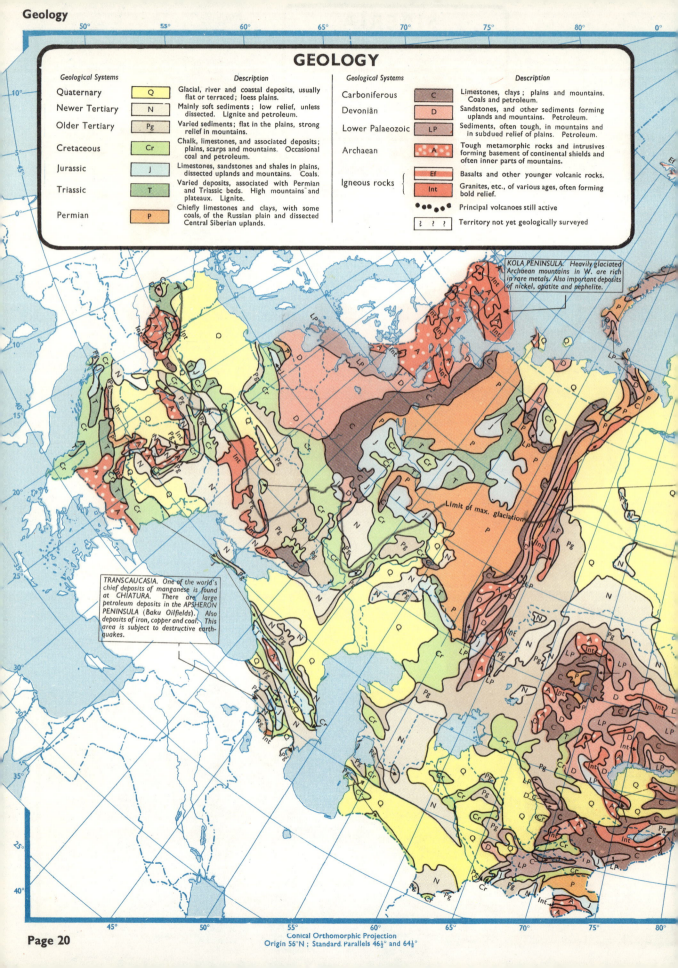

Geological Systems		Description
Quaternary	Q	Glacial, river and coastal deposits, usually flat or terraced; loess plains.
Newer Tertiary	N	Mainly soft sediments; low relief, unless dissected. Lignite and petroleum.
Older Tertiary	Pg	Varied sediments; flat in the plains, strong relief in mountains.
Cretaceous	Cr	Chalk, limestones, and associated deposits; plains, scarps and mountains. Occasional coal and petroleum.
Jurassic	J	Limestones, sandstones and shales in plains, dissected uplands and mountains. Coals.
Triassic	T	Varied deposits, associated with Permian and Triassic beds. High mountains and plateaux. Lignite.
Permian	P	Chiefly limestones and clays, with some coals, of the Russian plain and dissected Central Siberian uplands.

Geological Systems		Description
Carboniferous	C	Limestones, clays; plains and mountains. Coals and petroleum.
Devonian	D	Sandstones, and other sediments forming uplands and mountains. Petroleum.
Lower Palaeozoic	LP	Sediments, often tough, in mountains and in subdued relief of plains. Petroleum.
Archaean	A	Tough metamorphic rocks and intrusives forming basement of continental shields and often inner parts of mountains.
Igneous rocks	Ef	Basalts and other younger volcanic rocks.
	Int	Granites, etc., of various ages, often forming bold relief.
● ● ● ● ●		Principal volcanoes still active
? ? ? ?		Territory not yet geologically surveyed

KOLA PENINSULA. Heavily glaciated Archaean mountains in W. are rich in rare metals. Also important deposits of nickel, apatite and nephelite.

Limit of max. glaciation

TRANSCAUCASIA. One of the world's chief deposits of manganese is found at CHIATURA. There are large petroleum deposits in the APSHERON PENINSULA (Baku Oilfields). Also deposits of iron, copper and coal. This area is subject to destructive earthquakes.

Conical Orthomorphic Projection
Origin 56″N; Standard Parallels 46½° and 64½°

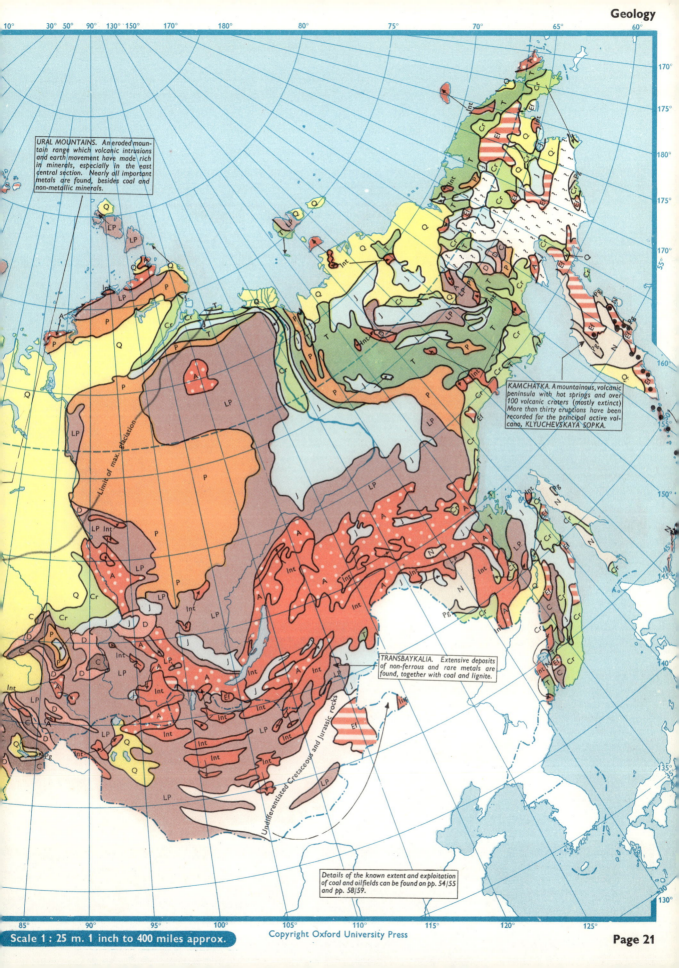

URAL MOUNTAINS. An eroded mountain range which volcanic intrusions and earth movement have made rich in minerals, especially in the east central section. Nearly all important metals are found, besides coal and non-metallic minerals.

KAMCHATKA. A mountainous, volcanic peninsula with hot springs and over 100 volcanic craters (mostly extinct) More than thirty eruptions have been recorded for the principal active volcano, KLYUCHEVSKAYA SOPKA.

TRANSBAYKALIA. Extensive deposits of non-ferrous and rare metals are found, together with coal and lignite.

Details of the known extent and exploitation of coal and oilfields can be found on pp. 54/55 and pp. 58/59.

Limit of max. glaciation

Undifferentiated Cretaceous and Jurassic rocks

Relief

ATLANTIC OCEAN

Limit of pack ice-average maximum (Spring)

North Sea

B. of Biscay

English Channel

Gulf of Bothnia

Gulf of Finland

Baltic Sea

Volga

Black Sea

Sea of Azov

Caspian Sea

Aral Sea

Lake Balkhash

Mediterranean Sea

Red Sea

Feet	
16,000	
10,000	
6000	
3000	
1500	
1000	
600	
300	
Sea Level	
Land Depr.	
100 fathoms	

Conical Orthomorphic Projection
Origin 56°N ; Standard Parallels 46½° and 64½°

Scale 1:25 m. 1

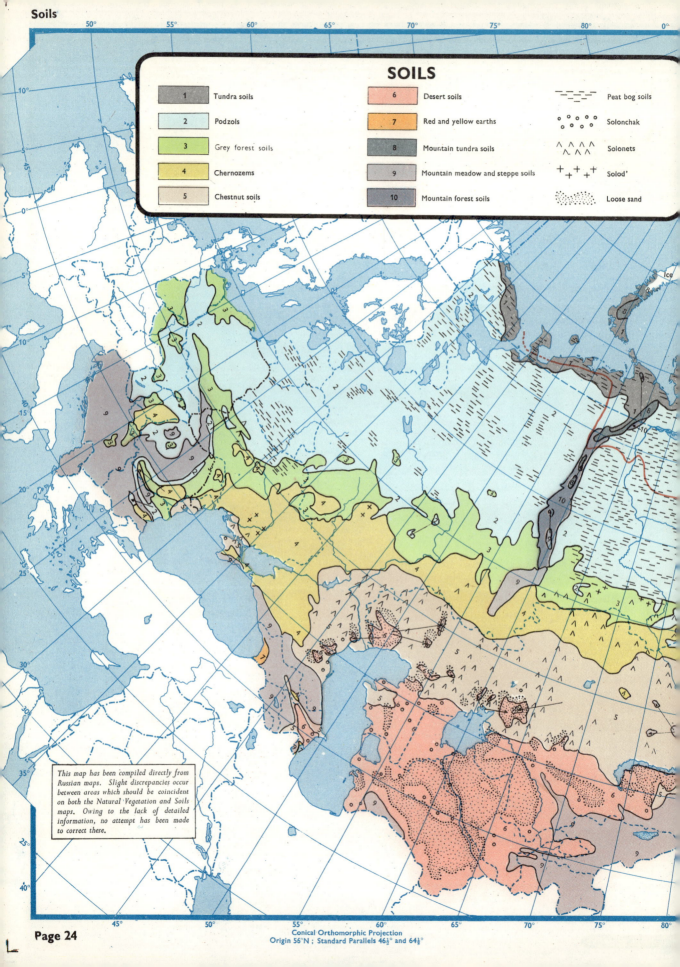

SOILS

1	Tundra soils	
2	Podzols	
3	Grey forest soils	
4	Chernozems	
5	Chestnut soils	
6	Desert soils	
7	Red and yellow earths	
8	Mountain tundra soils	
9	Mountain meadow and steppe soils	
10	Mountain forest soils	

Peat bog soils
Solonchak
Solonets
Solod'
Loose sand

This map has been compiled directly from Russian maps. Slight discrepancies occur between areas which should be coincident on both the Natural Vegetation and Soils maps. Owing to the lack of detailed information, no attempt has been made to correct these.

Conical Orthomorphic Projection
Origin 56°N ; Standard Parallels 46½° and 64½°

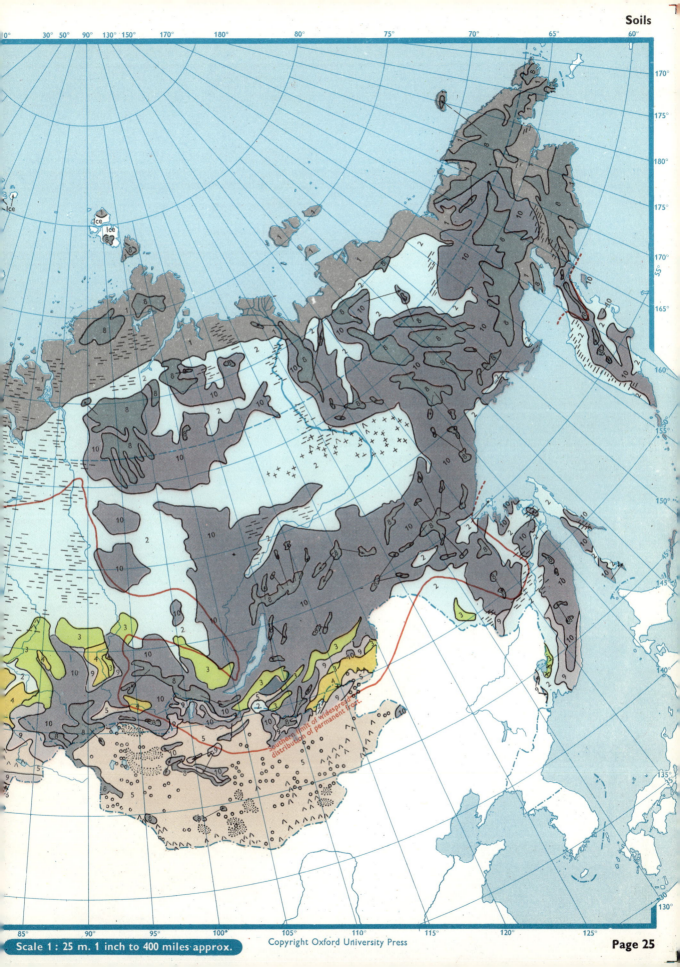

Scale 1 : 25 m. 1 inch to 400 miles approx.

Southern limit of widespread distribution of permanent frost.

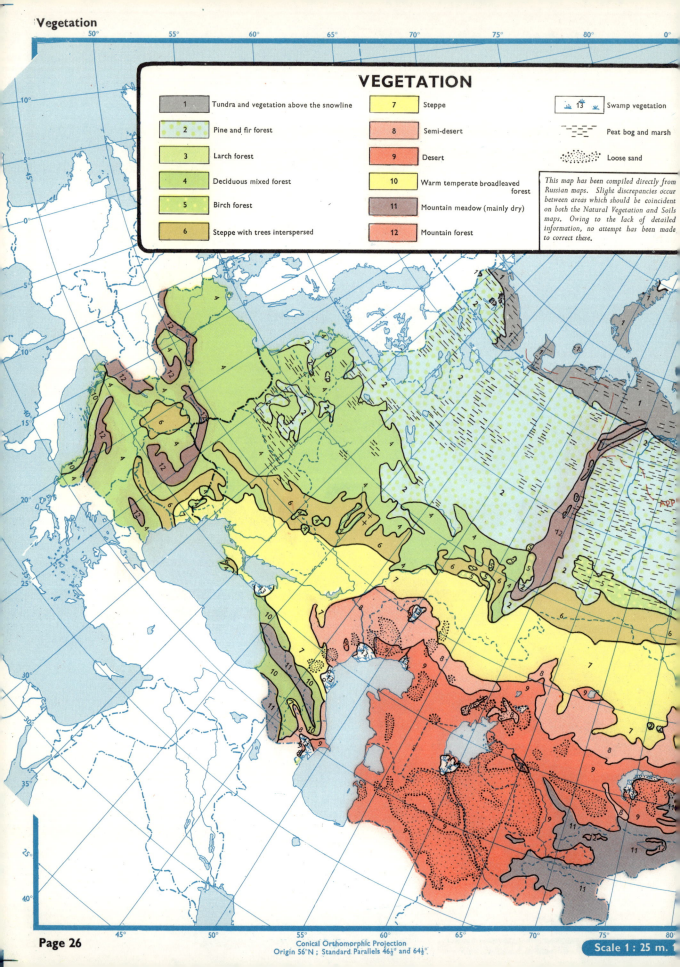

VEGETATION

1	Tundra and vegetation above the snowline	
2	Pine and fir forest	
3	Larch forest	
4	Deciduous mixed forest	
5	Birch forest	
6	Steppe with trees interspersed	

7	Steppe
8	Semi-desert
9	Desert
10	Warm temperate broadleaved forest
11	Mountain meadow (mainly dry)
12	Mountain forest

13	Swamp vegetation
	Peat bog and marsh
	Loose sand

This map has been compiled directly from Russian maps. Slight discrepancies occur between areas which should be coincident on both the Natural Vegetation and Soils maps. Owing to the lack of detailed information, no attempt has been made to correct these.

Conical Orthomorphic Projection
Origin 56°N ; Standard Parallels 46½° and 64½°

Scale 1 : 25 m.

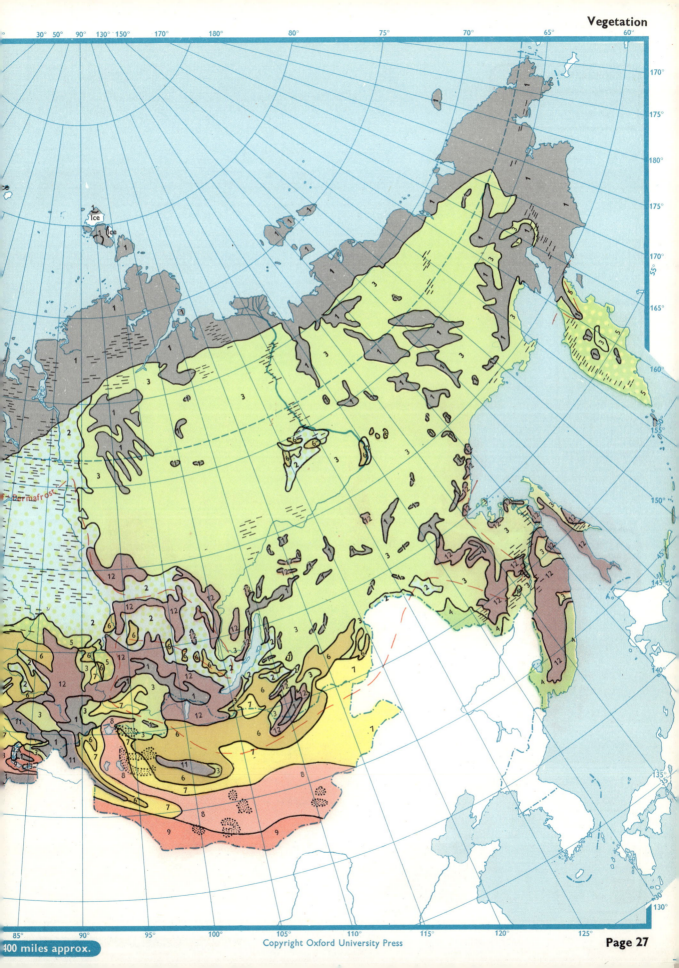

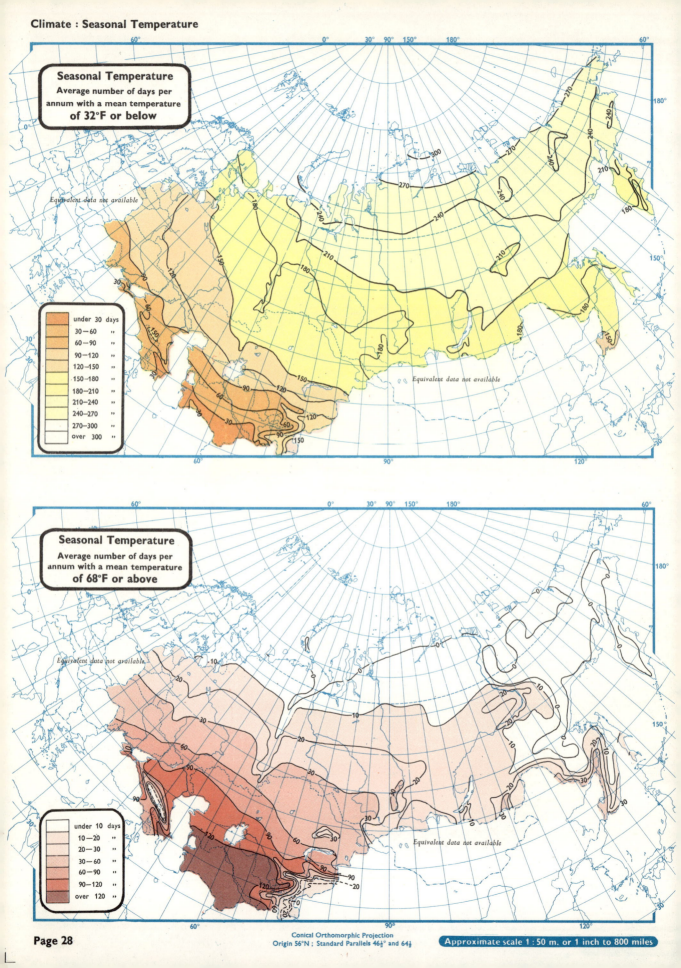

Seasonal Temperature
Average number of days per annum with a mean temperature of 32°F or below

Equivalent data not available

Equivalent data not available

	under 30 days
	30 — 60 ,,
	60 — 90 ,,
	90 — 120 ,,
	120 — 150 ,,
	150 — 180 ,,
	180 — 210 ,,
	210 — 240 ,,
	240 — 270 ,,
	270 — 300 ,,
	over 300 ,,

Seasonal Temperature
Average number of days per annum with a mean temperature of 68°F or above

Equivalent data not available

Equivalent data not available

	under 10 days
	10 — 20 ,,
	20 — 30 ,,
	30 — 60 ,,
	60 — 90 ,,
	90 — 120 ,,
	over 120 ,,

Conical Orthomorphic Projection
Origin 56°N ; Standard Parallels 46½° and 64½°

Approximate scale 1 : 50 m. or 1 inch to 800 miles

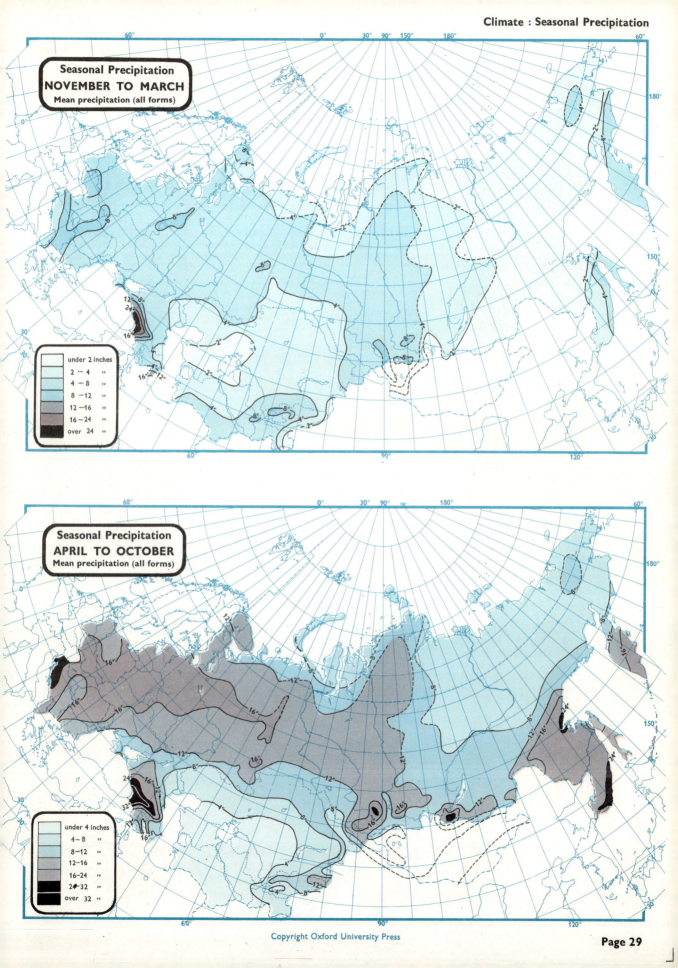

Seasonal Precipitation
NOVEMBER TO MARCH
Mean precipitation (all forms)

	under 2 inches
	2 — 4 ,,
	4 — 8 ,,
	8 — 12 ,,
	12 —16 ,,
	16 —24 ,,
	over 24 ,,

Seasonal Precipitation
APRIL TO OCTOBER
Mean precipitation (all forms)

	under 4 inches
	4 — 8 ,,
	8 —12 ,,
	12 —16 ,,
	16 —24 ,,
	24 —32 ,,
	over 32 ,,

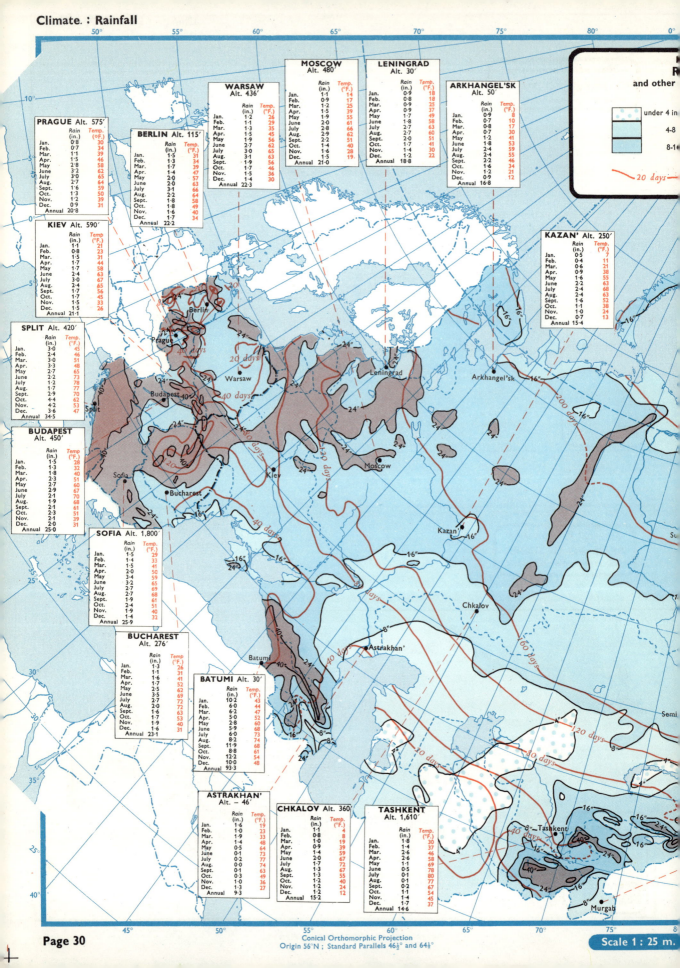

PRAGUE Alt. 575'

	Rain (in.)	Temp. (°F.)
Jan.	0·8	30
Feb.	0·7	34
Mar.	1·1	39
Apr.	1·5	46
May	2·8	58
June	3·2	62
July	3·0	65
Aug.	2·7	64
Sept.	1·6	59
Oct.	1·3	50
Nov.	1·2	39
Dec.	0·9	31
Annual	20·8	

KIEV Alt. 590'

	Rain (in.)	Temp. (°F.)
Jan.	1·1	21
Feb.	0·8	23
Mar.	1·5	31
Apr.	1·7	44
May	1·7	58
June	2·4	63
July	3·0	67
Aug.	2·4	65
Sept.	1·7	56
Oct.	1·7	45
Nov.	1·5	33
Dec.	1·5	26
Annual	21·1	

WARSAW Alt. 436'

	Rain (in.)	Temp. (°F.)
Jan.	1·2	26
Feb.	1·1	29
Mar.	1·2	35
Apr.	1·5	45
May	1·9	56
June	2·7	62
July	3·0	65
Aug.	3·1	63
Sept.	1·8	56
Oct.	1·7	46
Nov.	1·5	36
Dec.	1·4	30
Annual	22·3	

BERLIN Alt. 115'

	Rain (in.)	Temp. (°F.)
Jan.	1·5	31
Feb.	1·3	34
Mar.	1·7	39
Apr.	1·4	47
May	2·0	57
June	2·0	63
July	3·1	66
Aug.	2·2	64
Sept.	1·8	58
Oct.	1·8	49
Nov.	1·6	40
Dec.	1·7	34
Annual	22·2	

MOSCOW Alt. 480'

	Rain (in.)	Temp. (°F.)
Jan.	1·1	14
Feb.	0·9	17
Mar.	1·2	25
Apr.	1·5	39
May	1·9	55
June	2·0	61
July	2·8	63
Aug.	2·9	62
Sept.	2·2	51
Oct.	1·4	40
Nov.	1·6	28
Dec.	1·5	19
Annual	21·0	

LENINGRAD Alt. 30'

	Rain (in.)	Temp. (°F.)
Jan.	0·9	18
Feb.	0·9	18
Mar.	0·9	25
Apr.	0·9	37
May	1·7	49
June	1·8	58
July	2·8	63
Aug.	2·7	60
Sept.	2·0	51
Oct.	1·7	41
Nov.	1·4	30
Dec.	1·2	22
Annual	18·8	

ARKHANGEL'SK Alt. 50'

	Rain (in.)	Temp. (°F.)
Jan.	0·9	8
Feb.	0·7	10
Mar.	0·8	17
Apr.	0·7	30
May	1·2	41
June	1·8	53
July	2·4	59
Aug.	2·4	55
Sept.	1·6	46
Oct.	1·6	34
Nov.	1·2	21
Dec.	0·9	12
Annual	16·8	

KAZAN' Alt. 250'

	Rain (in.)	Temp. (°F.)
Jan.	0·5	7
Feb.	0·4	11
Mar.	0·6	21
Apr.	0·9	38
May	1·6	55
June	2·2	63
July	2·4	68
Aug.	1·8	63
Sept.	1·6	52
Oct.	1·1	38
Nov.	1·0	24
Dec.	0·7	13
Annual	15·4	

SPLIT Alt. 420'

	Rain (in.)	Temp. (°F.)
Jan.	3·0	45
Feb.	2·4	46
Mar.	3·0	51
Apr.	2·7	65
May	2·2	73
June	1·2	78
July	1·7	77
Aug.	2·9	70
Sept.	4·4	62
Oct.	4·2	53
Nov.	3·6	47
Dec.	34·5	
Annual	34·5	

BUDAPEST Alt. 450'

	Rain (in.)	Temp. (°F.)
Jan.	1·5	28
Feb.	1·3	32
Mar.	1·8	40
Apr.	2·3	51
May	2·9	60
June	2·9	67
July	2·1	70
Aug.	1·9	68
Sept.	2·3	61
Oct.	2·3	51
Nov.	2·1	39
Dec.	2·0	31
Annual	25·0	

SOFIA Alt. 1,800'

	Rain (in.)	Temp. (°F.)
Jan.	1·5	29
Feb.	1·4	33
Mar.	1·5	41
Apr.	2·0	50
May	3·4	59
June	3·2	65
July	2·7	69
Aug.	2·7	68
Sept.	1·9	61
Oct.	2·4	51
Nov.	1·9	40
Dec.	1·4	32
Annual	25·9	

BUCHAREST Alt. 276'

	Rain (in.)	Temp. (°F.)
Jan.	1·3	26
Feb.	1·1	31
Mar.	1·6	41
Apr.	1·7	52
May	2·7	62
June	3·5	69
July	2·7	72
Aug.	2·0	72
Sept.	1·6	63
Oct.	1·7	53
Nov.	1·9	40
Dec.	1·6	31
Annual	23·1	

BATUMI Alt. 30'

	Rain (in.)	Temp. (°F.)
Jan.	10·2	43
Feb.	6·0	46
Mar.	6·2	47
Apr.	5·0	52
May	2·8	60
June	6·0	73
July	5·9	74
Aug.	8·2	74
Sept.	11·9	68
Oct.	8·8	61
Nov.	12·2	54
Dec.	10·0	48
Annual	93·3	

ASTRAKHAN' Alt. – 46'

	Rain (in.)	Temp. (°F.)
Jan.	1·6	19
Feb.	1·0	23
Mar.	1·9	33
Apr.	1·4	48
May	0·5	64
June	0·1	73
July	0·2	77
Aug.	0·0	74
Sept.	0·1	63
Oct.	0·3	49
Nov.	1·0	36
Dec.	1·3	27
Annual	9·3	

CHKALOV Alt. 360'

	Rain (in.)	Temp. (°F.)
Jan.	1·1	4
Feb.	0·8	8
Mar.	1·0	19
Apr.	0·9	39
May	1·4	59
June	2·0	67
July	1·7	72
Aug.	1·3	67
Sept.	1·3	55
Oct.	1·2	40
Nov.	1·2	24
Dec.	1·2	12
Annual	15·2	

TASHKENT Alt. 1,610'

	Rain (in.)	Temp. (°F.)
Jan.	1·8	30
Feb.	1·4	37
Mar.	2·6	46
Apr.	2·6	58
May	1·1	69
June	0·5	78
July	0·1	80
Aug.	0·1	77
Sept.	0·2	67
Oct.	1·1	54
Nov.	1·4	45
Dec.	1·7	37
Annual	14·6	

R
and other

under 4 in

4-8

8-16

20 days

Conical Orthomorphic Projection
Origin 56°N ; Standard Parallels 46½° and 64½°

Scale 1 : 25 m.

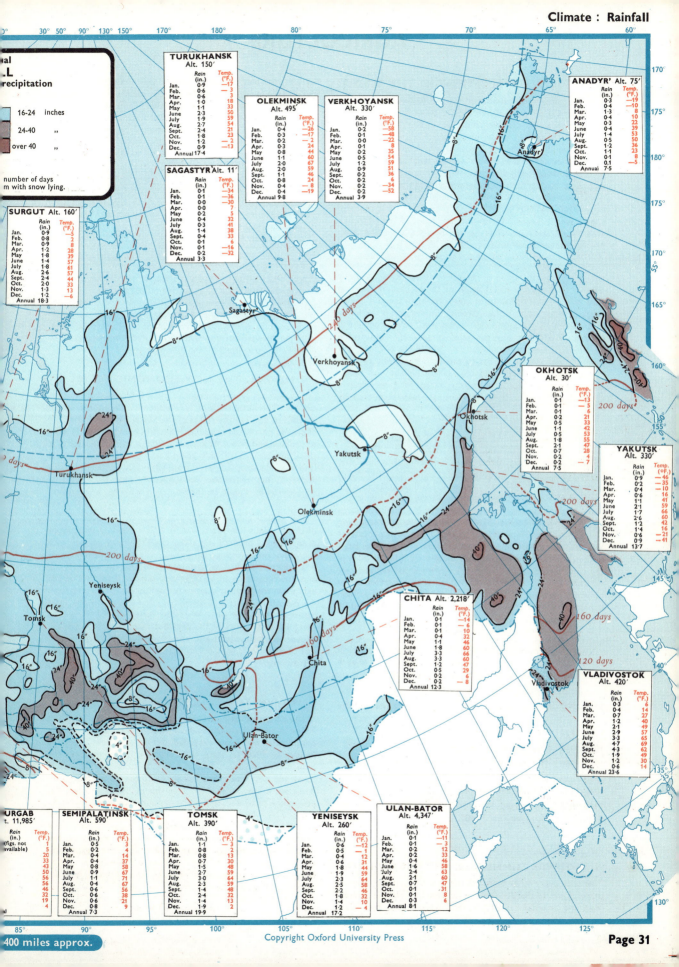

Annual precipitation

- 16–24 inches
- 24–40 ,,
- over 40 ,,

number of days with snow lying.

TURUKHANSK
Alt. 150'

	Rain (in.)	Temp. (°F.)
Jan.	0·9	−17
Feb.	0·6	−3
Mar.	0·6	3
Apr.	1·0	18
May	1·1	33
June	2·3	50
July	1·9	59
Aug.	2·7	54
Sept.	2·4	21
Oct.	1·8	23
Nov.	1·2	−2
Dec.	0·9	−13
Annual	17·4	

OLEKMINSK
Alt. 495'

	Rain (in.)	Temp. (°F.)
Jan.	0·4	−26
Feb.	0·3	−17
Mar.	0·2	−2
Apr.	0·3	24
May	0·8	44
June	1·1	60
July	2·0	67
Aug.	2·0	59
Sept.	1·1	46
Oct.	0·8	24
Nov.	0·4	8
Dec.	0·4	−19
Annual	9·8	

VERKHOYANSK
Alt. 330'

	Rain (in.)	Temp. (°F.)
Jan.	0·2	−58
Feb.	0·1	−48
Mar.	0·0	−22
Apr.	0·1	8
May	0·2	35
June	0·5	54
July	1·2	59
Aug.	0·5	51
Sept.	0·2	36
Oct.	0·2	6
Nov.	0·2	−34
Dec.	0·2	−52
Annual	3·9	

SAGASTYR Alt. 11'

	Rain (in.)	Temp. (°F.)
Jan.	0·1	−34
Feb.	0·1	−36
Mar.	0·0	−30
Apr.	0·0	7
May	0·2	5
June	0·4	32
July	0·3	41
Aug.	1·4	38
Sept.	0·4	33
Oct.	0·1	6
Nov.	0·1	−16
Dec.	0·1	−32
Annual	3·3	

ANADYR' Alt. 75'

	Rain (in.)	Temp. (°F.)
Jan.	0·3	−19
Feb.	0·4	−10
Mar.	1·3	−9
Apr.	0·4	10
May	0·3	22
June	0·4	39
July	1·3	53
Aug.	0·5	50
Sept.	1·2	36
Oct.	1·1	23
Nov.	0·1	8
Dec.	0·1	−5
Annual	7·5	

SURGUT Alt. 160'

	Rain (in.)	Temp. (°F.)
Jan.	0·9	−5
Feb.	0·8	2
Mar.	0·9	8
Apr.	1·2	28
May	1·8	39
June	1·4	57
July	1·8	61
Aug.	2·6	57
Sept.	2·4	44
Oct.	2·0	33
Nov.	1·3	13
Dec.	1·2	−6
Annual	18·3	

OKHOTSK
Alt. 30'

	Rain (in.)	Temp. (°F.)
Jan.	0·1	−13
Feb.	0·1	−5
Mar.	0·1	6
Apr.	0·2	21
May	1·1	33
June	1·1	42
July	0·5	53
Aug.	1·8	55
Sept.	2·1	47
Oct.	0·7	28
Nov.	0·2	4
Dec.	0·2	−7
Annual	7·5	

YAKUTSK
Alt. 330'

	Rain (in.)	Temp. (°F.)
Jan.	0·9	−46
Feb.	0·2	−35
Mar.	0·4	−10
Apr.	0·6	16
May	1·1	41
June	2·1	59
July	1·7	66
Aug.	2·6	60
Sept.	1·2	42
Oct.	1·4	16
Nov.	0·6	−21
Dec.	0·9	−41
Annual	13·7	

CHITA Alt. 2,218'

	Rain (in.)	Temp. (°F.)
Jan.	0·1	−14
Feb.	0·1	−6
Mar.	0·1	10
Apr.	0·4	32
May	1·1	46
June	1·8	60
July	3·3	66
Aug.	3·3	60
Sept.	1·2	47
Oct.	0·5	29
Nov.	0·2	6
Dec.	0·2	8
Annual	12·3	

VLADIVOSTOK
Alt. 420'

	Rain (in.)	Temp. (°F.)
Jan.	0·3	6
Feb.	0·4	14
Mar.	0·7	27
Apr.	1·2	40
May	2·1	49
June	2·9	57
July	3·3	65
Aug.	4·7	69
Sept.	4·3	62
Oct.	1·9	47
Nov.	1·2	30
Dec.	0·6	14
Annual	23·6	

MURGAB
Alt. 11,985'

	Rain (in.)	Temp. (°F.)
Jan.	1 (figs. not available)	3
Feb.	5	4
Mar.	20	10
Apr.	33	37
May	43	58
June	50	67
July	56	71
Aug.	56	56
Sept.	46	56
Oct.	32	38
Nov.	19	21
Dec.	4	9
Annual		

SEMIPALATINSK
Alt. 590'

	Rain (in.)	Temp. (°F.)
Jan.	0·5	3
Feb.	0·2	4
Mar.	0·4	14
Apr.	0·4	37
May	0·8	58
June	0·9	67
July	1·1	71
Aug.	0·4	65
Sept.	0·6	56
Oct.	0·6	38
Nov.	0·6	21
Dec.	0·8	9
Annual	7·3	

TOMSK
Alt. 390'

	Rain (in.)	Temp. (°F.)
Jan.	1·1	−3
Feb.	0·8	2
Mar.	0·4	13
Apr.	0·7	30
May	1·5	48
June	2·7	59
July	3·0	64
Aug.	2·3	58
Sept.	1·4	48
Oct.	2·4	32
Nov.	1·4	13
Dec.	1·9	2
Annual	19·9	

YENISEYSK
Alt. 260'

	Rain (in.)	Temp. (°F.)
Jan.	0·6	−12
Feb.	0·5	−1
Mar.	0·4	12
Apr.	0·8	31
May	1·8	44
June	2·7	59
July	2·3	64
Aug.	2·5	58
Sept.	1·4	46
Oct.	1·8	32
Nov.	1·4	10
Dec.	1·2	−4
Annual	17·2	

ULAN-BATOR
Alt. 4,347'

	Rain (in.)	Temp. (°F.)
Jan.	0·1	−11
Feb.	0·1	−3
Mar.	0·2	12
Apr.	0·3	33
May	0·4	46
June	1·6	58
July	2·4	63
Aug.	2·1	60
Sept.	0·7	47
Oct.	0·3	31
Nov.	0·1	8
Dec.	0·3	6
Annual	8·1	

400 miles approx.

Copyright Oxford University Press

Page 31

AGRICULTURE

HISTORICAL DEVELOPMENT IN THE U.S.S.R.

By far the greatest limitation on the expansion of Soviet agriculture is climate. While most parts of European Russia enjoy reasonably favourable conditions, in Siberia and Central Asia extreme cold in winter and drought in summer are the rule. It will be seen from the maps that in Asiatic Russia crops are confined (1) to a relatively narrow strip across southern Siberia, widening only here and there along the river valleys and (2) to a wider strip in Eastern Siberia as the Pacific Ocean is approached, and (3) to some parts of Kazakhstan and the other Asian republics where there is shelter and sufficient rainfall, or where human intervention has increased the cultivable area by irrigation and drainage, and by conquering the ever-present danger of soil erosion.

The black earth region of central European Russia and the northern Ukraine provides some of the richest agricultural land in the world. Rainfall is relatively high and the annual range of temperature is not extreme. This region is the home of Russian agriculture, and exported wheat both to the rest of the Empire and to foreign countries from the time of Ivan the Terrible. The extension of the sown area to the podzolic soils further north came later, but by the eighteenth century most of European Russia south of the Gulf of Finland was under cultivation or used for stock rearing, except for those areas in the east and north-east which were still mainly covered by forest.

In 1913 about 367 million hectares* (910 million acres) were classed as agricultural land, though only about 105 million hectares (260 million acres) were under cultivation. The figure of 367 million hectares represents rather less than 17% of the total area of the country. About 152 million hectares were comprised in large estates belonging to the royal family, the nobles and the monasteries; the remaining 215 million hectares were divided up into over 20 million peasant holdings averaging a little over 10 hectares (25 acres) apiece, but with a considerably larger proportion under cultivation than in the large estates, which included great areas of forest. Roughly 90% of the total area cultivated was under cereal crops (wheat, rye, oats and barley), and about 80 million tons of cereals were harvested; there were at the same time about 35 million horses, 60 million cattle, 120 million sheep and goats, and 20 million pigs.

Since the Revolution there have been six distinct phases in the attitude of the Soviet Government to agriculture, and each had a marked influence on the level of production.

1917–1922 The period of war communism, civil war and the crop failure of 1921.

1922–1928 The period of the New Economic Policy.

1928–1933 The forced collectivization.

1933–1941 The period of stabilization on the collective principle.

1941–1946 The war and post-war period.

1946–1955 The period of expansion under the fourth and fifth Five-Year Plans.

1917–1922 WAR COMMUNISM AND THE CIVIL WAR

As with every other branch of the Russian economy, agricultural output fell sharply after the Revolution. The big private, royal and ecclesiastical estates were expropriated, but there were too few people with the necessary technical and managerial skill to take over; some of the expropriated land was allocated to peasants, or later to returned soldiers and settlers from the towns, but a considerable amount of land went temporarily out of production. Crops were requisitioned by the authorities, but great difficulty was found in getting the growers to declare the full amount harvested, and the prices paid were excessively low. Deliveries to the towns fell catastrophically, and by the middle of 1918 there was famine in many of the larger cities. This set up a social tension between the town-dwellers and the agrarian population which has never been fully relieved. As early as 1918, measures were taken to expropriate the *kulaks* or wealthier peasants, who were accused of deliberately trying to sabotage the Revolution by withholding the food they produced.

Numerous experiments were tried to ease the situation, including the setting up of collective farms, but without much success. In 1921 the peasants in many parts of the country were in open revolt, and an exceptionally dry and hot summer made matters worse. The New Economic Policy was originally put forward early in that year, but did not become fully effective in agriculture till the following year.

1922–1928 THE NEW ECONOMIC POLICY

The new policy ended requisitions and restored the market system. The Rural Code of 1922 suspended the redistribution of land, recognized the rights of the peasant to the ownership of his equipment and produce, and left him free to decide what to produce and how. At the same time it was made clear that these measures were only temporary; private ownership of land was declared to be " transitory and destined to disappear ", but there was no suggestion that the tempo of collectivization should be forced.

This radical change of policy arrested the agrarian decline of the previous five years, and by 1928 output of most kinds of agricultural produce was above the 1913 levels, though with the important exception of wheat. The total cultivated area was greater (113 million hectares against 105 million in 1913), and it reflected the trend of general economic policy that the area sown with industrial crops had almost doubled, rising from 4·5 million hectares to 8·6 million. The peasants were again accused of sabotage when wheat deliveries in 1927 were little more than half those of the previous year. It was this fall in deliveries which gave rise to one of the most ambitious, if also one of the most costly, economic experiments ever tried in a big country.

1928–1933 THE COLLECTIVIZATION PERIOD

The collective farm (*kolkhoz*) and the state farm (*sovkhoz*) had both been tried experimentally during the war communism period, and had been extended during the New Economic Policy; in 1926–27, state and collective farms produced 1·3 million tons of cereals (out of a national total of 77·8 million tons), and delivered nearly half of this to the authorities. State farms were used largely for experimental work in areas not previously cultivated, and were organized on lines similar to those of manufacturing industry. Collective farms were introduced in areas where the peasants had been more than usually uncooperative. The land was pooled and the peasant lost the right to regard any part of it as his own, while the great extension of the area to be managed as a single unit was intended to make possible economies in planning, crop rotation and the use of buildings and equipment.

* 1 hectare=2·471 acres.

Some of the earlier experiments had been tolerably successful. The policy initiated in 1928 had a treble aim: (1) to extend the collective principle to a much larger sector of agriculture; (2) to suppress the section of the population most hostile to the regime; (3) to release peasants to work in the factories. The peasants were forced in large numbers into the collectives, in spite of strong resistance, and the last of the *kulaks* were eliminated altogether. Between October 1929 and March 1930 the number of peasant holdings collectivized rose rapidly from 4·1% to 58%. Thereafter collectivization was slowed down considerably, and became at least nominally voluntary. The number of peasant households in the collective farms increased from 416,000 in 1928 to nearly 6 million in 1930, over 13 million in 1931, and nearly 19 million by 1938; the number of private holdings declined from 24·5 million in 1928 to 1·3 million in 1938. In 1939 the size of a private holding was limited by decree to 1½ hectares, and in irrigated land to ¼ hectare.

The initial effect of collectivization was to increase the area sown and the size of the harvest. But by 1931 the peak was past and the resistance of the peasants was making itself felt. Between 1931 and 1933 over 6·5 million hectares went out of cultivation, and between 1930 and 1932 the cereal harvest dropped from 83·5 million tons to under 70 million. The 1932 season was one of near-famine. Thereafter a considerable improvement took place, until in 1937* the output of cereals passed 100 million tons for the first time; the livestock population was still, however, smaller than before the Revolution, or in the peak year of 1929. Of about 95 million tons of cereals produced in 1938, 86 million (90%) came from collective farms, under 9 million from state farms, and only some 430,000 tons from private holdings.

1941–1946 THE WAR AND POST-WAR PERIOD

During the first year of war the Germans occupied nearly half of the total sown area of the U.S.S.R., including the districts which produced 38% of the cereal crop and 84% of the output of sugar beet. Livestock losses included 7 million horses, 17 million cattle, 27 million sheep and goats and 20 million pigs. At the time of the greatest German advance, about 88 million people, or nearly half the population, were living in enemy-occupied territory. This loss of population was on a scale comparable to the loss of the food producing areas, and was the main reason why the rest of the Soviet Union escaped serious famine.

It was not possible to extend agriculture eastwards to the same extent as manufacturing industry. Large capital schemes of irrigation and drainage would have been needed, but could not be carried out because of the needs of the war industries; agricultural machinery was short for the same reason, and there was a seriously inadequate supply of fertilizers. The non-occupied areas thus emerged from the war little changed, and for the first three or four years after the war the main problem was to restore devastated agricultural land and buildings rather than to strike out in new directions. In 1945, total cultivated area was still only 105 million hectares, the cereals crop no more than 67 million tons, and the number of cattle down to 47 million. The period of greatest difficulty thus came after the war rather than during it, and was made worse by the exceptionally dry summer of 1946.

1946–1955 THE FOURTH AND FIFTH FIVE-YEAR PLANS

The 1946–1950 Plan aimed at a total cultivated area of 158·5 million hectares (included the newly acquired territories), but the result fell short by over 7 million hectares. The restoration of the war-damaged lands was complete by 1947–48, and the lands taken over from pre-war Poland, Czechoslovakia and Romania had been fitted in to the Soviet pattern. Including the new territories, the actual extent of cultivated land in 1950 was 147·4 million hectares, or 44% more than in 1946; the cereal crop was 120·1 million tons, or nearly twice as much as in 1946.

* See footnote to Yield of Cereals in the U.S.S.R.

Under the fifth Five-Year Plan (1951–55) the sown area was to be raised to 169 million hectares, apparently not including the virgin lands referred to below. By 1953, a total of 159·2 million hectares was actually under cultivation. The target of 181·6 million tons of grain for 1955 was much too high, being based on the yields obtained in the exceptionally favourable summer of 1950. In the later years of the Plan, the emphasis was shifted from increasing the yield per hectare to extending the total area sown.

In 1953–54 an enormous programme was launched for the ploughing up of new land (pp. 34/35), and teams of specially recruited workers from the cities were moved to the new areas. According to official Soviet sources, about 17·2 million hectares were actually ploughed up in 1954, including some 8 million hectares in Kazakhstan, and about 3·6 million hectares of new land were planted with cereals. There is no independent check of accuracy of these figures, but it is known that, in spite of disappointing yields, the amount of wheat harvested was too great for the available storage space. Reports indicated that by the end of May 1955 over 14 million hectares had been sown in the current season. Yields were again expected to be low.

Both plans concentrated on three main activities—the control of drought, increased mechanization, and the bringing of electricity to the farms. Drought control was planned partly through irrigation schemes, partly through an ambitious programme of planting shelter belts of forest trees, involving the afforestation of 6·15 million hectares by 1965. According to official Soviet statements, 1·35 million hectares had been planted by the end of 1950, but in 1953 the whole project seems to have been dropped after the death of Stalin, who had tried to force it through without adequate preparation.

The mechanization programme was intended to provide 325,000 tractors between 1946 and 1950. No estimate was given for the years 1951–53, but between April 1954 and April 1957 the number supplied was to be 500,000. Actual deliveries were as follows (all in 15 h.p. units).

1946–50	536,000
1951	137,000
1952	131,000
1953	139,000
1954	137,000

Between 1946 and 1950 93,000 combine harvesters were delivered. Output of these machines was then considerably increased, and deliveries in the years 1951–54 reached 172,000.

The programme of electrification fell far short of the plan (pp. 52/53). By 1950, 70,000 collective farms were to be supplied with electric power, and the capacity of rural power stations was to be raised to 1·5 million kw.; in practice the programme was nearly fulfilled in the Urals, in Moscow and Leningrad Oblasts and in Armenia, but fell far short elsewhere. The power station capacity achieved was no more than 770,000 kw. As late as the beginning of 1954, only 21,000 collective farms were supplied with electricity, but it must be remembered that their size had been increased by amalgamation in 1953–54. The plan for 1955 provided for the electrification of a further 1,500 collective farms bringing the total up to about 25,000.

The following tables show, first, how the fluctuations of agricultural output corresponded to the successive stages in the development of Soviet agrarian policy, and secondly how over a period of forty years cereal production has declined in importance and that of industrial crops has risen. Industrial crops include cotton and other vegetable fibres, rubber plants such as *kok sagyz* and *tau sagyz*, many of the oilseeds, tobacco, and a few others used in the chemical industry. The fibres and rubber-bearing plants are dealt with on pp. 48/49. A considerable proportion of the potato crop is used for the production of industrial alcohol, especially for the manufacture of synthetic rubber, but potatoes are excluded from the definition of industrial crops as used in the tables.

It is particularly noteworthy that the numbers of livestock maintained are considerably more elastic than the production of crops. Especially in the period of forced collectivization, peasants often preferred to kill their animals rather than have them confiscated or bought at low prices.

Continued on page 36

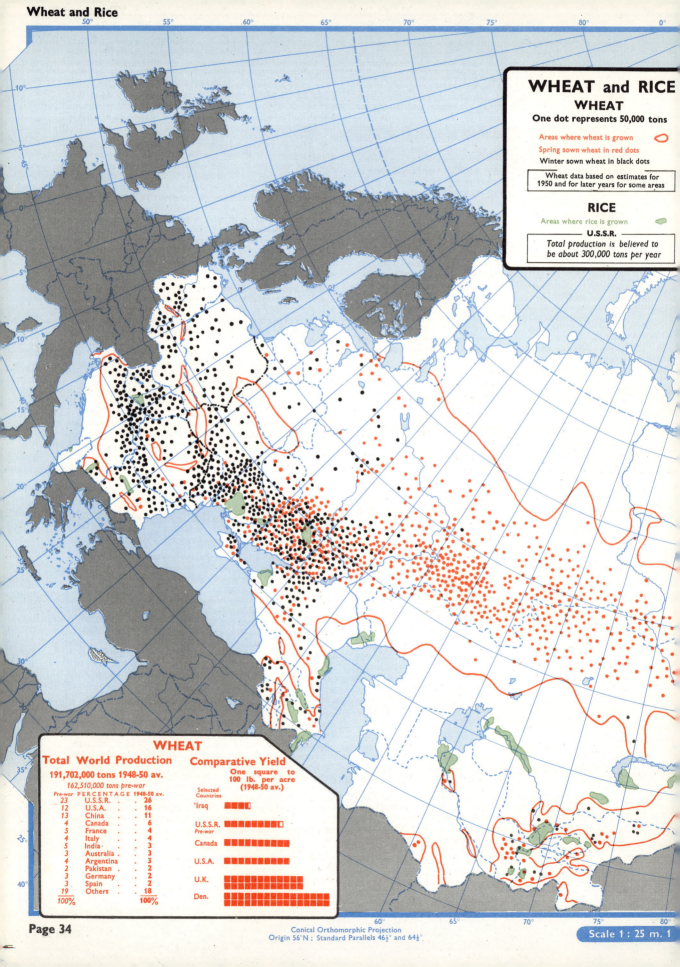

WHEAT and RICE

WHEAT

One dot represents 50,000 tons

Areas where wheat is grown

Spring sown wheat in red dots

Winter sown wheat in black dots

Wheat data based on estimates for 1950 and for later years for some areas

RICE

Areas where rice is grown

U.S.S.R.

Total production is believed to be about 300,000 tons per year

WHEAT

Total World Production

191,702,000 tons 1948-50 av.

162,510,000 tons pre-war

Pre-war PERCENTAGE		1948-50 av.
23	U.S.S.R.	26
12	U.S.A.	16
13	China	11
4	Canada	6
5	France	4
4	Italy	4
5	India	3
3	Australia	3
4	Argentina	3
2	Pakistan	3
3	Germany	2
3	Spain	2
19	Others	18
100%		100%

Comparative Yield

One square to 100 lb. per acre (1948-50 av.)

Selected Countries

'Iraq

U.S.S.R. Pre-war

Canada

U.S.A.

U.K.

Den.

Conical Orthomorphic Projection
Origin 56°N ; Standard Parallels 46½° and 64½°

Scale 1 : 25 m. 1

POTATOES
One dot represents 100,000 tons

World Production
230,099,000 tons 1948-50 av.
229,359,000 tons pre-war

Pre-war	PERCENTAGE	1948-50 av.
32	U.S.S.R.	31
14	Germany	15
16	Poland	13
7	France	6
4	U.S.A.	5
2	U.K.	4
4	Czechoslovakia	3
1	Netherlands	2
20	Others	21
100%		100%

Scale 1 : 60m approx.

No data for the Mongolian People's Republic

New Wheatlands
"Land where wheat production is to be increased or introduced."
Source : Priroda, No. 4, Moscow 1954

0   950 mls

Moscow

Tyumen'   Tomsk   Irkutsk
Ufa
Kuybyshev   Omsk
                    Barnaul
          Kokchetav
Aktyubinsk
          Karaganda

0 miles approx.

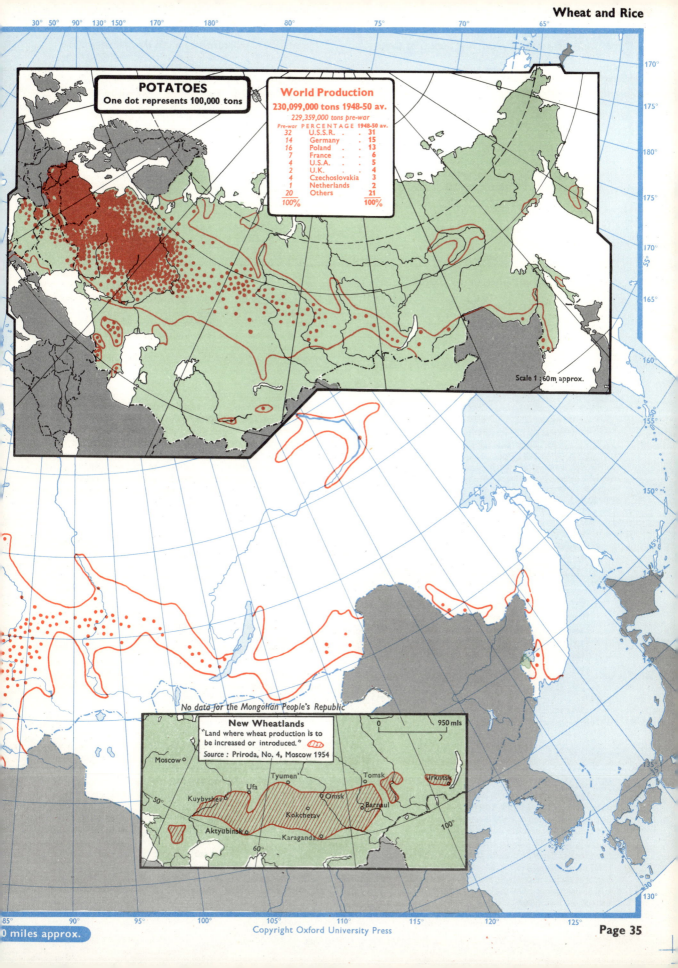

Cultivated Area, Cereals and Industrial Crops in the U.S.S.R.

(in million hectares)

	Total cultivated area	Cereals	%	Industrial crops	%
1913	105·0	94·4	89·8	4·5	4·4
1922	77·7	66·2	85·3	4·0	5·2
1928	113·0	92·2	81·6	8·6	7·6
1933	129·7	101·6	77·4	11·9	9·2
1938	136·9	102·4	75·0	11·0	8·1
1945	104·3	85·5	82·0	7·8	7·5
1950	147·4	103·6	70·2	12·4	8·4
1954	165·8	112·0	67·8	...	...
1955 Plan	169·0	107·0	63·4	13·9	8·2

Note : The areas given are within the frontiers of Russia as in each year.

Yield of Cereals in the U.S.S.R.

	Total yield (millions of tons)	Yield per Hectare (in tons)
1913	80·1	0·85
1922	56·3	0·85
1928	73·3	0·79
1933	89·8	0·88
1938	95·0	0·93
1945	67·0	0·79
1950	125·2	1·22
1954(est.)	130·0	1·16
1955 Plan	181·6	1·69

Note: From 1933 to 1953 total yield is calculated on " biological crop " estimates, on the basis of a theoretical yield per square metre multiplied by the area sown, less an arbitrary deduction of 10% for losses during harvest. Actual losses were later admitted to be nearer 20% to 25%, and the yield calculation is certainly too high. The final result of this calculation is to give a total yield estimate between 15% and 20% greater than actual yield. In 1953 and later years official Soviet figures make a larger allowance for losses.

Number of Cattle, Sheep, Goats and Pigs in the U.S.S.R.

(million head)

	Cattle	Sheep & Goats	Pigs
1916 July	60·6	121·2	20·9
1922 ,,	45·8	91·1	12·1
1928 ,,	70·5	146·7	26·0
1933 ,,	38·6	50·6	12·2
1938 ,,	63·2	81·3*	22·8*
1945 Dec.	47·0	69·4	10·4
1950 ,,	57·2	99·0	24·1
1954 Oct.	64·9	117·5†	51·0

* 1937   † Sheep only.

COLLECTIVE AND STATE FARMS

In 1950 there were about 260,000 collective farms and 4,000 state farms. The average size of a collective farm (*kolkhoz*) was said to be about 1,600 hectares (just under 4,000 acres) but individual collectives ranged from as little as 20 hectares to as much as 25,000. Since then a great many collectives have been amalgamated into much larger units, till in May 1955 the total number was given as about 91,000, averaging some 5,230 hectares each, and covering about 90% of the cultivated area of the Soviet Union. On the average, 31·5% of the land in the collectives was under cultivation, 26·9% pasture or hay, and 41·6% forest or waste.

The principle of collectivization is that the ownership of the land and buildings is vested in the collective, and though the individual has at least the theoretical right to leave, he cannot retain ownership of any of the land, even though it may previously have been his. Each *kolkhoz* is obliged to deliver a proportion of its produce to the appropriate marketing organization; the rest remains the property of the collective, some being allotted to the individual worker, some available for sale to him in return for cash which he has received in the form of a wage.

Until the spring of 1955 the *kolkhoz* could be directed what to produce, through a regional association on which each is represented, and which is also in touch with the higher administrative authorities of the State and of the Communist Party. A decree of March 1955 decentralized control and left the decisions to be taken by each *kolkhoz*. In 1955 about 30,000 organizers were sent out from the towns to direct the organization of the collectives, and in most cases they were appointed as chairmen of the *kolkhoz* committees. While some collective farms are mixed, there is also a great deal of specialization, which tends to be the rule in areas where certain crops thrive particularly well, such as sugar beet (pp. 38/39) in parts of the Ukraine, or which constitute the only areas where particular crops can be grown (e.g. cotton in the Uzbek S.S.R.). Great efforts are made, however, to avoid the possible perils of monoculture, and technical advice is made available from research institutes and elsewhere.

The *kolkhoz* usually owns only a small proportion of the machinery it employs. The larger or more specialized implements are provided through machine and tractor stations, and on an average there is one to every 20 or 30 collective farms. These stations, which date from before the main period of collectivization, were originally designed to provide machinery on a co-operative and economical basis to the smaller holdings which existed at that time. They are organized in much the same way as any other industrial undertaking of the state, and special rules govern the relations between them and the collective farms for which they work.

There is no collective principle in the State farm (*sovkhoz*). The land is owned by the State, and the work is organized on industrial lines under a manager, the workers are ordinary employees who receive a wage, and an appreciable proportion of the workers employed consists of forced labour. Since 1947 *sovkhoz* employees have been entitled to a half-hectare plot for their own use. A further difference is that the *sovkhoz* has its own machinery and does not rely on the machine and tractor stations. The average number of employees is about 300, and the average size of farm is about 30,800 hectares, of which about 5,400 hectares are cultivated; the rest is forest or waste. The proportion of cultivated land ranges from 70% in the few state farms in the black earth region to as little as 1·4% in the Far East.

For some years Soviet policy seemed to be turning against the state farm in favour of the collective, and the existing state farms became more and more specialized in the production of cereals. With the launching of the virgin lands programme in 1954 the *sovkhoz* came back into favour, and by May 1955 424 state farms had been set up in the new areas, bringing the total for the whole U.S.S.R. to 5,140.

A third type of organization which has been tried since the war, but is believed to have been abandoned, is the agricultural town—a very large farm or group of farms with the workers living in a single settlement enjoying all the normal town facilities and provided with transport to take them to their work. The object of this experiment was to assimilate as far as possible the conditions of agricultural work to those of manufacturing industry, but great difficulties of organization were met and the cost seems to have been unduly high.

AGRICULTURE IN EASTERN EUROPE

Eastern Europe shows great diversity in the character and history of its agriculture. The climate (pp. 30/31) ranges from the cold, windy winters and relatively warm summers of the Polish plains, where rainfall is distributed throughout the year, to the sub-tropical conditions of the Dalmatian coast and some inland valleys. The whole northern part is flat, and separated from the Danubian plain by the Carpathian range, while further south the country becomes more mountainous, so that there are great tracts where large-scale farming is impossible.

In much of Poland, Hungary and Romania *latifundia* farming was the rule for many centuries, the estates belonging to the royal houses, the nobles and sometimes the church. Every reformist movement included land reform in its programme as a political necessity—though sometimes, as in Romania, the reforms once carried out led to a sharp fall in efficiency and tended to create the same disadvantages as were being fought at the same time in southern countries.

In the Czech lands of Austria-Hungary the very large estate was seldom found, and in the former Turkish territories of the modern Yugoslavia, Bulgaria and Albania the characteristic organization was based on complex laws of inheritance which led to the fragmentation of holdings every time an owner died and his sons claimed the right of succession. At one time in Serbia the worst disadvantages of this constant breaking-up were avoided by the *zadruga* system under which the land was farmed in family units, but the *zadruga* was a dying institution even before the outbreak of the second World War. In contrast to the *latifundia* countries, where the problem was to break up large estates and settle new peasant owners on them, in the Balkans the commoner problem was how to group together into working units farms which were otherwise uneconomically small.

In either case the peasant owners suffered from lack of capital, and reform usually included the setting up of agrarian banks which lent them money at high rates of interest, so that after a few years of increasing peasant indebtedness the banks often replaced the former landowners. The second great disadvantage was that the shortage of equipment led peasants to keep their families on the land as additional labour—a trend intensified by the absence of alternative employment—with the result that practically every East European country except Germany and at least the western parts of Czechoslovakia suffered from agrarian over-population. Peasant discontent was one of the main reasons why, after the war, there was so little demand, in most areas, for the return of the old political systems.

The post-war agrarian revolution in each country was in two stages; (1) under widely-based " progressive " governments which still retained the land reform outlook of earlier days; and (2) the fully communist phase in which land reform was replaced by collectivization. In East Germany and Yugoslavia the communisation period began almost immediately; in Bulgaria, Romania and Albania during 1946; in Poland after the elections in February 1947; in Hungary after the final defeat of the Smallholders' Party Government in June 1947; and in Czechoslovakia only following the *coup d'état* in February 1948.

The collectivization experiments in each country were based more or less directly on the Soviet pattern. Apart from the reorganization of farms into large units, one of the most urgent tasks was to provide the collectives with machinery. The Soviet system of machine tractor stations was taken as a model. State farms were also set up in some areas, notably in Poland,

and in other areas various types of compulsory co-operative, short of full collectivization, were introduced as a temporary expedient. In every case, collectivization was resisted by the peasants, just as it had been in Russia, and agricultural production failed to expand according to plan. By 1952 an agrarian crisis affected virtually the whole of Eastern Europe.

During 1953 the policy of collectivization was slowed down or stopped in most parts of Eastern Europe. In Bulgaria, where the policy had gone furthest, there was no marked change, but in Hungary and Romania the process was put into reverse. This change in policy was apparently a joint decision by the Cominform countries, and so did not affect Yugoslavia, where in any case there had been little further collectivization since 1949.

POLAND

In Poland there are four types of collective farm in addition to a number of state farms; the differences depend on the amount of work the peasant is obliged to give and on the amount of land he is allowed to retain for his own use. State farms were set up soon after the second World War, but there was little collectivization before 1948. The 1951–55 Plan aimed at bringing 30% of the total cultivated area of the country into one form or another of socialized farming, but the slowing down of collectivization in 1953 and 1954 made it unlikely that this proportion would be reached.

State and Collective Farms in Poland

	State Farms		Collective Farms	
	Area cultivated	% of all cultivated	Area cultivated	% of all cultivated
	('000 ha)	land	('000 ha)	land
1949	1,350	9	55	0·5
1950	1,500	10	450	3·0
1951	1,650	10·8	729	4·6
1952	1,870	12	986	6·4
1953				
June	...	...	1,450	7·0
Dec.	...	...	1,500	7·2
1954 Sept.	...	...	1,800	9·0
1955 Plan	1,935	12·2	2,800	17.3

CZECHOSLOVAKIA

Before the change of government in February 1948 and the setting up of a fully communist system, there had been no attempt at collectivization, though some experiments in co-operative organization had been carried out. From 1948 to 1953 collectivization was pushed ahead rapidly, and by April 1953 44% of all arable land in the country was incorporated in collective and co-operative farms—the highest proportion of any East European country except Bulgaria. As in the neighbouring countries, the result was a fall in output and an agrarian crisis; the Government was forced to reverse its previous policy and to allow peasants to leave the collectives. By the end of 1954 the area of the collectives and co-operatives had been reduced by a third from that of early 1953.

Collective and Co-operative Farms in Czechoslovakia

	Number	Area ('000 ha)	% of all cultivated land
1952 October	8,600	2,120	44
1953			
June	8,300	1,930	40
Dec.	7,400	1,606	33
1954 Nov.	6,850	1,460	30

Continued on page 40

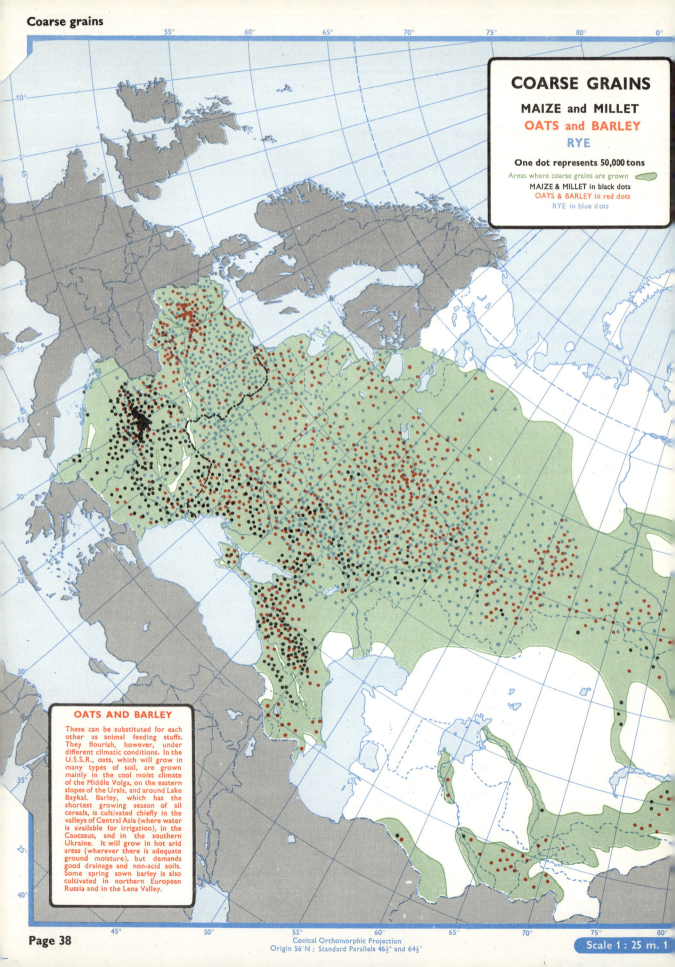

COARSE GRAINS

MAIZE and MILLET
OATS and BARLEY
RYE

One dot represents 50,000 tons

Areas where coarse grains are grown
MAIZE & MILLET in black dots
OATS & BARLEY in red dots
RYE in blue dots

OATS AND BARLEY

These can be substituted for each other as animal feeding stuffs. They flourish, however, under different climatic conditions. In the U.S.S.R., oats, which will grow in many types of soil, are grown mainly in the cool moist climate of the Middle Volga, on the eastern slopes of the Urals, and around Lake Baykal. Barley, which has the shortest growing season of all cereals, is cultivated chiefly in the valleys of Central Asia (where water is available for irrigation), in the Caucasus, and in the southern Ukraine. It will grow in hot arid areas (wherever there is adequate ground moisture), but demands good drainage and non-acid soils. Some spring sown barley is also cultivated in northern European Russia and in the Lena Valley.

Conical Orthomorphic Projection
Origin 56°N ; Standard Parallels 46½° and 64½°

Scale 1 : 25 m. 1

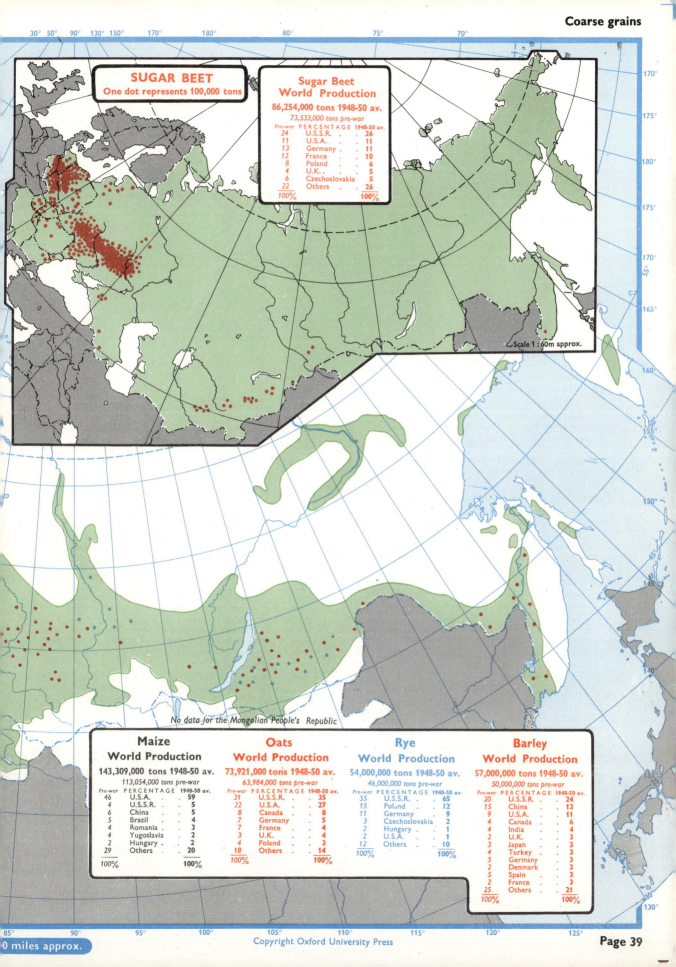

SUGAR BEET
One dot represents 100,000 tons

Sugar Beet World Production

86,254,000 tons 1948-50 av.

73,533,000 tons pre-war

Pre-war	PERCENTAGE	1948-50 av.
24	U.S.S.R.	26
11	U.S.A.	11
13	Germany	11
12	France	10
8	Poland	6
4	U.K.	5
6	Czechoslovakia	5
22	Others	26
100%		100%

Scale 1 : 60m approx.

No data for the Mongolian People's Republic

Maize World Production		**Oats** World Production		**Rye** World Production		**Barley** World Production	
143,309,000 tons 1948-50 av.		73,921,000 tons 1948-50 av.		54,000,000 tons 1948-50 av.		57,000,000 tons 1948-50 av.	
113,054,000 tons pre-war		63,984,000 tons pre-war		46,000,000 tons pre-war		50,000,000 tons pre-war	
Pre-war	1948-50 av.	Pre-war	1948-50 av.	Pre-war	1948-50 av.	Pre-war	1948-50 av.
46 U.S.A. 59		31 U.S.S.R. 35		55 U.S.S.R. 65		20 U.S.S.R. 24	
4 U.S.S.R. 5		22 U.S.A. 27		15 Poland 12		15 China 13	
6 China 5		8 Canada 8		11 Germany 9		9 U.S.A. 11	
5 Brazil 4		7 Germany 5		3 Czechoslovakia 2		4 Canada 6	
4 Romania 3		7 France 4		2 Hungary 1		4 India 4	
4 Yugoslavia 2		3 U.K. 4		2 U.S.A. 1		2 U.K. 3	
2 Hungary 2		4 Poland 3		12 Others 10		3 Japan 3	
29 Others 20		18 Others 14				4 Turkey 3	
						5 Germany 3	
						2 Denmark 3	
						5 Spain 3	
						2 France 3	
						25 Others 21	
100% 100%		100% 100%		100% 100%		100% 100%	

0 miles approx.

EAST GERMANY

Little attempt was made at collectivization before 1950, and at the end of 1952 there were still only 1,305 collective farms comprising no more than 3% of the cultivated area of the country. During the first half of 1953 there was a vigorous burst of collectivization, but in the second half of the year the area collectivized fell sharply. A new period of collectivization began in the summer of 1954; at the end of the year 14% of the land was collectively owned and managed.

Collective Farms in East Germany

	Area cultivated ('000 ha)	% of all cultivated land
1952 Dec.	160	3
1953		
March	475	7
June	798	12
Dec.	715	11
1954 Dec.	921	14

HUNGARY

In 1948 the total number of agricultural holdings in Hungary was given as 1,092,000, of which 423,000 belonged to small-holders, 587,000 to medium farmers, and 82,900 to *kulaks*. By 1952 the *kulaks*, though not eliminated, owned only 5% of the land. Medium farmers still held 36% and smallholders 25%; the remaining 34% (25% of arable land) was collectivized.

A decree of July 1953 permitted peasants to leave the collectives, and a sharp fall in the number of collectives, the area collectively owned, and the number of households incorporated, was the immediate result. Unlike Poland and East Germany, in Hungary the policy of disbanding the collective farms continued well into 1954.

Collective Farms in Hungary

End year	No. of collectives	Households incorporated	Area ('000 ha)	% of all cultivated land
1948	380	10,000	28	0.5
1949	1,520	46,000	148	2.6
1950	2,620	140,000	285	5.0
1951	4,652	350,000	1,030	18.2
1952	5,315	413,000	1,400	24.6
1953	4,677	263,700	1,120	20.0
1954 April	4,500	...	978	18.0

BULGARIA

In Bulgaria state and collective farms together covered 63% of the total arable area by the end of 1952. No reversal of policy occurred in 1953 similar to that in the other East European countries. The average arable area per collective was given in 1953 as 2,550 hectares.

Collective Farms in Bulgaria

	Number	Area cultivated ('00 ha)	% of all cultivated land
1947	549	...	3.7
1950	2,587	1,578	43
1953	2,747	5,000	61

ROMANIA

Two distinct types of socialized agriculture have been introduced in Romania. One is the collective farm on the Soviet model, though a great deal smaller; in 1954 the average size was only 415 hectares. The second is the joint tilling association, in which the peasants retain limited rights of ownership in the land, but combine for crop planning, the use of machinery and marketing. The setting up of new collectives virtually ceased after the spring of 1953, but between then and the end of 1954 a further 700 joint tilling associations were established.

Collectives and Joint Tilling Associations in Romania

	Number		Area ('000 ha)		% of all cultivated land	
	Collectives	J.T.A.	Collectives	J.T.A.	Collectives	J.T.A.
1952 Dec.	1,800	1,800	...	...	...	...
1953 June	2,000	2,000	982		12	
1954						
June	2,048	2,344	850	272	10	3.2
Dec.	...	2,700	...	...	...	...

The degree of socialization varies widely from one part of the country to another. In the summer of 1954, collectives and joint tilling associations covered 29.6% of the cultivated land in Constanţa province, 4.9% in Oradea, 4.1% in Iaşi, and only 3% in Craiova.

In June 1954 there were 220 machine and tractor stations.

ALBANIA

As late as April 1953 the number of collectives was stated to be only 115, incorporating 6,291 families and covering 21,776 hectares, or 6% of the farmland of the country. No later information is available.

YUGOSLAVIA

The "peasant working co-operative" of post-war Yugoslavia is hardly a collective farm in the strict Russian sense, though nearer to it than to the old peasant co-operative. It was first introduced in the Vojvodina in 1945–46, and efforts were made to extend the co-operative principle to the whole country. This proved impossible, however, partly because of peasant resistance, and partly because large-scale farming cannot be carried out in the mountain areas.

After the break with the Cominform in June 1948 collectivization was greatly intensified; but after the end of 1949 there was little further change until 1952 when the policy was put into reverse. By the beginning of 1955 the total number of collectives had been reduced to under 1,000.

Collectivization in Yugoslavia

	Number of Co-operatives	Households incorporated	Area ('000 ha)
1945 Dec.	31	...	...
1946 ,,	545	25,062	122
1947 ,,	779	40,590	214
1948 ,,	1,318	60,158	370
1949			
,, Mar.	3,046	166,287	880
,, June	4,535	226,087	1,260
,, Sept.	5,246	255,733	1,420
,, Dec.	6,625	340,739	1,870
1950 Dec.	6,968	418,659	2,250
1951 June	6,994	429,784	2,625
1952 June	6,908	...	...
1954 Dec.	990	...	890

AREAS AND PRODUCTION IN THE U.S.S.R.

CEREAL CROPS

One of the most striking economic developments in the Soviet Union has been the extension of the area in which cereals, and especially wheat, can be grown. From the original home of wheat in the black earth region of Central Russia and the Ukraine, its cultivation has spread north to the latitude of Leningrad in European Russia, south into large areas of Kazakhstan, and east along the narrow temperate belt of Siberia. Wheat is also grown in sheltered districts and in some river valleys to the north of the main Siberian belt. A hardy variety has been bred which ripens in 100 days from sowing, and is grown in such difficult areas as that of Yakutsk, where the short summer is fairly hot and sunny even though the frost-bound winter lasts for nearly nine months of the year.

Cultivation of coarse grains tends to vary with the number of livestock kept, but for some years these grains have tended to form a smaller proportion of all cereal crops grown. Until 1954 only barley was being grown over a greater area than in pre-war days.

Starting with the 1955 season, the Soviet authorities launched an ambitious campaign to raise the area under maize from 3½ million hectares in 1954 to 28 million hectares by 1960; for 1955 alone, 16 million hectares were planned. The reasons for this campaign were said to be that higher yields to the hectare could be got from maize than from other crops, that the leaves and stems make more and better silage, and that some of the waste can be turned to industrial use, especially for the manufacture of cellulose. The extra grain obtained was to be used chiefly as stock-feed, and especially for pigs. A common disadvantage of maize-growing on a large scale is that it demands more labour in harvesting than other grains, but it was claimed that this difficulty had been overcome by planting in small squares which could be harvested by a specially designed machine.

The areas chiefly affected by the maize programme are in the Ukraine and Central Asia, but some is to be grown as far north as the Komi A.S.S.R. and the southern part of Arkhangel'sk Oblast. Practically none is to be planted east of the Urals. The land needed for maize is to be found at the expense of the other coarse grains, and also of sugar beet, a policy which will intensify the shortage of sugar already recorded in 1954 and 1955.

Areas under Cereal Crops in the U.S.S.R.

	1931 %†	1952 ('000 ha)
Wheat	38·5	48,000
Rye	27·9	29,100
Oats	18·3	15,300
Barley	7·2	9,070
Maize	4·2	3,660
Others*	3·6	2,380

* Mainly millets, sorghums and buckwheat; rice excluded.
† % of total cereal area.

Output of Five Principal Grains in the U.S.S.R.
(Actual Yield Estimates)
(in thousand metric tons)

	1931	1935–39 av.	1950	1952
Wheat	20,500	33,000	29,500	35,000
Rye	22,000	22,000	23,000	23,000
Oats	11,100	18,000	11,000	11,000
Barley	5,300	9,000	7,000	8,000
Maize	4,800	4,000	3,500	3,500

ROOT CROPS IN THE U.S.S.R.
POTATOES

Apart from their importance as a food for human beings and animals, potatoes are extensively grown in the Soviet Union as an industrial crop. They are used for the production of alcohol for all purposes, but especially for the synthetic rubber industry, which tends to be concentrated in or near the main potato-growing areas.

From 1934 to 1938 the average area under potatoes was about 7·0 million hectares, with an average annual production of about 74 million tons. The area has been considerably enlarged since the war by the incorporation into the Soviet Union of potato-growing districts in the Baltic Republics, Poland and East Prussia, and by the extension east and north of potato cultivation. In 1950 the area under potatoes grown for food was given as 8·3 million hectares, and output 87 million tons. Under the fifth Five-Year Plan, by 1955 the area was to reach 10 million hectares.

SUGAR BEET

Almost all the sugar consumed in the U.S.S.R. is derived from sugar beet. Cane cultivation is confined to a few small areas in the south where the climate approaches the sub-tropical.

Until about 1936 sugar was imported to supplement inadequate home production. Between 1933 and 1938 the area under sugar beet was reduced to make room for other crops, but technical improvements increased the yield so much that in 1938 it was even possible to export a little. During the war some of the best beet growing lands were lost, and output of sugar fell heavily, but the 1938 level of production was regained by 1950.

Under the fifth Five-Year Plan the area of sugar beet was planned to increase to 1,550,000 hectares by 1955. This figure was exceeded in 1953. In 1954 imports of sugar were resumed, and in the first half of 1955 orders for 700,000 tons were placed abroad. The probable reason for this return to imports in spite of the increased area under sugar beet is that the crop is now of greater industrial importance than before. Substantial quantities are believed to be used for producing alcohol for the manufacture of synthetic rubber and plastics. Consumption of sugar by the public has also increased.

Sugar Beet: Area and Production in the U.S.S.R.

	Area (in '000 hectares)	Production (in '000 tons)
1931	1,395	14,800
1938	1,182	24,400
1948	1,149	19,800
1950	1,154	24,000
1951	1,198	27,000
1952	1,198	27,500
1953	1,570	...

VEGETABLE OILS IN THE U.S.S.R.

Sunflower seed is the largest source of vegetable oils in the Soviet Union, followed by cottonseed and linseed. Hemp, rape and sesame have been cultivated for long periods, but on a much smaller scale. In the last three decades there has been an expansion of soya cultivation, especially in the Far East, and of groundnuts in a few areas. The tung tree has also been introduced, and is economically important around Batumi.

OILSEEDS: ACREAGES AND PRODUCTION
(areas in '000 hectares; production in '000 metric tons)

SUNFLOWER SEED

	1938	1949	1950	1951	1952
Area	3,140	3,440	3,550	3,640	3,700
Production	2,027	2,300	2,400	2,200	2,400

COTTON SEED

	1938–9	1948–9	1950–1	1951–2	1952–3
Production	1,660	1,200	1,870	1,910	1,950

LINSEED

It would appear that in recent years the U.S.S.R. has become the world's second largest producer of linseed, coming after the U.S.A. but ahead of Argentina.

	1938	1949	1950	1951	1952
Production	800	650	650	670	630

Continued on page 44

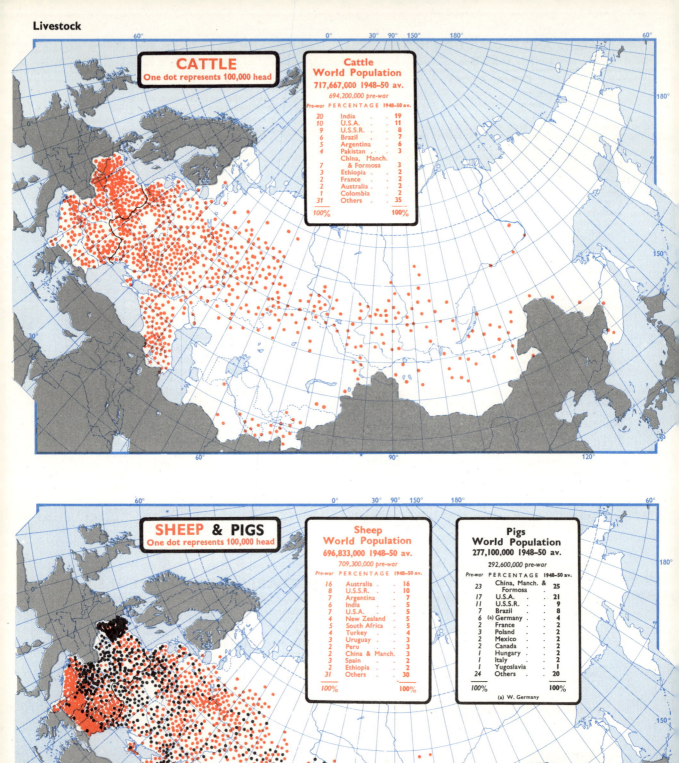

CATTLE
One dot represents 100,000 head

Cattle
World Population
717,667,000 1948–50 av.
694,200,000 pre-war

Pre-war	PERCENTAGE	1948–50 av.
20	India	19
10	U.S.A.	11
9	U.S.S.R.	8
6	Brazil	7
5	Argentina	6
4	Pakistan	3
7	China, Manch. & Formosa	3
3	Ethiopia	2
2	France	2
2	Australia	2
1	Colombia	2
31	Others	35
100%		100%

SHEEP & PIGS
One dot represents 100,000 head

Sheep
World Population
696,833,000 1948–50 av.
709,300,000 pre-war

Pre-war	PERCENTAGE	1948–50 av.
16	Australia	16
8	U.S.S.R.	10
7	Argentina	7
6	India	5
7	U.S.A.	5
4	New Zealand	5
5	South Africa	5
4	Turkey	4
3	Uruguay	3
2	Peru	3
2	China & Manch.	3
3	Spain	2
2	Ethiopia	2
31	Others	30
100%		100%

Pigs
World Population
277,100,000 1948–50 av.
292,600,000 pre-war

Pre-war	PERCENTAGE	1948–50 av.
23	China, Manch. & Formosa	25
17	U.S.A.	21
11	U.S.S.R.	9
7	Brazil	8
6	(a) Germany	4
3	France	2
3	Poland	2
2	Mexico	2
2	Canada	2
2	Hungary	2
1	Italy	2
1	Yugoslavia	1
24	Others	20
100%		100%

(a) W. Germany

Conical Orthomorphic Projection
Origin 56°N ; Standard Parallels 46½° and 64½°

Approximate scale

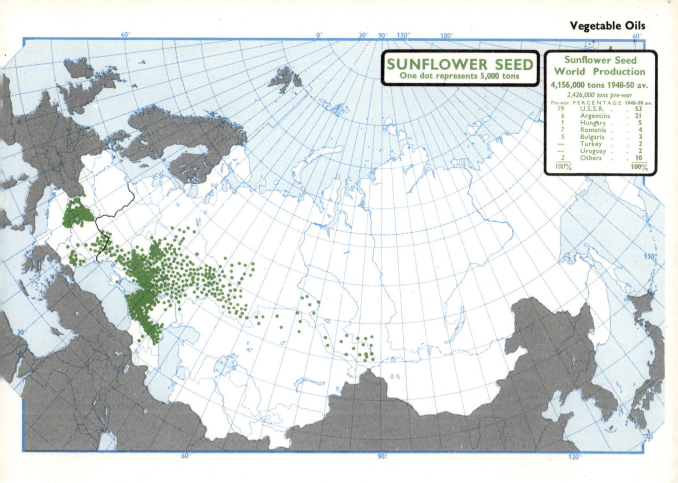

SUNFLOWER SEED
One dot represents 5,000 tons

Sunflower Seed World Production
4,156,000 tons 1948-50 av.
2,426,000 tons pre-war

Pre-war	PERCENTAGE	1948-50 av.
79	U.S.S.R.	53
6	Argentina	21
1	Hungary	5
7	Romania	4
5	Bulgaria	3
—	Turkey	2
—	Uruguay	2
2	Others	10
100%		100%

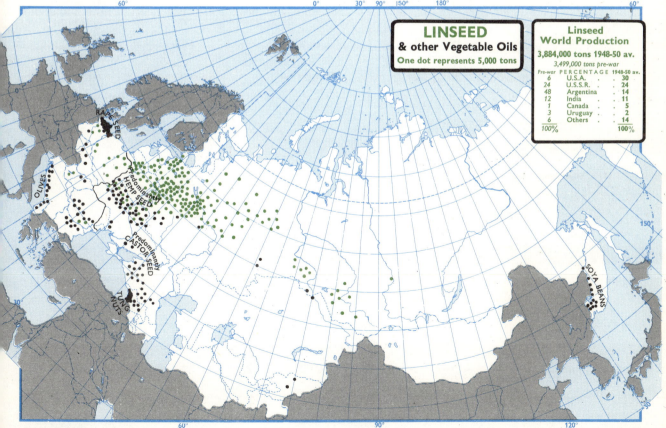

LINSEED
& other Vegetable Oils
One dot represents 5,000 tons

Linseed World Production
3,884,000 tons 1948-50 av.
3,499,000 tons pre-war

Pre-war	PERCENTAGE	1948-50 av.
6	U.S.A.	30
24	U.S.S.R.	24
48	Argentina	14
12	India	11
1	Canada	5
3	Uruguay	2
6	Others	14
100%		100%

RAPE SEED

OLIVES

Predominantly HEMP SEED

Predominantly CASTOR SEED

TUNG NUTS

SOYA BEANS

OTHER OILSEEDS

Information about the production of other oilseeds is scanty. The latest available data, mostly based on estimates, may be briefly summarized.

Soya Beans	Production 1952 about 165,000 tons.
Rape Seed	Pre-war area 69,000 hectares.
Hemp Seed	Production about 250,000 tons a year.
Castor Seed	Production 100,000—120,000 tons a year.

AREAS AND PRODUCTION IN EASTERN EUROPE

The following tables show the area and production of the more important crops for representative pre-war years and for years since 1948 where the figures are available. All areas are in thousand hectares, and production in thousand metric tons. For years before 1939, the figures relate to the pre-war frontiers.

POLAND

		1938	1948	1950	1955 Plan
Wheat	Area	1,760	1,382	1,495	
	Production	2,048	1,621	1,854	2,280
Rye	Area	5,790	5,150	5,280	...
	Production	6,445	6,300	6,500	6,900
Oats	Area	2,280	1,756	1,720	...
	Production	2,500	2,403	2,125	2,800
Barley	Area	1,200	860	840	...
	Production	1,368	1,010	1,077	1,960
Potatoes	Area	2,750*	2,530†	2,650	...
	Production	38,014*	28,828†	36,835	39,750
Sugar beet	Area	224*	252†	275	...
	Production	5,962*	5,229†	6,377	7,800

* 1934–38 average. † 1948–49 average.

EAST GERMANY
(Estimated figures)

		1948	1950	1952
Wheat	Area	447	505	...
	Production	926	815	...
Rye	Area	1,290	1,290	...
	Production	1,790	2,950	...
Oats	Area	561	530	...
	Production	780	1,140	...
Barley	Area	250	260	...
	Production	420	...	...
Potatoes	Area	810	810	...
	Production	9,932	14,645	...
Sugar beet	Area	205	210	215
	Production	3,545	4,871	5,550

CZECHOSLOVAKIA

		1938	1950	1952
Wheat	Area	900	800	...
	Production	1,790	1,540	...
Rye	Area	980	473†	...
	Production	1,676	1,223†	...
Oats	Area	740	...	...
	Production	1,333	972†	...
Barley	Area	660	595†	...
	Production	1,210	1,034†	...
Maize	Area	93	...	...
	Production	217	239†	...
Potatoes	Area	715*	575†	...
	Production	9,635*	6,780†	...
Sugar beet	Area	163*	197†	210
	Production	4,664*	4,954†	2,045

* 1934–38 average. † 1948–50 average.

HUNGARY

		1938	1950	1952
Wheat	Area	1,620	1,460	...
	Production	2,690	2,040	...
Rye	Area	620	650†	...
	Production	727	775†	...
Oats	Area	228	205†	...
	Production	300	265†	...
Barley	Area	458	...	...
	Production	645	674†	...
Maize	Area	1,180	...	...
	Production	2,545	...	...
Potatoes	Area	290*	...	...
	Production	2,133*	...	...
Sugar beet	Area	46*	106†	115
	Production	960*	1,283†	1,375

ROMANIA

		1938	1948	1950	1952
Wheat	Area	3,830	2,425	...	...
	Production	4,820	2,540	...	...
Rye	Area	455	138	...	...
	Production	464	66	...	...
Oats	Area	675	512	...	...
	Production	487	265	...	...
Barley	Area	1,282	680	...	...
	Production	850	280	...	...
Maize	Area	5,050	4,320†	...	...
	Production	5,342	5,280†	...	...
Sugar beet	Area	30*	...	66§	93
	Production	398*	...	823§	775

* 1934–38 average. † 1947. § 1948–50 average.

BULGARIA

		1938	1948	1950	1952
Wheat	Area	1,400	1,260†	...	...
	Production	2,150	1,525	...	...
Rye	Area	182	162†	...	...
	Production	200	95†	...	...
Oats	Area	135	156†	...	...
	Production	117	77†	...	...
Barley	Area	220	191†	...	...
	Production	344	131†	...	...
Maize	Area	685	760	...	...
	Production	776	895	...	...
Sugar beet	Area	1*	...	37§	37
	Production	115*	...	385§	325

* 1934–38 average. † 1947. § 1948–50 average.

YUGOSLAVIA

		1938	1948	1950	1952
Wheat	Area	2,130	1,880	1,790	1,830
	Production	2,750	2,525	1,825	1,675
Rye	Area	256	250	257	295
	Production	227	251	218	225
Oats	Area	355	347	390	335
	Production	323	344	195	215
Barley	Area	415	318	325	315
	Production	410	353	266	257
Maize	Area	2,600	2,370	2,200	2,280
	Production	4,712	4,071	2,085	1,470
Potatoes	Area	270*	...	222†	237
	Production	1,498*	...	1,516†	1,128
Sugar beet	Area	27*	...	86†	76
	Production	509*	...	1,148†	512

* 1935–38 average. † 1948–50 average.

ALBANIA

No reliable information is available on areas or production. In both 1938 and 1947 there were approximately 100,000 hectares under maize (the chief crop), and production was about 140,000 tons. Total production of food grains was planned to reach 284,000 tons in 1955. The 1950–55 plan aimed at a sixteen-fold increase in output of sugar beet, to 100,000 tons in the latter year.

VEGETABLE OILS IN EASTERN EUROPE

In Eastern Europe linseed is more important than sunflower, which is extensively grown only in Yugoslavia; Poland is the largest producer of linseed in Europe, and the seventh largest in the world. Groundnuts are grown on a very small scale in Bulgaria and Yugoslavia.

OILSEEDS: AREAS AND PRODUCTION IN EASTERN EUROPE

(areas in '000 hectares; production in '000 metric tons)

SUNFLOWER SEED

		1938	1949	1950	1952	1952
Bulgaria	Area	190	...	...	...	...
	Production	116	...	...	...	...
Hungary	Area	7	290	...	...	...
	Production	31	199	...	...	...
Romania	Area	200	500	...	...	...
	Production	195	300	...	...	...
Yugoslavia	Area	20	129	114	102	93
	Production	25	152	68	93	50

COTTON SEED

		1938–9	1948–9	1950–1	1951–2	1952–3
Bulgaria	Production	13	13	14	14	14
Romania	Production	2	9	11	16	16

LINSEED

		1938	1949	1950	1951	1952
E. Germany	Production	14	18	...	...	...
Poland	Production	69	63	69	83	...

OTHER OILSEEDS

Information about the production of other oilseeds is scanty. The latest available data, mostly based on estimates, may be briefly summarized.

SOYA BEANS

Yugoslavia: Production 1952 about 2,000 tons.

RAPE SEED

Poland:	Production 1952 about 85,000 tons.
Yugoslavia:	„ „ „ 5,000 „
E. Germany:	„ 1949 „ 110,000 „
Czechoslovakia:	„ 1948 „ 4,000 „
Hungary:	Pre-war production about 11,000 tons a year.
Bulgaria:	„ „ „ 21,000 „ „
Romania:	„ „ „ 52,000 „ „

FORESTRY AND TIMBER IN THE U.S.S.R. AND EASTERN EUROPE
(pp. 46/47)

About half of the entire area of the U.S.S.R. is covered by forest, and there are extensive forests in the Carpathians and other parts of Eastern Europe. Taken together, these vast wooded regions form rather more than a third of the forest area of the whole world.

Coniferous forest greatly exceeds deciduous in area. The latter is found only in parts of Eastern Europe (where oak and beech are the commonest trees), in a belt across southern Russia, in a narrow band between the Siberian plain and the mountains of Central Asia, and in parts of the Far East. Oak, beech, sycamore, elm and chestnut are the most widespread of the broadleaved trees; the distribution of ash and walnut is more limited, though limes extend well to the north into the coniferous areas.

Of the conifers, the European larch (*Larix decidua*) is found furthest north, but in eastern Siberia it yields place to *Larix sibirica*. A variant of the white fir, *Abies sibirica*, is widespread, as are the Scots pine (*Pinus sylvestris*), *Pinus cembra* and, in the Far East, *Pinus koraiensis*. The Norway spruce (*Picea abies*) covers large areas in western Russia and in the Carpathians. In the mountain areas of Central Asia, south of the deciduous belt, *Abies sibirica* is found again, and there are huge forests of *Picea schrenkiana* at heights between 4,500 and 10,000 feet. Mediterranean-type forest occurs in the lower Caucasus and on the Black Sea littoral.

THE TIMBER INDUSTRY

Timber is one of the major exports of the Soviet Union, though its full development has been hampered by lack of machinery and by transport difficulties. Some of the sawmilling centres are very large. For instance, Arkhangel'sk exists almost solely as a sawmilling and timber-exporting town, and others in northern Russia, such as Igarka and Kotlas, depend on timber for the whole of their activity. The total forest area of the U.S.S.R. is officially estimated at 743 million hectares (2,869,000 square miles), though only a small proportion of this is suitable for exploitation.

WOODPULP AND PAPER

The enormous supplies of softwood available have made the U.S.S.R. and Poland into major producers of woodpulp and paper, in which both have a considerable export trade. In 1952 the U.S.S.R. produced about 2 million tons of woodpulp, of which a little over half was said to be chemical and the rest mechanical. Consumption of paper and board was rather under 2 million tons, and of newsprint about 400,000 tons. In recent years wood distillation has become an increasingly important part of the chemicals industry.

Production of Sawn Softwood, Woodpulp and Paper in the U.S.S.R.

	Sawn softwood (million cubic metres)	Woodpulp ('000 tons)	Paper ('000 tons)
1938	33	...	831*
1950	36	...	1,194
1954	50	2,000†	1,763

* 1937 † 1952 Slightly over half was chemical pulp.

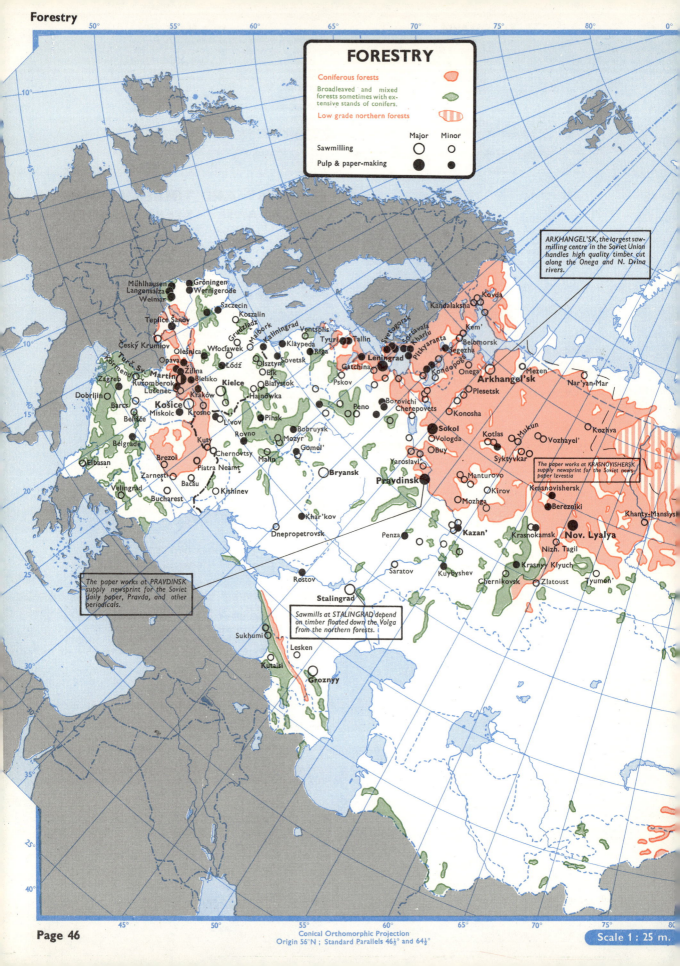

FORESTRY

Coniferous forests

Broadleaved and mixed forests sometimes with extensive stands of conifers.

Low grade northern forests

Major   Minor

Sawmilling

Pulp & paper-making

ARKHANGEL'SK, the largest saw-milling centre in the Soviet Union handles high quality timber cut along the Onega and N. Dvina rivers.

The paper works at KRASNOYISHERSK supply newsprint for the Soviet news-paper Izvestia

The paper works at PRAVDINSK supply newsprint for the Soviet daily paper, Pravda, and other periodicals.

Sawmills at STALINGRAD depend on timber floated down the Volga from the northern forests.

Mühlhausen
Langensalza
Weimar
Gröningen
Wernigerode
Szczecin
Koszalin
Teplice Sanov
Grudziadz
Milbork
Kaliningrad
Ventspils
Český Krumlov
Włocławek
Klaypeda
Tyuri
Tallin
Svetogorsk
Sortavala
Kovda
Kandalaksha
Kem'
Belomorsk
Segezha
Olesnica
Olsztyn
Sovetsk
Riga
Kharlu
Pitkyaranta
Onega
Mezen
Opava
Łódź
Leningrad
Kondopoga
Arkhangel'sk
Nar'yan-Mar
Martin
Žilina
Bielsko
Gatchina
Plesetsk
Zagreb
Ruzomberok
Kielce
Białystok
Pskov
Kozhva
Dobrljin
Kraków
Hainowka
Borovichi
Konosha
Barcs
Košice
Miskolc
Krosno
Peno
Cherepovets
Kotlas
Mukun
Vozhayel'
Belišče
L'vov
Peno
Sokol
Syktyvkar
Belgrade
Kuty
Rovno
Malin
Vologda
Buy
Mozhga
Krasnovishersk
Elbasan
Brezoi
Chernovtsy
Gomel'
Yaroslavl'
Manturovo
Berezniki
Velingrad
Zarnesti
Piatra Neamt
Bryansk
Kirov
Khanty-Mansiysk
Bacău
Kishinev
Pravdinsk
Nov. Lyalya
Bucharest
Khar'kov
Krasnokamsk
Nizh. Tagil
Dnepropetrovsk
Penza
Kazan'
Mozhga
Krasnyy Klyuch
Sukhumi
Lesken
Saratov
Kuybyshev
Chernikovsk
Zlatoust
Tyumen'
Kutaisi
Rostov
Groznyy
Stalingrad

Conical Orthomorphic Projection
Origin 56°N ; Standard Parallels 46½° and 64½°

Scale 1 : 25 m.

Ust'-Kamchatsk

Igarka

Yakutsk

Aleksandrovsk
Poronaysk
Dolinsk
Soviet Harbour
Korsakov
Yuzh.-Sakhalinsk

Peleduy

Mogochin
Yeniseysk
Birakan
Khabarovsk
Kansk
Bratsk
Blagoveshchensk
Krasnoyarsk
Tayshet
Zima
Iman
Novosibirsk
Chita
Lesozavodsk
Barnaul
Irkutsk
Ulan-Ude
Khapcheranga
Vladivostok
ipalatinsk

Approx. limit of 'Permafrost'

TEXTILES, FIBRES AND RUBBER

The development of the Soviet textile and rubber industries is closely connected with general agricultural policy. Cotton in particular is one of the most important of the " new crops ", and much energy has gone into the opening up of new areas in the south where it can be grown on an irrigated or non-irrigated basis. Also among the chief new crops are the two rubber-bearing plants, *kok sagyz* and *tau sagyz*, while in many areas of European Russia large acreages of potatoes are planted each year to provide the alcohol from which synthetic rubber is manufactured; the two together have made the U.S.S.R. less dependent of outside supplies of natural rubber. Flax remains a principal crop in several areas. The Soviet Union is still not self-sufficient in wool, although many areas are suitable for sheep-rearing. Constant efforts are being made to increase the wool supply.

TEXTILE FIBRES
FLAX

Flax has been extensively cultivated in Russia for hundreds of years. It is suited to the relatively wet and cool areas of central and north central Russia and Poland, and thrives both on podzolic and alluvial soils. The chief disadvantage of flax as a textile fibre is that it involves a great deal of labour in gathering and treatment before it can be converted into thread, and efforts to mechanize the various processes have not been entirely successful. It is widely grown as a source of linseed, though the best seed varieties do not produce the best fibre, and the timing and methods of harvesting are such that it is difficult to get both crops from the same plant.

Before 1913 there were about 1 million hectares under fibre flax, mostly west of the Urals, though some experimental growing had been started in Siberia. The yield in those days was about 330 kilos of fibre per hectare, a figure which was never reached again till 1948. (In Northern Ireland yield in 1948–1952 averaged 680 kg per hectare). In the years immediately before the second World War the area planted was twice that in 1913, but yields were down to 265 kg per hectare. The area in 1947 was little more than half that of 1938, but a steady increase followed, and yields also improved considerably. In 1952 the Soviet Union accounted for 84% of the world's area under fibre flax and for about 77% of world production of fibre.

Area and Production of Flax in the U.S.S.R.

	('000 acres)	('000 ha)	Fibre ('000 tons)
1913	2,470	1,000	330
1934–48 av.	5,073	2,054	545
1947	2,803	1,135	264
1950	4,199	1,700	610
1952	4,742	1,920	710

COTTON

Some cotton has been grown in Russia at least since the beginning of the 19th century, when it was cultivated on a small scale in the more accessible parts of Central Asia. As late as 1913, however, over half the country's needs were still being met by imports. After the Revolution no progress was made in expanding production for several years, and serious efforts began only with the first Five-Year Plan (1928–32).

Successful cotton-growing depends on a combination of climatic and physical conditions: abundant water, plentiful hot sunshine during the growing and ripening season, and freedom from intense winter cold. In the whole Soviet Union, such conditions are found only in Uzbekistan and in river valleys elsewhere in Central Asia. Development under the first two Five Year Plans went on several parallel lines. The most suitable areas were the scene of great irrigation works, harnessing the waters of the Amu Darya and Syr' Darya rivers and trying to reproduce " nilotic " conditions in areas where the winters are too cold to be ideal. Elsewhere, cotton was introduced in the southern Ukraine, Crimea and the North Caucasian valleys, which all suffered from the disadvantage of insufficient water.

Nevertheless, despite the special difficulties encountered, production rose steadily and imports were reduced from over 100,000 tons in 1929 to less than half that amount during 1934–36; in 1937 there was a net exportable surplus of some 40,000 tons, although the area planted was rather less than in several earlier years.

During the war, areas were reduced in order to make use of the land for grain crops, and the peak production of 1940 does not seem to have been equalled till 1949. Under the fourth Five-Year Plan the main concentration was less on increasing the area planted than on raising the yield.

Area and Production of Cotton in the U.S.S.R.

	('000 acres)	('000 ha)	Unginned weight ('000 tons)
1913	1,699	688	740
1934–38 av.	5,000	2,024	2,600
1946	3,220	1,300	1,600
1950 plan	4,200	1,700	3,100
1953	6,175	2,500	4,300

Note: The best estimates of production are given in terms of unginned cotton—i.e., including the seeds, husks, small twigs, etc. A rough conversion to fibre weight may be made by dividing the unginned weight by 4.

WOOL

A high proportion of Russian sheep rearing has been traditionally for meat production rather than wool. In most of the country the breeds most commonly favoured approximate to the English hill rather than the down varieties, so that the clip is both small in quantity and of poor quality. Cross-bred sheep were introduced during the 18th century, and in the drier and hotter southern areas the fat-tailed types prevail, but apart from a few specialized zones such as that east of Moscow there is little breeding of the heavy-fleeced varieties. The merino is still bred in places, and in central Asia the *karakul* is of great economic importance, though the value of the latter is in its skin rather than its fleece.

For most years both before and since the Revolution sheep and goats have been grouped together in the published figures of livestock. The numbers rose sharply after about 1921, and in 1928 reached a peak which has not been equalled since. After that year, the forced collectivization caused a more marked reduction in livestock than in other branches of agriculture. By the middle of 1934 the number of sheep and goats was barely a third of the 1928 figure. Direct comparisons are, however, made more difficult by the fact that from 1934 onwards the figures available are for 1st January (before the breeding season) whereas for earlier years they are for 1st July (after breeding is over). (Livestock map, p. 42).

Production of Raw Wool in the U.S.S.R.

	(in metric tons)
1933	54,900
1938	137,600
1941 Plan	189,400
1953	230,000
1955 Plan	450,000

HEMP

Of the coarser fibres, only hemp is extensively cultivated; there are two varieties, " northern " and " southern " and the former is the more important in the U.S.S.R. As with flax, a proportion of the hemp cultivated is grown for its seed rather than for fibre.

Pre-war estimates show that there were about 688,000 hectares under fibre hemp, and annual production of fibre was over 200,000 tons. After the war both the area planted and the tonnage produced were smaller; an estimate for 1952 mentions 370,000 hectares and 145,000 tons of fibre, equivalent to about 40% of world output of " true " hemp.

MINOR VEGETABLE FIBRES

A number of less important fibres are produced from a variety of plants, of which most have either been introduced from outside the U.S.S.R. or have been developed from relative scarcity. Among the most valuable of these is *kenaf*, a hard fibre of Indian origin used chiefly in making sacking and other coarse cloths. *Kendyr'* is another hard fibre, though less coarse than *kenaf* or the jute which the latter resembles. It was originally found in Russian Central Asia, and is used in making durable cloths and rope. *Ramie* is a variety of nettle introduced into Russia from China during the nineteenth century. It yields a very fine, soft, tough fibre which can easily be mistaken for silk when made up into yarn or cloth.

SILK

In some parts of Southern Russia and Central Asia silk cultivation has been established for many centuries, and formed the basis of one of the chief traditional industries of the region. It is still economically important in the Caucasus and Black Sea areas and in valleys of the Central Asian mountain system. Some of the cloths produced are of very high quality.

RUBBER

There is no part of the Soviet Union where climatic conditions are suitable for the growing of *Hevea brasiliensis*, the world's principal source of natural rubber, and until recent years virtually all the rubber consumed was imported. A synthetic rubber industry grew up in the decade before the second World War, using (in the main) industrial alcohol distilled from fermented potatoes. In some areas potatoes (pp. 34/35) are specially grown for this purpose rather than for human consumption, and the synthetic rubber industry tends to be located in or near such areas.

Considerable quantities of natural rubber are, however, produced from several rubber-bearing plants. About 1930 it was noticed that a type of dandelion, *Taraxacum kok sagyz*, found in the Tien Shan mountains yielded a latex-like substance when bruised. It was discovered later that a related plant, *tau sagyz*, had similar properties. Both plants were grown experimentally in the drier parts of European Russia south and south-west of the Urals, and *kok sagyz* was acclimatized with relative ease. In 1939 about 26,000 hectares were planted with these plants, and nearly 5,000 tons of natural latex were produced. By 1943 the area had been raised to 243,000ha to make up for the loss of imports during the war; output is uncertain.

Experiments have also been tried with the guayule, *Parthenium argentatum*, a small shrub whose original habitat was in northern Mexico. Guayule is now grown on quite a considerable scale, but is less important as a source of rubber than *kok sagyz* or *tau sagyz*.

THE TEXTILE INDUSTRY
U.S.S.R.

As early as the beginning of the 18th century, the wool industry tended to concentrate in the Moscow region; but it was not till the introduction of the new and heavier types of machinery from England after 1780, and the use of cotton as a principal textile fibre, that the concentration became really marked. There was a sharp increase in imports of machinery after 1800. Cotton cloths began to be produced, almost entirely from imported cotton, on a commercial scale in the first years of the 19th century and gradually replaced both wool and linen in the cheaper markets. After 1830 the expansion of production was almost as rapid as in England half a century earlier, and marked the first and most decisive phase of Russia's industrial revolution. High protective tariffs were the rule, coupled sometimes with absolute prohibition of the import of certain goods. Even the highly diversified textile industry could not fully resist the traditional Russian tendency towards monopolistic organization.

Before the Revolution of 1917, Moscow and Ivanovo were the principal textile centres, each surrounded by a number of smaller manufacturing towns. St. Petersburg (Leningrad) was also of some importance. Cotton was far ahead of wool as the leading fibre, and the greater part of the raw cotton used was still being imported.

After the Revolution there was at first still further concentration of the industry into the central industrial area; Ivanovo Oblast especially developed into one of the densest textile areas in Europe, though Moscow Oblast remained the biggest producer. In accordance with ideas on industrial organization

then current, some of the new factories set up in the 1920s were of vast size. Rather later a new policy was introduced. Textile factories were established in the new cotton growing districts around Tashkent; in intermediate areas such as the neighbourhood of Saratov; and in various parts of Siberia.

About a third of Soviet spinning capacity and nearly a quarter of the weaving capacity were destroyed during the second World War. During reconstruction some of the destroyed capacity was not restored in the same places as previously. Under the 1946–1950 Plan further development was carried out in the Urals and in Siberia, and there was also further expansion in Central Asia. An official Soviet estimate stated, however, that in 1950, 52% of the cotton cloth output of the Soviet Union came from Vladimir, Ivanovo and Kostroma Oblasts, which also produced about 70% of all the linen cloth manufactured.

Production of Cotton and Woollen Cloth in the U.S.S.R.

(Million metres)

	Cotton	Wool
1930	2,515	101
1940	3,880	120
1945	1,674	57
1950	3,800	155
1954	5,500	242

Production of Linen, Silk and Rayon in the U.S.S.R.

(Millions of square metres)

	Linen	Silk and Rayon
1940	270	78
1950	420	130
1952	257	225
1953	288	400
1954	295	516

TEXTILES IN EASTERN EUROPE

The main concentrations of the East European textile industry are in Bohemia and western Moravia, the southern part of East Germany and in the Łódź area of Poland. Other locations are widely scattered, and all the countries of the region produce textiles, which enter into their foreign trade comparatively little except in so far as they are exported to the Soviet Union in exchange for raw cotton. Practically all the cotton used is imported, though some is grown in Bulgaria, Romania, the Hungarian plain and Yugoslav Macedonia. Poland is the largest producer of flax and Bulgaria of wool.

Textile Fabric Production in Eastern Europe

(Millions of square metres)

	Cotton			Wool			Synthetic Fibres		
	1937	1953	1954	1937	1953	1954	1937	1953	1954
Poland	288	499	524	37·7	70·5	71·2	23·0	68·0	74·8
Hungary	...	209	227	20·2	16·4†	16·6	...	...	...
Romania	...	192§	205§	...	28·9	30·0	...	15·2‡	16·8‡
Bulgaria	34*	119	...	5·2*	9·8	...	...	...	...
Albania	...	50	59	...	1	...	...	...	...
Yugoslav.	...	...	...	...	...	...	...	...	...
Czech.	...	...	...	17·2¶	28·6¶	...	...	...	...
E. Ger.	All fabrics, 1953, 391 million square metres.								

* 1939   † 1952   ‡ Includes silk   ¶ '000 tons   § incl. viguna

Note: In 1949 the production of linen in Poland was 39·3 million square metres; figures for later years are not available. In 1950 rayon output was given as 7,820 tons.

Production of Certain Textile Fibres in Eastern Europe, 1952

(Tons)

	Cotton	Flax	Hemp	Rayon (filament)
Poland	—	50,000	6,000	12,200
E. Germany	—	...	...	16,400
Czechoslovakia	—	14,000	5,000	7,250
Hungary	—	2,000	7,000	680
Romania	8,167	6,000	30,000	1,360
Bulgaria	7,260	—	6,000	...
Yugoslavia	...	10,000	30,000	...

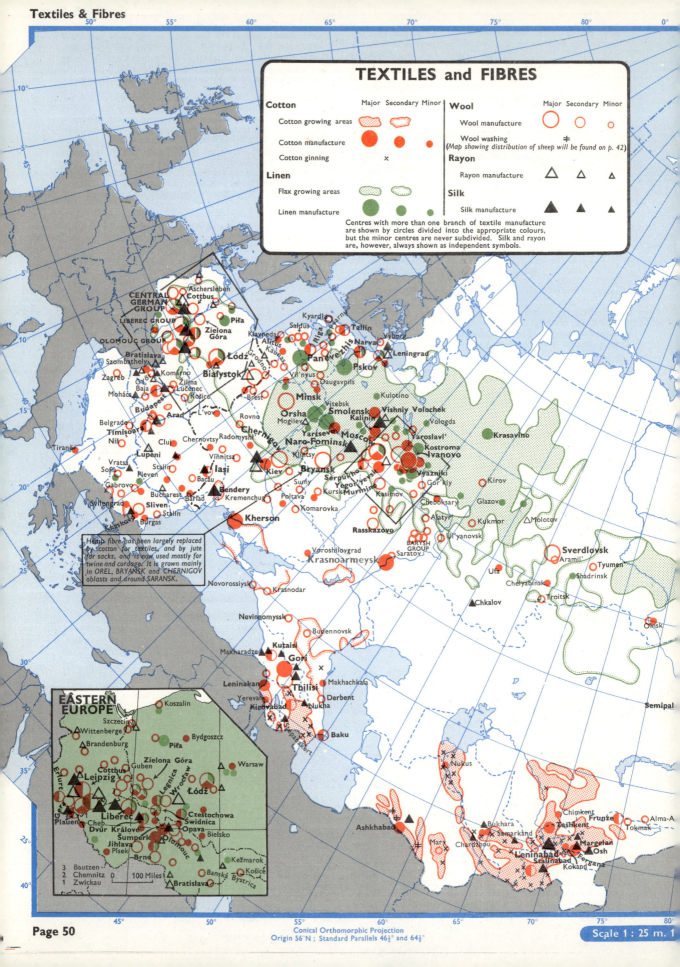

Conical Orthomorphic Projection
Origin 56°N ; Standard Parallels 46½° and 64½°

Scale 1 : 25 m. 1

RUBBER

Major Secondary Minor

Synthetic rubber manufacture ○ ○ ○

District where rubber-
bearing plants are grown

Leningrad
Yaroslavl'
Kursk Yefremov
Voronezh Tambov Kazan'
Kadiyevka
Krasnoyarsk
Yerevan
Temir-Tau
Sumgait

Alcohol derived from potatoes, which
are grown over much of European
Russia and Eastern Europe, is used
extensively in synthetic rubber
manufacture.

Kok Sagyz and similar rubber-bearing
plants, which will grow in temperate
regions, are of some importance in the
U.S.S.R. Nevertheless, the U.S.S.R.
relies largely on the production of
synthetic rubber, and imports natural
rubber (Hevea) from S.E. Asia.

Scale 1 : 60m approx.

Birobidzhan

Krasnoyarsk Kansk
△Kemerovo
Tulun
sk
Barnaul
Biysk
Ulan-Ude
Khadkhal
Ulan-Bator

MOSCOW DISTRICT

Shcherbakov Yaroslavl' Kostroma
Kineshma
Yur'yevets
Kalyazin Ivanovo Shuya
Gor'kiy
Vyazniki
Murom
Greater 2
Moscow 3
Kuntsevo 1
Yegor' Arzamas
yevsk Kasimov
Serpukhov Murmino
Zaraysk
MOSCOW 1 Noginsk
DISTRICT 3 Orekhovo-Zuyevo
0 100 Miles 2 Pavlovskiy Posad

00 miles approx.

COAL AND ELECTRIC POWER

COAL AND LIGNITE IN THE U.S.S.R.

The potential coal and lignite reserves of the U.S.S.R. are so vast that there will be ample supplies of fuel for many decades. An estimate made in 1937 gave total reserves as 1,441,000 million tons of coal (anthracite, bituminous and sub-bituminous) and 211,000 million tons of lignite, which together would represent about 20% of the reserves believed to exist in the world. Proved reserves, or those whose existence was then fairly definitely established, were substantially less—perhaps 630,000 million tons of coal and 118,000 million tons of lignite—but there was reason to believe that very large deposits would be found in Siberia. Two areas shown on Soviet maps as containing unexplored coal-bearing strata cover a considerable part of north and north-eastern Siberia; in the 1937 estimate a total of 782,000 million tons of coal and lignite was included for these areas, though the figure was based on nothing more positive than geological indications of the presence of coal.

At the 1953 rate of consumption of over 300 million tons a year (perhaps 250 million tons of coal and 70 million tons of lignite) these total estimated reserves should last for 2,100 years, and the measured reserves for over 60 years. Consumption is increasing rapidly, however, and a more reasonable assessment of prospects would suggest that the Donets Basin or Donbass will be the first of the major fields to be exhausted, in perhaps 60 to 100 years, and that the main centres of production will shift to the Pechora Basin, to Karaganda, and even more to the Kuzbass, the only one of the big new basins which is also the centre of heavy industry. The metallurgical industries of the Urals will probably come to depend increasingly on coal from Karaganda.

The oldest centre of coal production is the western part of the Donbass, which in 1913 was the source of 87% of the coal consumed in Russia; until after 1926, this was almost the only supply of good coking coal, which accounts for the early growth of the metallurgical industry there. The eastern part of the basin was first exploited in the 1920s, and anthracite began to be mined in considerable quantities; later, mining was extended to Rostov Oblast, where the main coalfield is geologically part of the Donbass. In 1937 it was estimated that of the reserves in the whole basin (including Rostov Oblast and the North Caucasus fields) about 30% consisted of anthracite, 16% sub-anthracite, 5% coking coal, and the remaining 49% bituminous coals of varying quality.

During the war the whole Donbass was occupied by the Germans and ceased to supply coal for Soviet consumption, but the installations were not destroyed and production continued on a restricted scale. The loss of the Donbass provided a stimulus to the development of some other areas. In the Urals, where coal has been mined for at least three quarters of a century, output more than doubled between 1940 and 1945. In the Kuzbass output increased by half in the same period; in 1945 this was the largest single producing area in the U.S.S.R. Output in the Moscow Lignite Basin also doubled, even though the western part of the basin was occupied by the enemy for a period. The Karaganda Basin was also developed during the war.

The estimate of reserves of 6,600 million tons for the Urals is probably too small. Two thirds of the known deposits are believed to consist of bituminous coals, including some anthracite in the west and south, coking coal in scattered small deposits, and considerable quantities of the poorer qualities; the more northerly, and those in Chelyabinsk Oblast, deposits are largely lignite. The lack of coking coal caused the metallurgical industries of the Urals during the 1930s to depend heavily on Kuzbass coal (see pp. 60/61).

In the Kuzbass, the existence of coal deposits was proved early in the present century, and by 1913 output reached 770,000 tons. The quality is high, and includes a large proportion of good coking types. The industrialization of the Kuzbass after the foundation of Stalinsk in 1931 stimulated exploitation, so that in 1937 nearly 18 million tons were produced, and in 1940 just under 25 million tons.

The most spectacular of the newly developed fields are Pechora and Karaganda. In the Pechora basin, production began only in the 1930s, and did not become significant till the war and the construction of the railway from Kotlas to Vorkuta. Between 1940 and 1945 output increased tenfold, and in 1950 was two and a half times the 1945 figure. The coal is entirely bituminous and of good quality, justifying the technical troubles in an area where climatic conditions are so unfavourable. Karaganda is an older field but here again full-scale exploitation was delayed by transport difficulties. Most of the coal is bituminous, but there is some lignite.

Coal and Lignite Production by Coal Basins in the U.S.S.R.

(in million tons)

	1913† Output	%*	1926 Output	%*	1937 Output	%*	1940 Output	%*	1945 Output	%*	1950 Output	%*	1954 Output	%*
Donbass (excl. Rostov)	25.3	87.0	17.9	69.3	68.7	54.0	83.7	50.2	28.6	19.2	85.0	32.5		
Rostov & North Caucasus	—	—	1.75	6.8	8.45	6.6	10.9	6.6	6.0	4.0	12.0	4.6		
Urals	1.22	4.2	1.57	6.1	8.08	6.3	12.0	7.3	25.8	17.2	35.0	13.4		
North Russia	—	—	—	—	0.57	0.4	0.4	0.2	4.0	2.7	10.0	3.8	Regional breakdown not available	
Karaganda	0.09	0.3	0.01	0.04	4.1	3.2	7.0	4.2	12.5	8.4	17.0	6.5		
Other Central Asian fields	0.13	0.4	0.13	0.5	0.92	0.7	1.7	1.0	1.9	1.4	4.0	1.5		
Kuzbass	0.77	2.6	1.78	6.9	17.7	13.9	23.7	14.3	32.3	21.7	43.0	16.4		
Cheremkhovo	0.59	2.0	0.54	2.1	4.08	3.2	5.95	3.6	6.7	4.5	8.4	3.2		
Transbaykal	0.25	0.9	0.22	0.8	1.74	1.4	2.55	1.5	2.9	2.0	3.6	1.4		
Other Siberian & Far Eastern fields	0.37	1.3	0.88	3.4	4.82	3.7	6.8	4.1	7.8	5.2	10.0	3.8		
Moscow Basin (lignite)	0.3	1.0	0.95	3.7	7.51	5.9	10.0	6.0	20.0	13.4	31.2	11.9		
Western Ukraine	—	—	—	—	0.3	0.2	0.4	0.2	—	—	1.2	0.5		
Transcaucasia	0.07	0.2	0.09	0.3	0.39	0.33	0.6	0.4	0.7	0.5	1.6	0.6		
Totals	**29.09**		**25.18**		**127.4**		**165.7**		**149.2**		**262.0**		**346**	
Totals, hard coal only	**28.5**		**22.8**		**113.0**		**145.0**		**113.0**		**209.2**		**...**	

† Excluding Polish coalfields, then in Tsarist Empire.   * percentage of total Soviet output.

COAL AND LIGNITE IN EASTERN EUROPE

All the countries of Eastern Europe produce coal or lignite, though only Poland has large deposits of the better quality bituminous coals, chiefly in Silesia. The principal bituminous deposits of Czechoslovakia are extensions southwards and westwards of the Silesian field, but generally the quality of coal found on the Czech side of the border is inferior. Elsewhere bituminous coal is rare and sporadic in distribution, but the lack of it is compensated by a relative abundance of lignite.

In recent years, Poland has been the fifth largest producer of hard coal in the world, after the U.S.A., U.S.S.R., Great Britain and West Germany. (In 1938 both France and Japan were above Poland on the list). Production was expanded very rapidly by the Germans during the war; output was doubled between 1938 and the end of 1940, and thereafter increased steadily year by year, though at a less spectacular pace. Both before and since the war, coal has played a large part in Poland's export trade. Though Silesia is by far the biggest coalfield, the qualities mined are inferior to those of the Wałbrzych field. The main source of lignite is in the former German lands, lying between Poznań and the River Oder.

Coal and Lignite Production in Eastern Europe

(in thousand tons)

		1935	1945	1950	1953	1954
Poland	Coal	28,542	27,364	78,099	88,560	91,296
	Lignite*	18		4,840	5,280	7,200
E. Ger.	Coal	5,042†	...	2,810	3,550	...
	Lignite	89,892†	...	138,200	175,300	183,800
Czech.	Coal	10,892	11,720	18,449	20,350	21,500
	Lignite	15,114	15,400	31,373	32,750	36,100
Hungary	Coal	823	708	13,200	2,100	22,000
	Lignite	6,840	3,576		19,200	
Bulgaria	Coal	93	231	5,711	8,420	8,600
	Lignite	1,566	3,558			
Romania	Coal	278	211	300	6,000	...
	Lignite	1,667	1,820	3,300		4,900
Albania	Lignite	20	16	42	...	
Yugoslav.	Coal	400	757§	1,154	925	982
	Lignite	4,035	6,047§	11,717	10,300	12,658

* Pre-war Poland till 1945; thereafter includes lignite production in former German territories.

† 1936.   § 1946; figures for 1945 were not published.

ELECTRIC POWER IN THE U.S.S.R.

In Tsarist times, Russia produced proportionately less electric power than any of the other leading countries. In 1913 the total generating capacity of the country was put at 1·1 million kw., and production at 1,900 million kwh., or 13·9 kwh. per head of the population. The efficiency of the plant was low; production was 1,727 kwh. per kw. of installed capacity, which represents an effective utilization of under 20%. All the plants were thermal, and roughly 60% of the power generated was from oil fuel, the rest being from relatively good quality coal.

Production fell sharply during the Revolution and its aftermath, and although generating capacity in 1921 was a little higher than in 1913 (1·2 million kw.), output was only 500 million kwh. The first hydro-electric plant in Russia had been constructed at Volkhov, near Leningrad, in 1918. In 1920 the so-called Goelro plan was launched, providing for the construction of 1·11 million kw. of new thermal capacity and 640,000 kw. hydro-electric. By 1928 capacity had risen to 1·9 million kw., and output to 5,000 million kwh. (of which 4% was hydro-electric), showing the greatly improved utilization figure of 2,630 kwh. per kw. At the same time thermal generation was modified to use coal dust, lignite and peat instead of good quality coal, so that costs were reduced. The plan was declared fully implemented by the end of 1930, though it would seem that the hydro-electric part of it was still unfulfilled.

Under the first and second Five-Year Plans (1928–37) more thermal plant using low grade fuels were built or projected and hydro-electric resources began to be properly explored. The third Five-Year Plan, from 1938 onwards, carried on development, and in particular started an ambitious series of dams, reservoirs and hydro-electric installations on the upper Volga.

Among the thermal plant constructed were two very big ones near Moscow (Elektrogorsk and Shatura) using local supplies of peat, and several in the Ukraine fired by anthracite dust from the Donbass. The best known project of this period was the Dneproges hydro-electric installation at Zaporozh'ye, which was brought into operation in 1932 and was raised to an eventual capacity of 550,000 kw.—the biggest in the world till the construction of the Grand Coulee in the U.S.A. The Svir'stroy plant north-east of Leningrad was completed a little later, and a number of lesser installations were built in various parts of European Russia. In 1940 total capacity had been raised to 11·3 million kw. and output to 48,300 million kwh., representing 4,274 kwh. per kw.; 10·5% of production was now hydro-electric.

The Upper Volga scheme was the first of a type which has become increasingly important since the war. The Uglich installations were opened in 1940. The following year the bigger Shcherbakov project was completed, which involved the flooding of over 2,000 square miles of land to form the Rybinsk reservoir, since that time the largest artificial lake in the world. Since 1945 another vast reservoir has been completed on the Don at Tsimlyansk with a plant of 160,000 kw. capacity, and even bigger ones are under construction on the Volga at Stalingrad and Kuybyshev. These schemes underline the close connection between hydro-electric development and irrigation, and a number of the larger projects recently completed or under construction are designed to serve both purposes.

The fourth Five-Year Plan (1946–50) projected an increase in total capacity of 22·4 million kw. with an output of 82,000 million kwh.; the actual amount of construction involved was greatly increased by the destruction in the war of some 5 million kw. of generating capacity (including Dneproges) equivalent to 44% of the 1940 total. The final performance was better than the plan; actual output in 1950 was 90,300 million kwh.

The 1951–55 Plan aimed at a total production of 162,500 million kwh. The projected installations at Kuybyshev (capacity 2,100,000 kw.) and Stalingrad (capacity 1,700,000 kw.) are to be the biggest in the world. Four more, at Gor'kiy, Mingechaur (completed 1954), Ust'-Kamenogorsk (completed 1953) and on the Kama river near Molotov (operating 1954), were to be of about 500,000 kw. each.

Even with all these developments, however, the U.S.S.R. remains relatively badly supplied with power. Production in 1953 was 133,000 million kwh., or rather more than 600 kwh. per head of the population, which compares with 2,770 kwh. in the U.S.A., 1,280 kwh. in Britain, 1,020 kwh. in West Germany, 950 kwh. in France, 680 kwh. in Italy, and 350 kwh. in Spain.

Electricity Capacity and Production in the U.S.S.R.

Year	Capacity (mn.kw.)	Production ('000 mn. kwh.)	Hydro-electricity (percentage)	Production per head of population (kwh.)	Production per kw. of installed capacity (kwh.	Effective utilization %
1913	1·1	1·9	—	14	1,727	19·9
1928	1·9	5·0	4	34	2,630	30·0
1934	6·3	21·0	...	133	3,333	38·0
1940	11·3	48·3	10·5	283	4,274	48·8
1950	20·1	90·3	18·3	440	4,492	51·3
1953	27·8	133·0	20†	600	4,785	54·6
1954	30·3	146·0	...	...	4,818	55·0
1955	(40·0)	(162·5)	(25)	(720)	(4,625)	(53·0)

† Estimate.

ELECTRIC POWER IN EASTERN EUROPE

Since 1945 electricity production has been increased by the most important producers in Eastern Europe, East Germany, Poland and Czechoslovakia. Each has a well developed grid system, though the East German network was dislocated after the war by dismantling, the division of Germany and the loss of Silesia to Poland. Many East German generating stations use lignite. Elsewhere in Eastern Europe electricity generation has also increased since 1945. In Yugoslavia a large hydro-electric scheme is in operation at Jablanica, and there are plans to build additional stations and to link them to the West German, Austrian and Italian grids.

Electric Power Production in Eastern Europe

(in million kwh.)

	1937	1954
E. Germany	13,900*	24,600
Poland	6,300	15,400
Czechoslovakia	4,100	13,600
Hungary	1,400†	4,800
Romania	1,200†	3,700
Bulgaria	280	1,800
Yugoslavia	933¶	2,381‡

* 1936 est.   † 1938.   ¶ 1934–8 av.   ‡ 1949–51 av.

Figures for Albania are not available

Production of Coal and Lignite in the U.S.S.R. & Eastern European Countries

■ 1 mn. metric tons of hard coals
□ 1 mn. metric tons of lignite
□ 1 mn. metric tons of coal, type unknown

Upper figure : Hard Coals
Lower figure : Lignite

Output in mn. metric tons

97	Donbass (including Rostov & N. Caucasus) (1950)
43	Kuzbass (1950)
35 ★	Urals (1950)
31	Moscow (1950)
22 ★	Siberian & Far Eastern fields (excluding Kuzbass) (1950)
17 ★	Karaganda (1950)
10	North Russia (Vorkuta) (1950)
4 ★	Central Asian coalfields excluding Karaganda (1950)
2 ★	Minor coalfields (1950)

Output in mn. metric tons

3 / 175	East Germany (1953)
89 / 5	Poland (1953)
20 / 33	Czechoslovakia (1953)
21 ★	Hungary (1953)
1 / 10	Yugoslavia (1953)
8 ★	Bulgaria (1953)
6 ★	Romania (1953)

★ No data on proportions of output of hard coals and Lignite

No data for Albania

EAST GERMANY is the world's largest producer of lignite

Many of the thermal electric stations of the MOSCOW area are fired by lignite or peat.

Development of the PECHORA coalfield followed the completi[on] Kotlas-Vorkuta railway in 1942

When completed, the STALINGRAD and KUYBYSHEV hydro-electric stations will be the largest in the world.

DNEPROGES hydro-electric plant before destruction in 1941 had a capacity of 550,000 kw, then the largest in the world. Since reconstruction its capacity has been raised to 651,000 kw.

The DONBASS, the principal Soviet coal producer, contains both good coking coals and anthracite.

EASTERN EUROPE

0   50 Miles

Conical Orthomorphic Projection
Origin 56°N ; Standard Parallels 46½° and 64½°

Scale 1 : 25 m. 1

OAL and ELECTRIC POWER

ucing Centres

	Major	Secondary	Minor
thracite			
uminous			
nite			

Fields

as coal & anthracite

extent of coal and lignite basins

where exploitation is uncertain or where deposits known not to be exploited are shown thus ?

ricity Generating Stations

	Major	Minor
Hydro-electric stations	▲	▲
Thermal stations	●	●

U.S.S.R. Coal & Lignite Reserves
(in thousand million tons ; 1937 estimate)

	Hard Coal	Lignite
Pechora-Vorkuta	36·5	—
Moscow Lignite Basin	—	12·4
Urals		6·6*
Donbass (excl. Rostov)	34·7	—
Rostov and North Caucasus	9·8	—
Western Ukraine	—	0·5
Georgia	0·36*	
Karaganda (incl. Ekibastuz)	33·4	1·3
Fergana	16·4*	
Kuzbass	375·0	—
Minusinsk Basin	19·0	—
Yenisey-Chulym	—	43·0
Kansk	—	42·0
Irkutsk (Cheremkhovo)	79·0	
Transbaykal	2·0*	
Bureya	26·0	—
Amur	—	2·6
Sakhalin	3·0	—
Minor Far Eastern fields	4·3	—
Total estimated (approx.)	630·0	118·0
Total measured (1939)	19·5*	
Total operational (1939)	2·5*	

* Type uncertain

The large coal and lignite basins of NORTH SIBERIA have been only partly explored. They are believed to contain immense reserves.

Noril'sk

Zyryanka

Sangar

Kangalasskiye Kopi

Verkh.-Vilyuysk

Kochumdek ?

Seligdar

Vanavara ?

YENISEY-CHULYM LIGNITE FIELD

Aleksandrovsk
Oktyabr'skiy
SAKHALIN COALFIELDS
Uglegorsk
Ugol'nyy
Sinegorsk

BUREYA COALFIELD
Sredniy Urgal
Chekunda

AMUR LIGNITE FIELD
Raychikhinsk

Tomsk
Nazarovo
Kansk
KUZBASS COALFIELD
Chernogorsk
Barnaul
MINUSINSK COALFIELD
Cheremkhovo
Irkutsk

TRANSBAYKAL COALFIELDS
Bukachacha
Chita
Baley
Ulan-Ude
Gusinoozersk

Kamenogorsk
ukhtarma

Dzun Bulak

Ulan-Bator
Nalaykha

Sayn Shanda

Artem
Suchan
Okeanskoye
Kraskino

KUZBASS

Tomsk
Anzhero–Sudzhensk
Barzas
Yashkino
Kemerovo
Novosibirsk
Toguchin
Leninsk-Kuznetskiy
Belovo
Listvyanskiy
Kiselevsk
Yuzhkuz-bassgres
Prokop'yevsk
Mezhdurechensk
Stalinsk
Osinniki
Barnaul
Shushtulep

0 Miles 50

DONBASS

7 Bokovo-Antratsit
9 Gundorovka (Donetsk)
4 Irmino
6 Krasnyy Luch
2 Makeyevka
1 Novo-Ekonomicheskoye
10 Novoshakhtinsk
5 Parizhskaya Kommuna
8 Sverdlovsk
3 Yenakiyevo
11 Zugres

0 Miles 50

Rubezhnoye
Kadiyevka
Gorlovka
Krasnodon
Stalino
Chistyakovo
Shtergres
Shakhty

00 miles approx.

PETROLEUM AND NUCLEAR FUELS

The Soviet Union is the third largest producer of petroleum in the world, after the United States and Venezuela. For a brief period at the beginning of this century the Tsarist Empire was at the head of the list, but in the troubled decades that followed the lead was lost.

HISTORY OF THE INDUSTRY IN THE U.S.S.R.

The first record of crude oil production in Russia was in 1833, from deposits near the surface in the present Groznyy oilfield, though full scale commercial production did not begin there till 1893. The small West Ukrainian field is the oldest within the present boundaries of the U.S.S.R. (and among the oldest in the world), though at the time of its first exploitation in 1860 it lay within the Austro-Hungarian Empire. The real start of the Russian petroleum industry was, however, in 1873, when the first wells were sunk at Sabunchi, Balakhany and Romany to the north of Baku. With the addition of Bibi-Eybat in 1882, and Surakhany and Binagady in 1903, the Baku field became the largest single producing area in the world; it was to provide the mainstay of the Russian industry up to the time of the Second World War. When Russian production reached its pre-revolutionary peak of $11\frac{3}{4}$ million tons in 1901, overwhelmingly the greater part came from the Baku region.

Before 1917, Groznyy was second in importance. A gusher was drilled in 1893, and by 1910 output from the original group of wells passed the million ton mark. The so-called New Groznyy field was discovered in 1913, and added another 400,000 tons a year. The Maykop (or Kuban') field in the northern Caucasus was opened in 1908, Cheleken Island (an offshore extension from Nebit-Dag) in 1909, and Emba in 1912. Before the Revolution there was also small production in the Fergana valley, and the Ukhta-Pechora field in the extreme north was known to exist but had not been exploited.

Production by Fields, 1908 and 1916

	1908		1916	
	('000 tons)	%*	('000 tons)	%*
Baku	7,570	89·5	7,820	81·0
Groznyy	850	10·5	1,560	16·1
Maykop	1	0·01	32	0·3
Emba	—	—	265	2·7
Cheleken	—	—	16	0·2

* percentage of Russian output

Note: Maykop showed a peak production of 151,000 tons in 1912, and Cheleken of 218,000 tons in 1911.

Production of oil before 1917 was entirely in the hands of private companies, with the Nobel group (which had acquired a controlling position in the early developments in the Baku area) predominant. In 1914 about 60% of the capital invested in the industry was foreign-owned, British investors alone holding some 86 million gold dollars, or 40% of the total of Russian and foreign investments.

The Revolution caused considerable disruption and total output of crude oil fell from over 9,000,000 tons in 1916 to 3,500,000 tons in 1920. During part of this period, the Baku field was in the short-lived independent Republic of Azerbaydzhan, but was reconquered by the Red Army in 1920. Production recovered slowly, but it was not till 1927 that the level of 1916 was reached, and the pre-war peak of 1901 was not passed until 1928.

No new fields had so far been opened up under the Soviets, though the eastern part of the Nebit Dag deposit began to be exploited soon after the Revolution. Great interest was, however, shown in the area between the Volga and the Urals, and in 1929 oil was proved at Verkhne-Chusovskiye Gorodki. The discovery was of enormous significance. During the next few years (the period of the first Five-Year Plan) exploration continued on a large scale; in 1931 further deposits were proved at Ishimbay, in 1933–34 production started there and at Krasnokamsk, in 1937–38 at Tuymazy and Buguruslan, so that by the end of the second Five-Year Plan (1937) a vast series of new oilfields was being opened up in the heart of the country. These fields together form what is collectively known as the Second Baku; by 1953 production there had passed that of the Baku area, and was approaching half the total for the whole Soviet Union.

The small Dagestan oilfield was explored in 1926, and production began in 1934. Further extensions of the Baku field took place, including a number of centres at some distance from the main field. On Sakhalin Island in the Far East, Japanese interests had secured concessions in Soviet territory and had been producing oil since 1921; after 1935 a number of new wells were opened by the Russians as well. Also during the period of the second Five-Year Plan (1933–37), the exploitation of the Fergana field was extended, and new centres were brought into production in the Surkhan Darya valley near the border of Afghanistan.

Production by Fields, 1938

	('000 tons)	% of total
Baku and Georgia	20,700	73·4
Groznyy	2,300	8·1
Kuban' (incl. Maykop)	2,000	7·1
Second Baku	1,600	5·6
Emba (Gur'yev)	500	1·8
Central Asia	300	1·1
Sakhalin	300	1·1
Dagestan	200	0·7
Nebit-Dag (incl. Cheleken)	200	0·7
Ukhta-Pechora	100	0·4
Total	**28,200**	**100**

In the autumn of 1939 the West Ukrainian oilfield became Soviet territory for the first time, but was lost again to the Germans in 1941. The German drive of 1941–42 brought them into the northern Caucasus, and the Kuban' field and part of the Groznyy field were also temporarily lost, and Baku itself was threatened for a time. The Second Baku was comparatively safe, as were the Central Asian and other fields; the result was that the war intensified the trend (actually expressed in the Third Five-Year Plan), to reduce the relative importance of Baku and increase that of the Volga–Urals area. Had the Third Five-Year Plan been carried out according to schedule, by 1942 Baku would have been supplying only 57% of the total, the Second Baku would have risen to 14·7%, Emba to 4·2%, Central Asia to 3·7%, while the other fields changed but little in relative position. Planned total output was 47,400,000 tons; but, with the country disrupted by war and two fields in enemy occupation, production in 1942 was under 32 million tons.

The Fourth Five-Year Plan (1946–50) was more soundly based on actual possibilities, and output for 1950 was planned at 35,440,000 tons; actual production in that year came to 37,600,000 tons. Under the Plan, Baku was to be down to 47·9% of the total, and the R.S.F.S.R. fields generally (including the Second Baku, Groznyy, Kuban', Sakhalin and minor fields) up to 40·9%. The Second Baku remained the scene of the most intensive activity; new wells were still being bored at Baku and elsewhere, however, and the damaged Ukrainian, Kuban' and Groznyy fields were being restored as rapidly as possible. Unfortunately, reliable figures to show the relative importance of the various fields in 1950 are not available, though estimates put Baku at about 45% (lower than planned) and the Second Baku at about 28%.

Little information is available about the other minor fields which were explored in Siberia during this period.

CURRENT PRODUCTION

In 1953, total oil production in the Soviet Union was about 52 million tons, and in 1954 about 59·5 million tons. The available evidence suggests that Baku is nearing exhaustion, and produces only some 17 million tons a year (33% of the 1953 total). In 1953 the output of the Second Baku was about half the total for the whole country, or 26 million tons. The other fields thus produced some 9 million tons between them. In 1954 output in the Second Baku was about 27 million tons, of which the Bashkir A.S.S.R. produced roughly 12 million tons, Kuybyshev Oblast 8 million tons, and the Tatar A.S.S.R. about 4 million tons. Production in the U.S.S.R. approximately doubled between 1947 and 1953, and there is likely to be still further expansion in the years to come.

CHARACTERISTICS OF SOVIET OIL PRODUCTION

The character of the deposits varies very considerably from area to area. In the Baku region there are 22 oil horizons capable of production, which account for the high and continual output since 1873, and for the relatively small number of wells. The deepest well sunk before 1950 went to 16,700 feet. In the Second Baku, which really consists of a number of separate oilfields spread over an area as large as the British Isles, there are usually only two or three productive horizons. One well at Tuymazy is 10,800 feet in depth, and the average depth for the whole area is 1950 was 6,560 feet. In most of the other oilfields, the depths reached are less.

The total number of wells in the Soviet Union is perhaps 25,000, which gives an annual output per well of over 2,000 tons. This average compares with about 430,000 tons per well in the Middle East, and 600 tons in the U.S.A.

The average specific gravity of Soviet crude oil is relatively high, at 0·922 (A.P.I. gravity 22·35, or 6·82 barrels per metric ton). The average for the U.S.A. is 0·850 (7·418 barrels per ton); the only areas which show a higher specific gravity than the Soviet Union are some of the small European fields, as in Austria and Albania, and the Diyarbakir field in Turkey.

RESERVES

The estimating of reserves is a very hazardous business, and figures as wide apart as 6,000 million and 20,000 million tons have been given. The Soviet Government claimed in 1952 total *known* reserves of oil and natural gas as 4,500 million tons, which would be adequate for 60 to 100 years of consumption at the present rate and at estimated future rates. These known reserves exclude the possibilities of great unexplored deposits in several parts of the country, and also exclude some which are known to exist but have not been measured.

REFINING CAPACITY

In 1910, a total of 8,230,000 tons of crude oil was refined, mainly near the centres of production. As late as 1937, when 25,700,000 tons of crude oil passed through the refineries, 77% of the total refinery capacity was in the Baku area, but in the years between 1937 and 1945 many new refineries were built in the Second Baku and elsewhere, and the emphasis shifted from refining on the spot to the conveying of crude oil through pipelines to be refined nearer the places where it would actually be used. Total refining capacity is not known, but is probably keeping pace with output. Nor is there any recent information on total cracking capacity, though in 1951 it was estimated at some 14 million tons a year.

NATURAL GAS AND OIL SHALE

Natural gas is found in association with most of the oil deposits, but is frequently pumped back into the wells to maintain pressure. In some cases, however, it is piped either to feed local installations or to large towns.

The Baku oilfield is rich in natural gas. The only centres where gas is found otherwise than in association with petroleum are in Kuybyshev and Saratov Oblasts, and at Dashava in the Western Ukraine. Estimates of production are unsatisfactory; output from the Saratov fields in 1947 was said to be 545 million cubic metres, but figures for the other fields are not available. Estimates of reserves are wholly unreliable, but there is no doubt that they are very large.

Oil shale is found mainly on the northern border of the Estonian SSR. Petroleum is recovered from it as a useful secondary source.

ROMANIAN PETROLEUM

Romania was the first country in the world where petroleum was produced on a commercial scale. The main, and oldest, oilfield was opened up near Ploeşti on the southern edge of the Carpathians in 1857, when some 2,000 barrels (roughly 290 metric tons) were extracted from two hand-dug wells. Mechanical drilling was introduced by stages from 1880 to 1887; by 1900 output reached 218,000 tons, and 1,915,000 tons in 1913. During the first World War there was a set-back, the 1913 level not being reached again till the middle 1920s;

thereafter a steady increase followed, to a peak of 8,700,000 tons in 1936.

For some years afterwards production was considerably down, but recovered with the development of new fields after 1945. In 1953 the 1936 record was passed for the first time.

Production of Petroleum in Romania
1936 and 1946–1954)
('000 metric tons)

1936	1946	1947	1948	1949	1950	1951	1952	1953	1954
8,600	4,250	3,800	4,300	4,700	5,300	6,500	8,400	9,500	10,200

OTHER EAST EUROPEAN COUNTRIES

Apart from the Soviet Union and Romania, there is also some output of petroleum in all the other East European countries except East Germany. Hungary developed its oil industry mainly during the war, and output has exceeded half a million tons in almost every year since 1944. Total output from Eastern Europe (excluding Romania but including Yugoslavia) in 1952 was about 1,100,000 tons.

Production of Petroleum in Eastern Europe, 1938, and 1946-54
('000 metric tons)

	1938	1946	1947	1948	1949	1950	1951	1952	1953	1954
Albania	127	122	183	244	305	132	305	100	200	235
Czech.	19	24	37	48	60	81	97	180	195	202
Hungary	43	660	570	484	510	512	510	520	850	1,194
Poland	507	117	126	135	153	178	185	210	245	...
Yugoslav.	1	29	34	37	64	110	147	151	172	216

There is also small production in Bulgaria

NUCLEAR FUELS

Details of the output of uranium and other radio-active minerals are completely lacking. A guess is that in 1945 about 10 tons of uranium were produced in the Soviet Union, East Germany, Czechoslovakia and Bulgaria together, and that by 1950 output had risen to perhaps 150 tons; but the actual figures may well have been higher than these. Jáchymov in Czechoslovakia probably accounted for the greater part of the output between 1945 and 1948, the mine having been worked since 1922 and fully exploited by the Germans during the war. It seems that this mine, where uranium is recovered from pitchblende, is now nearing exhaustion, and the nearby German mines at Annaberg and Schneeberg are now probably the main producers in this area.

In the Soviet Union itself, the oldest mine is at Tyuya Muyun in the Kirgiz SSR, where 534 tons of uranium-vanadium ore are known to have been produced in 1925–26. Other uranium-vanadium deposits of unknown value are scattered over much of Central Asia. In the Lake Baykal region there are betafite deposits, where uranium is associated with calcium, columbium and tantalum, all of which are mined at Slyudyanka.

Thorium is mined at Tarak and Kazachinskoye in Krasnoyarsk Kray, and radium is recovered from radio-active water at Ukhta in North Russia.

The probability is that none of these deposits compares in size with those of the Belgian Congo, Canada or Australia. Current Soviet policy on the development of atomic energy would suggest confidence in the supply of the minerals, but it is impossible to say what reserves exist or for how long they will last.

NUCLEAR ENERGY

The Soviet Union was one of the first countries in the world to experiment in the harnessing of nuclear energy to produce electricity. A 5,000 kw. plant was brought into operation in the Moscow area in 1954, and it was reported that a plant of 20,000 kw. capacity was to be inaugurated in 1956. Other plant, some of them considerably larger, were said to be under construction in several parts of the country, especially in the Lake Baykal area.

There is an arrangement between the Soviet Union and the other Cominform countries under which information on nuclear development is exchanged. In 1955 the U.S.S.R. exported at least one breeder-reactor to Czechoslovakia.

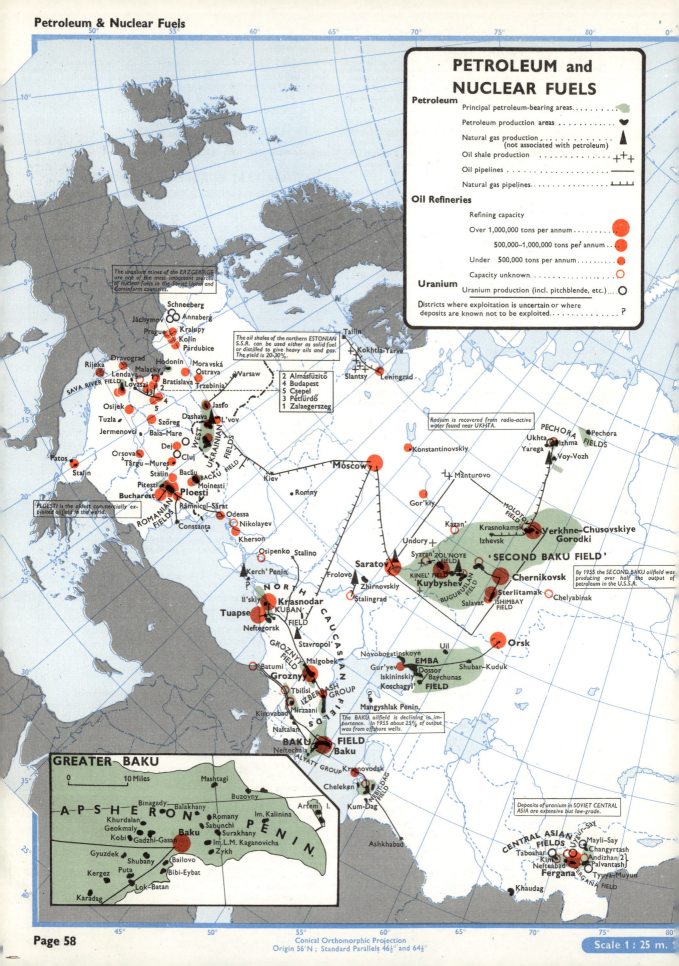

PETROLEUM and NUCLEAR FUELS

Petroleum

Principal petroleum-bearing areas . . . .

Petroleum production areas . . . .

Natural gas production . . . .
(not associated with petroleum)

Oil shale production . . . .

Oil pipelines . . . .

Natural gas pipelines . . . .

Oil Refineries

Refining capacity

Over 1,000,000 tons per annum . . . .

500,000–1,000,000 tons per annum . . .

Under 500,000 tons per annum . . . .

Capacity unknown . . . .

Uranium

Uranium production (incl. pitchblende, etc.) . . .

Districts where exploitation is uncertain or where
deposits are known not to be exploited . . . . ?

*The uranium mines of the ERZGEBIRGE
are one of the most important sources
of nuclear fuels in the Soviet Union and
Cominform countries.*

*The oil shales of the northern ESTONIAN
S.S.R. can be used either as solid fuel
or distilled to give heavy oils and gas.
The yield is 20-30%.*

2	Almásfüzitő
4	Budapest
5	Csepel
3	Pétfürdő
1	Zalaegerszeg

*Radium is recovered from radio-active
water found near UKHTA.*

*PLOESTI is the oldest commercially ex-
ploited oilfield in the world.*

'SECOND BAKU FIELD'

*By 1955 the SECOND BAKU oilfield was
producing over half the output of
petroleum in the U.S.S.R.*

*The BAKU oilfield is declining in im-
portance. In 1955 about 25% of output
was from offshore wells.*

GREATER BAKU

0   10 Miles

APSHERON PENIN.

*Deposits of uranium in SOVIET CENTRAL
ASIA are extensive but low-grade.*

Conical Orthomorphic Projection
Origin 56°N ; Standard Parallels 46½° and 64½°

Scale 1 : 25 m.

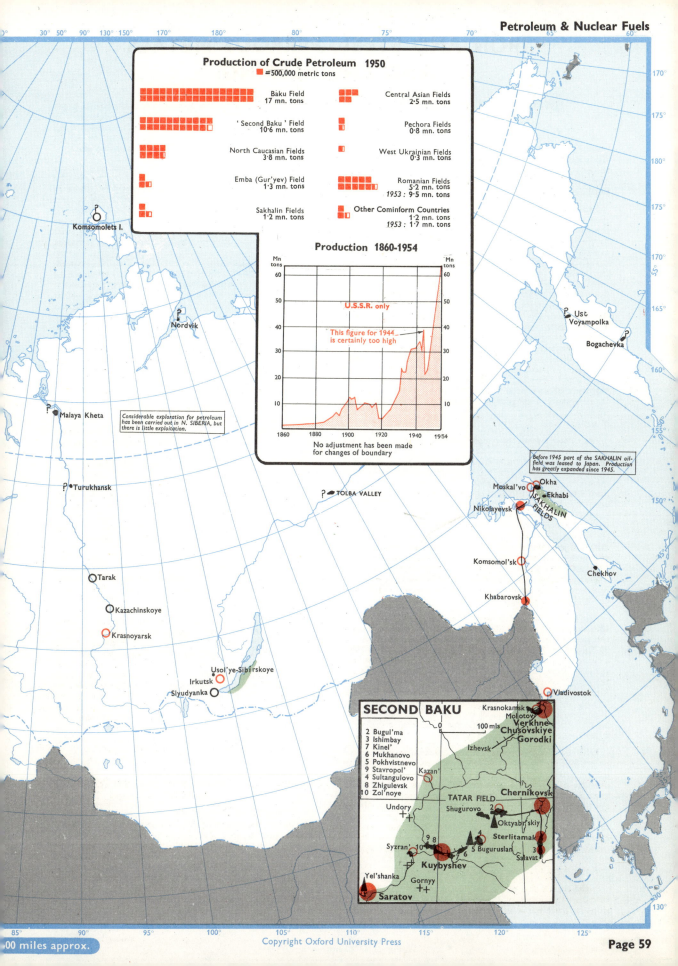

Production of Crude Petroleum 1950

■ = 500,000 metric tons

Baku Field
17 mn. tons

Central Asian Fields
2·5 mn. tons

'Second Baku' Field
10·6 mn. tons

Pechora Fields
0·8 mn. tons

North Caucasian Fields
3·8 mn. tons

West Ukrainian Fields
0·3 mn. tons

Emba (Gur'yev) Field
1·3 mn. tons

Romanian Fields
5·2 mn. tons
1953: 9·5 mn. tons

Sakhalin Fields
1·2 mn. tons

Other Cominform Countries
1·2 mn. tons
1953: 1·7 mn. tons

Production 1860-1954

Mn tons

U.S.S.R. only

This figure for 1944
is certainly too high

No adjustment has been made
for changes of boundary

Considerable exploration for petroleum
has been carried out in N. SIBERIA, but
there is little exploitation.

Komsomolets I.

Nordvik

Ust
Voyampolka

Bogachevka

Malaya Kheta

TOLBA VALLEY

Before 1945 part of the SAKHALIN oilfield was leased to Japan. Production
has greatly expanded since 1945.

Turukhansk

Moskal'vo Okha
Ekhabi
SAKHALIN
FIELDS

Nikolayevsk

Chekhov

Tarak

Komsomol'sk

Kazachinskoye

Khabarovsk

Krasnoyarsk

Usol'ye-Sibirskoye

Irkutsk

Slyudyanka

Vladivostok

SECOND BAKU

Krasnokamsk
Molotov
Verkhne
Chusovskiye
Gorodki

2 Bugul'ma
3 Ishimbay
7 Kinel'
6 Mukhanovo
5 Pokhvistnevo
9 Stavropol'
4 Sultangulovo
8 Zhigulevsk
10 Zol'noye

0   100 mls

Izhevsk

Kazan'

Chernikovsk

TATAR FIELD
Shugurovo   2
Oktyabr'skiy
Sterlitamak

Undory

9 8
Syzran' 10   5 Buguruslan
Kuybyshev   7 6

Salavat

Yel'shanka   Gornyy

Saratov

00 miles approx.

IRON AND STEEL, FERROUS METALS
AND ENGINEERING

HISTORY OF THE INDUSTRY
IN THE U.S.S.R.

In Russia, as in other countries, the first use of iron was in prehistoric times when primitive tools and weapons were fashioned out of more or less pure metal. In Russia too, as in the countries of western Europe and elsewhere in the world, new techniques of smelting and alloying made possible the beginnings of industrialization.

Small local industries based on iron casting and hand working grew up in medieval times, and were later developed to some economic importance under the Tsars from Ivan the Terrible to the Empress Catherine. The small industries were located especially in the Moscow area and in the middle Volga valley, using iron ore from central Russia, and local supplies of charcoal for smelting; the shift to the Urals did not come till the 18th century. In 1718, towards the end of the reign of Peter the Great, about 27,000 tons of pig-iron were produced in the Empire, chiefly from ore mined in the Urals, and for several decades Russia was by far the largest producer of iron in the world. Peter visited England to study British industrial methods, taking back with him numerous technicians from the west to help in his vigorous programme of industrial and military development, and during the second half of his reign over twenty up-to-date smelters for iron and other metals were constructed. Moscow, Nizhniy Novgorod (Gor'kiy), Tula and Mozhaysk became major industrial centres, and in the north St. Petersburg (Leningrad) was being built. Each had a considerable engineering industry, especially for the manufacture of arms.

After Peter's death the tempo slackened, though there was a partial revival under Catherine the Great (1762–1796). Contemporary development in England was much more rapid, and in 1800 both England and Russia produced approximately the same amount of pig iron (about 160,000 tons). During the next fifty years British output increased more than tenfold, while Russian production barely doubled; and in 1870 Russia had fallen to seventh place with an output of 370,000 tons. The six principal producers were Britain (nearly 6 million), U.S.A. (1,770,000), Germany (1,260,000), France (1,180,000), Belgium (570,000) and Austria (410,000 tons). Russian output in that year was barely more than 3% of the world total. Between 1870 and the start of the first World War there was a considerable increase, to a peak of 4,623,000 tons in 1913, and Russia had risen again to fifth place in the world after the U.S.A., Britain, Germany and France.

It is hard to say exactly when steel was first produced. In 1866 output was about 10,000 tons, and remained at roughly the same level for the next ten years. There was a rise to 50,000 tons in 1877, 100,000 tons in 1878 and 290,000 in 1880, a figure which was not reached again till 1890. After that year the increase was more rapid, to 4,750,000 tons in 1913 and the pre-revolutionary peak of 4,900,000 tons two years later, when Russia was the fourth largest producer in the world after the U.S.A., Germany and Britain. Output of steel exceeded that of pig-iron for the first time in 1908, partly because exports of pig-iron had been reduced, but chiefly because there were larger quantities of scrap metal available.

The growth of coal production over the same period is traced on pp. 52/53. The comparatively slow growth of the Russian iron and steel industry was due not to lack of fuel, but to the shortage of capital and technical skill, and to a political and social atmosphere inimical to rapid industrial progress. Unlike the petroleum industry (pp. 56/57) iron and steel was almost entirely run by Russian interests, with relatively little foreign capital. Foreign technicians remained, but their number decreased as the opportunity for employment in their own countries widened. Nevertheless, the tradition of the Scottish engineer in Russian industry survived up to the Revolution, and even for a time after it.

In the chaotic days of the Revolution and the wars of intervention, most forms of industrial activity slowed down or stopped; there was actually no pig-iron produced in 1918, and output of steel fell almost to nothing. When the Soviets began to organize the metallurgical and engineering industries they were faced with an acute shortage of managers, technicians and skilled workers. Output of steel did not pass the pre-revolution peak till 1929, and pig-iron not till the following year.

The first Five-Year Plan (1928–1932) was designed especially to develop the heavy industries of the Soviet Union, and iron and steel were expanded on lines different from those of Tsarist days. Even before 1917, Moscow had ceased to be a main centre, and the greater part of the output of iron and steel came from the Urals and the Ukraine, near the sources of iron ore, while the engineering industry was showing a tendency to drift to the same, or nearby, areas to lessen transport costs. These trends were intensified under the Soviet régime. The Urals and the Donbass were developed at remarkable speed, and further east the Kuzbass was opened up as a new centre for heavy industry. A new development, too, was the concentration of a whole series of metallurgical and engineering operations into a single vast plant or *kombinat*; the first, initiated at Magnitogorsk in the eastern Urals in 1931, was followed the next year by the founding of a huge works at Stalinsk (formerly Kuznetsk) on the Tom' river in Siberia. By the end of the five years covered by the Plan, the full effect had still not been felt. Output of steel in 1932 was 5,920,000 tons, and of pig-iron 6,180,000 tons. Even though this was a slump year in the west, it was still significant that the Soviet Union was the world's second largest producer of both pig-iron and steel.

During the first two Five-Year Plans, the development of the Urals area and the Kuzbass proceeded on parallel lines. The Urals are rich in iron ore but short of coking coal, which is plentiful in the Kuzbass. The Plans aimed at making the two areas complementary, at least until the iron ore reserves known to exist in the Kuzbass could be exploited. No similar problem arose in the Ukraine (the other great metallurgical area), where both good coal and iron ore are mined, and where development during the first two Plans was equally rapid. On the eve of the war, in 1939, the Soviet Union as a whole produced over 15 million tons of pig-iron and nearly 19 million tons of steel. Western recovery from the slump had, however, pushed Russian production of each back to third place, Germany now being slightly ahead.

During the war the industrial region of the Ukraine was lost to the Germans, and many installations were destroyed to prevent them from being used by the invaders. This was the chief factor in the sharp decline in production, which reached a low level of 4,800,000 tons of pig-iron and 8,100,000 tons of steel in 1942. By 1943 further developments east of the Urals were helping to redress the balance, and in 1948 output again reached the high levels of 1939. The great expansion of such centres as Sverdlovsk, Nizhniy Tagil, Serov and Alapayevsk to the east of the hills, and Chelyabinsk to the south, took place mainly during the war. After the war Novotroitsk was constructed as a smaller Magnitogorsk, with all the processes concentrated into a single large plant; the first blast furnace came into operation in October 1954. Most of the other new centres are more specialized, and no more huge *kombinats* are planned; current Soviet thought apparently regards this wartime decentralization as preferable to the experiments in large scale organization tried under the first two Plans.

CURRENT OUTPUT OF IRON AND STEEL IN THE
U.S.S.R.

Since 1947 Soviet iron and steel production has grown at a phenomenal rate, steel especially increasing by between 3 and 4 million tons a year. No figures of output by producing regions are available after 1951, but the following table indicates that the importance of the Urals and the Kuzbass is still growing. To-day the Kuzbass is itself a major producer of iron ore, so that the 1,200 mile rail haul from the Urals is no longer essential to the development of this area. The figures available do not, however, distinguish between the Urals and the Kuzbass.

Output of Pig Iron* and Crude Steel in the U.S.S.R., 1937-1954

*Including ferro-alloys

Production of Pig Iron and Ferro-Alloys in Eastern Europe

(in '000 metric tons)

	1938	1950	1953	1954
Czechoslovakia	1,233	2,050	2,790	2,800
Poland	963*	1,490	2,299	2,700
Hungary	335	482	760	820
Bulgaria	0·13†	4·2	10·6¶	...
Romania	133	335	456	432
E. Germany	74	337	1,068	1,400
Yugoslavia	59	212	270	367

† 1939   ¶ 1952

* Pre-war figures for Poland exclude production in the parts of Silesia then in Germany.

Production of Crude Steel in Eastern Europe

(in '000 metric tons)

	1938	1950	1953	1954
Czechoslovakia	1,873	3,190	4,430	3,000*
Poland	1,441	2,515	3,597	4,000
Hungary	648	1,022	1,500	1,490
Romania	276	558	719	630
E. Germany	1,695	995	2,177	1,700*
Yugoslavia	227	428	515	616

* Rolled steel

Output of Pig Iron and Crude Steel by Regions in the U.S.S.R.

	1937 Pig Iron ('000 tons)	%*	1937 Crude Steel ('000 tons)	%*	1949 Pig Iron ('000 tons)	%*	1949 Crude Steel ('000 tons)	%*	1951 Pig Iron ('000 tons)	%*	1951 Crude Steel ('000 tons)	%*
Ukraine¶	9,200	63·5	9,300	52·7	7,000	42·2	7,000	30·0	11,000	49·9	11,300	36·1
U.S.S.R. in Europe†	1,200	8·3	3,300	18·7	900	5·4	3,000	12·9	1,000	4·5	3,500	11·2
U.S.S.R. in Asia§	4,100	28·2	5,100	28·6	8,700	52·4	13,300	57·1	10,100	45·6	16,500	52·7
Totals	**14,500**	**100%**	**17,700**	**100%**	**16,600**	**100%**	**23,300**	**100%**	**22,100**	**100%**	**31,400**	**100%**

¶ Including Rostov Oblast and Crimea   † Excluding Urals   § Including Urals   * percentage of total

IRON AND STEEL IN EASTERN EUROPE

Iron and steel have been produced for over a century in several areas of Eastern Europe, especially in Silesia, Moravia and Bohemia, and in the former provinces of Brandenburg and Saxony-Anhalt in East Germany. The metallurgical industries of Slovakia, Hungary and Romania are newer, having undergone their main development since the second World War—though the Slovak and Hungarian industries expanded considerably while the war was still in progress. In all these countries there was a sharp fall during the period of revolution and dislocation after 1945, and in East Germany some plant was removed as reparations. Since 1948 German industry has been re-equipped and by 1952 production exceeded the 1938 level. All of them, however, now include the expansion of iron and steel production in their current plans for industrial development. In 1953 the whole group of Eastern European countries (including Yugoslavia) produced 4·7% of the world's pig iron and 5·4% of its steel.

IRON ORE IN THE U.S.S.R.

The U.S.S.R. is well supplied with iron ore. Estimates made in 1938 showed measured reserves of nearly 4,000 million tons and "total" reserves of roughly 8,300 million tons; the first of these alone would be adequate for 80 years' consumption at the present rate. Ultimate reserves are without doubt much higher than either of these estimates, though a figure of 267,000 million tons given by a Soviet scientist is too high. However, the figure of "total" reserves quoted above excludes several large deposits now known to exist in the northern Urals, in Kazakhstan and in Transbaykalia.

With such vast reserves available; there is no need at present to use inferior ores, and only a very small proportion (2% to 3%) of the ore mined is of less than 42% Fe content. The highest qualities are up to 68% (Krivoy Rog and Bakal), and the average for the whole country ranges from 55% to 57%.

The Krivoy Rog deposit in the Ukraine is the best known, and for many years was the largest single source of supply. It is also the biggest deposit of hematite. The other main Ukrainian deposit, at Kerch', is larger in terms of ore but of relatively poor quality with a high phosphorus content. Between them, Krivoy Rog and Kerch' account for rather more than half the annual iron ore production of the Soviet Union.

In the Urals, limonite ($2Fe_2O_3.3H_2O$) is the most prevalent ore, especially in the northern and central parts of the range. The quality improves from north to south (though little is yet known about the important deposits at Serov in the north) and the highest Fe percentages are reached at Bakal and in the Komarovo-Zigazinskiy field. These are said to be the best ores in the U.S.S.R., and contain relatively high proportions of manganese and silicon. Magnetite (Fe_3O_4) is the principal ore in the southern Urals (hence the name of Magnitogorsk—"magnetic mountain"), but is also found further north. At Kushva and Nizhniy Tagil magnetite occurs in conjunction with hematite (Fe_2O_3), which is otherwise rather rare in the Urals.

The Kuzbass ores suffer from the grave disadvantage of a rather high zinc content; the iron content ranges from 43% to 53%. The ore has been fairly fully worked out in the northern part of the deposit, but there are still reserves of up to 100 million tons known to exist in the area of Tashtagol, which is now the chief centre. East of the Kuzbass, large deposits are known to exist in the Abakan area, but until the railway between Abakan and Stalinsk is completed it will not be possible to exploit them fully.

Further east, the Baykal area is likely to become of much greater importance, though here again communications are far from adequate for full development. The Bratsk and Nizhne-Ilimsk deposits are known to contain well over 300 million tons of good quality magnetite, and the area is still incompletely explored. The other Siberian deposits are still of minor economic significance, though it may be that in time full exploration and improved communications will transform the picture.

Continued on page 64

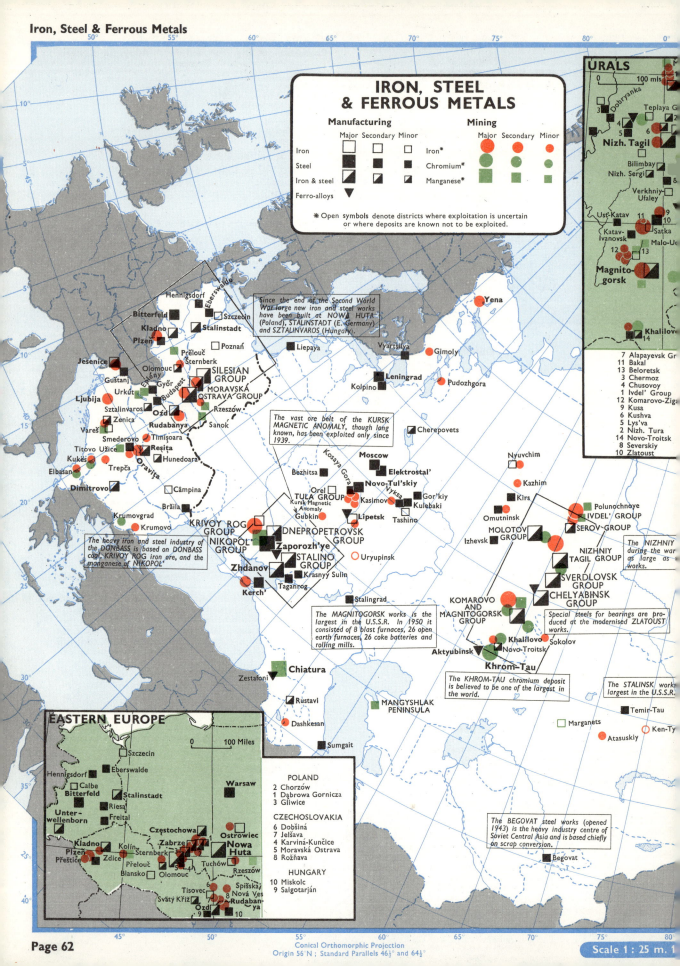

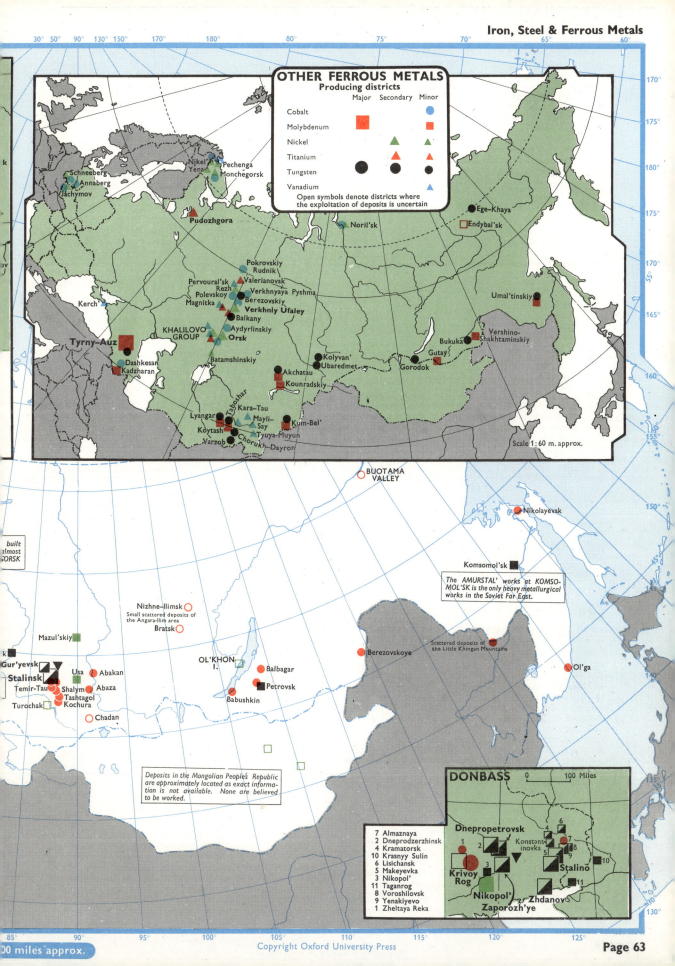

OTHER FERROUS METALS
Producing districts

	Major	Secondary	Minor
Cobalt			●
Molybdenum	■	■	■
Nickel		▲	▲
Titanium	▲	▲	▲
Tungsten	●	●	●
Vanadium			▲

Open symbols denote districts where the exploitation of deposits is uncertain

Scale 1 : 60 m. approx.

Nikel Yena · Pechenga · Monchegorsk
Schneeberg · Annaberg · Jáchymov
Pudozhgora
Kerch
Ege–Khaya
Endybal'sk
Noril'sk
Umal'tinskiy
Pokrovskiy Rudnik
Pervoural'sk · Valerianovsk
Rezh · Verkhnyaya Pyshma
Polevskoy · Berezovskiy
Magnitka · Verkhniy Ufaley
Balkany
KHALILOVO GROUP · Aydyrlinskiy · Orsk
Batamshinskiy
Tyrny–Auz
Dashkesan · Kadzharan
Vershino–Shakhtaminskiy
Bukuka · Gutay
Kolyvan' Ubaredmet · Gorodok
Akchatau
Kounradskiy
Tabochar · Kara–Tau
Lyangar · Mayli–Say
Köytash · Tyuya–Muyun · Kum–Bel'
Varzob · Chorukh–Dayron

BUOTAMA VALLEY
Nikolayevsk
Komsomol'sk

The AMURSTAL' works at KOMSOMOL'SK is the only heavy metallurgical works in the Soviet Far East.

built almost GORSK

Nizhne–Ilimsk
Small scattered deposits of the Angara–Ilim area
Bratsk
Mazul'skiy
OL'KHON I.
Berezovskoye
Scattered deposits of the Little Khingan Mountains
Ol'ga
Gur'yevsk
Stalinsk
Usa · Abakan
Temir–Tau · Abaza
Shalym
Tashtagol · Kochura
Turochak
Chadan
Balbagar
Petrovsk
Babushkin

Deposits in the Mongolian Peoples' Republic are approximately located as exact information is not available. None are believed to be worked.

DONBASS

0   100 Miles

7 Almaznaya
2 Dneprodzerzhinsk
4 Kramatorsk
10 Krasnyy Sulin
6 Lisichansk
5 Makeyevka
3 Nikopol'
11 Taganrog
8 Voroshilovsk
9 Yenakiyevo
1 Zheltaya Reka

Dnepropetrovsk
Konstantinovka
Krivoy Rog
Nikopol' Zaporozh'ye
Stalino
Zhdanov

00 miles approx.

In the European U.S.S.R. the Yena deposit is very large, being estimated in 1938 at approaching 1,000 million tons of magnetite with Fe content up to 48 %. Further south, the titaniferous magnetite deposit at Pudozhgora is relatively small (under 100 million tons), but of great interest because of the composition of the ore, which includes only 26·5 % iron, but up to 10 % titanium, 1·5 % manganese, 27 % SiO₂, and a trace of vanadium. Like Yena, however, this deposit is still under-exploited because of its remoteness. Elsewhere in European Russia, the most noteworthy deposit is at Uryupinsk with over 700 million tons of rather low quality magnetite. Great interest is also being shown in the curious feature of central European Russia known as the Kursk Magnetic Anomaly (pp. 20/21 & 62/63). The first mine was sunk in 1939 at Gubkin.

Elsewhere the Dashkesan magnetite deposit in Azerbaydzhan is the most important at present exploited on a large scale. Several deposits in Central Asia, notably near Karaganda, are likely to become more significant as communications improve.

Production of Iron Ore by Regions in the U.S.S.R.

	1940 ('000 tons)	%*	1950 ('000 tons)	%*
Ukraine¶	18,700	62·5	21,000	52·5
Other U.S.S.R. in Europe†	1,400	4·7	1,400	3·5
U.S.S.R. in Asia§	9,800	32·8	17,600	44·0
Total	29,900	100·0	40,000	100·0

¶ Incl. Rostov Oblast and Crimea   † Excl. Urals   § Incl. Urals
* percentage of total

IRON ORE IN EASTERN EUROPE

All the East European countries except Albania and Bulgaria produce iron ore. Czechoslovakia is the largest producer, but the ores average less than 42 % Fe content, and some are very low grade quartzites. The Romanian ores average 40 %, but reserves are small, being estimated at 40 million tons in 1948.

Production of Iron Ore in Eastern Europe

(in '000 metric tons)

	1938	1946	1950	1953	1954
Czechoslovakia	1,420	1,116	1,575	2,300	...
Poland	872	395	793	1,345	1,600
Hungary	370	133	368	.	...
Romania	139	107	395	675	700
E. Germany	...	...	400	1,299	...
Yugoslavia	607	394	826	795	...

Note: For the years 1946 and 1950 the figures for Yugoslavia are based on official data. The others are unofficial estimates.

FERROUS METALS IN THE U.S.S.R. AND EASTERN EUROPE

The Soviet Union is abundantly provided with manganese and chromium and is the world's largest producer of both, adequately provided with nickel, titanium and vanadium and notably short of cobalt, molybdenum and tungsten. " Shortage " is a relative term since the nature of the metallurgical industry is determined largely by the materials available, and the standards that apply in, say, the U.S.A. do not necessarily have the same validity in the U.S.S.R. In alloying metals, various substitutions are possible. The only serious disadvantage the Soviet Union is likely to suffer is the fact that the three metals of which supplies are shortest—cobalt, molybdenum and tungsten—are the three most usually substituted for one another in high speed steels.

MANGANESE

The two principal Soviet deposits of manganese ore, at Nikopol' and Chiatura, contain between them almost a third of world reserves. Nikopol' (reserves estimated in 1938 at 522 million tons) is the largest deposit in the world. Reserves at Chiatura are estimated at 211 million tons. Total reserves of the Soviet Union may be nearly 800 million tons, sufficient for 160 to 200 years at present rates of consumption.

Manganese is valuable as a deoxidiser in steel production, and also assists in removing sulphur; and as an alloy metal in steels of high tensile strength and resistence to abrasion.

Minor uses are in electric batteries and in the chemicals industry, but these account for less than 5 % of consumption. The principal ores are pyrolusite (MnO₂) and psilomelane (H₄MnO₅). The first of these is often found in association with iron ore, and is not then usually separated from the iron in smelting; where it is absent, or present only in small quantity, extra manganese is added to produce ferro-manganese, which becomes the basis of manganese steel.

Before the war, the U.S.S.R. had a large exportable surplus of manganese, and in some years was the world's largest exporter; in 1937 about a million tons of ore were shipped abroad. During the war the Nikopol' mine was lost to the Germans, and exports almost stopped, while Soviet industry was forced to rely on the resources of Chiatura and a few smaller deposits. Exports were resumed in 1945, though on a smaller scale than previously. Between 1948 and 1953 exports to western countries were insignificant, but in 1954 shipments began again, especially to the United Kingdom and France.

In Eastern Europe, Czechoslovakia is a significant producer, though the ores are of low quality (below 30 % Mn content). Hungary, Romania and Yugoslavia show appreciable production, and there is also a very small output in Bulgaria, which reached 3,000 tons in 1937.

Production of Manganese Ore in the U.S.S.R. and Eastern Europe

	1938 ('000 tons)	%*	1946 ('000 tons)	%*	1950 ('000 tons)	%*	1953 ('000 tons)	%*
U.S.S.R.	2,283	45·4	2,600	58·2	3,900	57·4	5,000	54·8
Czechoslovakia	105	2·0	...	...	168	2·5	240	2·6
Hungary	22	0·4	15	0·3	...	...	...	...
Romania	60	1·1	19	0·4	25	0·4	...	...
Yugoslavia	4	0·1	8	0·2	13	0·2	10	0·1

* percentage of world total

CHROMIUM

Chromium, important as an alloy metal since the invention of stainless steel, is used wherever corrosion is likely to occur, and also increases the resistance of steel to high pressures, temperatures and speeds. Cast iron and steel are often electroplated with chromium.

Until 1934 the Soviet Union imported ferro-chrome, but this was for technical reasons rather than because of any shortage of chromium. The U.S.S.R. and Turkey are the largest producers of chromite in the world, with roughly equal output.

The first large deposit of chromite to be exploited was at Sarany, which produced some 10,000 tons of ore in 1926, when total production was over 30,000 tons; the rest came from scattered smaller deposits in the Urals. The quality of the Urals ore tends to be low, and the industrial use of chromium did not develop fully till after the discovery of the vast Khrom Tau deposit in 1937. This deposit contains over 15 million tons of chromite, and is probably the largest in the world.

In Eastern Europe, only Yugoslavia is a major producer, though Albania is said to have produced 52,000 tons of " chrome ore " in 1950. Chromite is found also in Bulgaria (1943 estimate 7,000 tons) and in Romania, where there has been little development.

Production of Chromite in the U.S.S.R. and Yugoslavia

	1938 ('000 tons)	%*	1946 ('000 tons)	%*	1950 ('000 tons)	%*	1953 ('000 tons)	%*
U.S.S.R.	...	...	...	...	457	20·2	610	17·2
Yugoslavia	50	4·6	93	8·1	115	5·1	127	3·4

* percentage of world total

NICKEL

Nickel is the main alloy metal used in stainless steels and other non-ferrous alloys where resistance to corrosion is important; it is also used for electro-plating, and in special alloys where low conductivity of heat and electricity are demanded. The principal ores are pentlandite, a compound sulphide of nickel and iron which is usually found in association with other metallic ores, especially copper; and the more complex garnierite, a hydrated silicate of nickel and manganese.

No nickel was produced in the Soviet Union before 1934, but to-day the country is self-sufficient. Known reserves are adequate for three decades of consumption at the present rate, and further large deposits are confidently believed to exist. The earliest centre of production was at Verkhniy Ufaley. Monchegorsk came into production in 1938, and Orsk and Noril'sk in 1940. After the war the U.S.S.R. acquired from Finland the very important mines near Pechenga, which helped to sustain production until the restoration of Monchegorsk, put out of action by German bombardment.

In Eastern Europe, only Germany produces nickel, and there the output is very small.

Production of Nickel Ore in the U.S.S.R.

(in terms of metal)

	('000 tons)†	%*
1938	2·6	2·3
1946	20·3	15·0
1950	29·0	20·0
1953	39·6	19·2

† estimate   * percentage of world total

MOLYBDENUM

Molybdenum supplies are inadequate for Soviet industry, and imports and rigorous substitution have been necessary. Molybdenum is an alloy metal used in high-speed steels; the pure metal is also used in filaments in lamps and thermionic valves. The main substitute metal for alloys is tungsten, of which supplies within the U.S.S.R. are also inadequate.

The chief source of molybdenum in pre-war days was the tungsten-molybdenum deposit at Tyrny Auz. This mine and its installations were destroyed in 1941 to prevent their falling into German hands, and the work of restoration seems to have been slow. The Umal'tinskiy mine supplied only a small part of Soviet needs, but more was obtained with great difficulty from the copper ores of Kounradskiy.

No molybdenum is known to exist in Eastern Europe, apart from an insignificant deposit in Yugoslavia and one in Romania which was largely worked before the war. Soviet consumption since the war has been at the rate of 1,000–2,000 tons of metal a year, but this has probably been chiefly imported from Manchuria and North Korea.

TUNGSTEN

Tungsten is derived chiefly from the mineral wolfram. Two mines were worked in the Chita area before 1917, and numerous other small deposits were brought into production subsequently. After 1935 Tyrny Auz became the main source, and total output for the whole country rose to perhaps 2,000 tons of concentrate a year. After the loss of Tyrny Auz, fresh reliance was placed on the Chita deposits, and the Gorodok mine was heavily worked and perhaps almost exhausted. Total reserves were estimated in 1945 at up to 30,000 tons (60% WO₃), which is adequate for only a few years' consumption unless supplemented by imports or substitution. Tungsten is used chiefly as a hardening agent in high-speed steels and cutting instruments.

There is a very small output of tungsten (in association with tin and bismuth) in Eastern Germany, but otherwise none in Eastern Europe. Soviet output figures are not available.

COBALT

Cobalt is employed to increase the magnetic qualities of steel, and also as a hardening agent, but its main uses are in the chemicals industry and as a pigment. The development of jet aircraft has increased the demand for cobalt steels which are resistant to very high speeds and temperatures.

Most of the cobalt produced in the U.S.S.R. is mined in association with nickel; its distribution closely follows that of the nickel deposits; recoveries are said, however, to be very low. The main sources are in the southern Urals and northern Kazakhstan. Total reserves have been estimated at 25,000 tons (metal content), which would suggest exhaustion in under twenty years.

A little cobalt is mined in East Germany, but there is none anywhere else in Eastern Europe. Soviet production figures are not published, and no reliable estimates exist.

TITANIUM AND VANADIUM

Prospects for the two remaining alloy metals are relatively favourable. Only a small proportion of titanium supplies is used in alloys, the greater part being consumed in the manufacture of pigments. Vanadium is a valuable alloy metal, especially as a means of reducing the liability of steel to fatigue. Titanium is found chiefly in the minerals ilmenite and rutile, though some Soviet production is from titaniferous magnetites. Vanadium is sometimes associated with mica and sometimes with titanium, but the main source in the U.S.S.R. is as a by-product from the iron ores mined at Kerch'.

Supplies of both metals seem to be adequate for at least several decades' consumption, though no production figures are available. Neither metal is found in Eastern Europe.

ENGINEERING

U.S.S.R.

Originally the engineering industry was closely associated with the production of iron and steel, and its location was determined by the location of smelting centres. Early in the nineteenth century this ceased to be wholly true, and engineering became more widely dispersed. New engineering centres were developed in areas where iron and steel were not produced but where the development of railways made the transport of heavy materials easier.

With the growth of the great metallurgical industries of the Urals and the Kuzbass, engineering also developed, using the metal which these industries produced. During the Second World War engineering in these districts was energetically developed for strategic as well as purely economic reasons. As large-scale agriculture was pushed eastward, new centres were needed for the manufacture of agricultural machinery; similarly the opening up of minerals deposits has made it desirable that mining machinery should be produced fairly near. Just as the development of the Donbass had stimulated the growth of engineering in towns like Kramatorsk which now supply large quantities of mining equipment, so the goldfields and other mineral industries of Eastern Siberia are equipped from relatively new engineering centres such as Irkutsk.

The gross value of engineering production rose from an index number of 100 in 1940 to 437 in 1954, little short of the 1955 planned level of 460. The index of agricultural machinery was stated to have risen from 100 in 1940 to 1,280 in 1954, including an increase in tractor production from 52,000 (15 h.p. units) in the earlier year to 137,000 in the later. Other production figures for 1954 include 5,027,000 kwh. steam turbines, 2,396,000 bicycles, 1,247,000 sewing machines, 2,890,000 radio sets and over 3 million electric irons. The index of output of oil refinery equipment (1940=100) stood at 720 in 1953.

EASTERN EUROPE

Before the Second World War only Eastern Germany, Bohemia and Upper Silesia possessed highly developed engineering industries. Hungary was an exporter of railway rolling stock; oil drilling machinery was made in Romania, and agricultural machinery at centres scattered over the whole area. Hungary, Romania, Bulgaria, Albania and Yugoslavia were, however, obliged to import most of their engineering requirements, chiefly from Germany.

The East German industry, which was famous for its precision engineering, was largely dismantled immediately after the war, but restarted in 1949; in 1953, output included 15,300 tractors, 14,000 commercial vehicles and 29,000 motor cars. Czechoslovakia has not only expanded the types of engineering traditionally practised, but has become a very large producer of electric motors (384,000 in 1953). Poland, after acquiring some of the German industries in Silesia, has greatly enlarged her production of rolling stock, machine tools, bicycles and radio sets (269,000 in 1953), and supplies most of her own needs in agricultural machinery and tractors. In Hungary and Romania the expansion has been decidedly less marked, and Albania still has little engineering. Yugoslavia is now a small exporter of agricultural machinery.

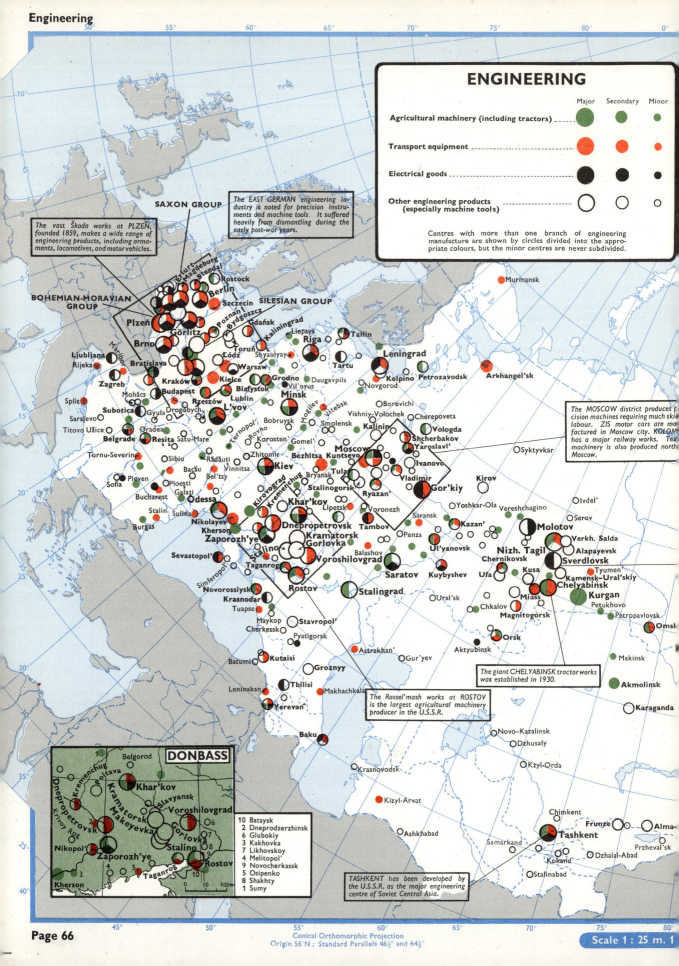

ENGINEERING

	Major	Secondary	Minor
Agricultural machinery (including tractors)			
Transport equipment			
Electrical goods			
Other engineering products (especially machine tools)			

Centres with more than one branch of engineering manufacture are shown by circles divided into the appropriate colours, but the minor centres are never subdivided.

The vast Škoda works at PLZEŇ, founded 1859, makes a wide range of engineering products, including armaments, locomotives, and motor vehicles.

The EAST GERMAN engineering industry is noted for precision instruments and machine tools. It suffered heavily from dismantling during the early post-war years.

The MOSCOW district produces p
cision machines requiring much skil
labour. ZIS motor cars are ma
factured in Moscow city. KOLOM
has a major railway works. Tex
machinery is also produced north
Moscow.

The giant CHELYABINSK tractor works was established in 1930.

The Rossel'mash works at ROSTOV is the largest agricultural machinery producer in the U.S.S.R.

TASHKENT has been developed by the U.S.S.R. as the major engineering centre of Soviet Central Asia.

DONBASS

10 Bataysk
2 Dneprodzerzhinsk
6 Glubokiy
3 Kakhovka
7 Likhovskoy
4 Melitopol'
9 Novocherkassk
5 Osipenko
8 Shakhty
1 Sumy

SAXON GROUP
BOHEMIAN-MORAVIAN GROUP
SILESIAN GROUP

Conical Orthomorphic Projection
Origin 56°N ; Standard Parallels 46½° and 64½°

Scale 1 : 25 m. 1

EASTERN EUROPE

EASTERN GERMANY
5 Aschersleben
3 Brandenburg
6 Dessau
7 Erfurt
9 Gera
4 Hennigsdorf
8 Jena
10 Plauen
2 Rathenow
1 Stendal
11 Zwickau

CZECHOSLOVAKIA
14 Česká Třebová
12 Cheb
13 Kladno

POLAND
19 Bytom
20 Chorzów
17 Opole
21 Pabianice
15 Racibórz
16 Świdnica
18 Zabrze

Rostock
Demmin
Wismar
Neubrandenburg
Magdeburg
Berlin
Halle
Leipzig
Döbeln
Dresden
Görlitz
Frankfurt
Zielona Góra
Poznań
Leszno
Wrocław
Chomutov
Prague
Plzeň
Domažlice
Kroměříž
Brno
Gottwaldov
Moravská Ostrava
Bratislava
Gdańsk
Koszalin
Kaliningrad
Chełmno
Elbląg
Szczecin
Bydgoszcz
Toruń
Gniezno
Warsaw
Żyrardów
Łódź
Kielce
Kraków
Rzeszów
Tĕšín/Cieszyn
Košice
Váç
Miskolc

0   50   100m

Magadan

Yakutsk

Nikolayevsk
Aleksandrovsk

Komsomol'sk
Yuzhno-Sakhalinsk
Nevel'sk

Bodaybo

Svobodnyy
Khabarovsk

Blagoveshchensk

Tomsk
Krasnoyarsk

Ust'-Kut

Cheremkhovo
Chita

Spassk-Dal'niy

Irkutsk
Ulan-Ude

Nakhodka
Vladivostok

Stalinsk
...ovka
...haul
Biysk

Rubtsovsk
...i palatinsk
Ust'-Kamenogorsk

Ulan – Bator

Large railway works at ULAN-UDE supply locomotives and rolling stock for the Trans-Siberian Railway.

...sh works at SVERDLOVSK ...avy machinery for the ...try in the Urals.

The new industrial section of KRAS-NOYARSK, developed since 1930, makes agricultural machinery, railway equipment, and mining machinery.

MOSCOW DISTRICT

4 Dmitrov
7 Elektrostal'
3 Klin
5 Kol'chugino
10 Kolomna
6 Kuntsevo
8 Murom
12 Ozery
9 Pavlovo
13 Plavsk
11 Serpukhov
2 Shuya
14 Skopin
1 Vichuga

Shcherbakov
Yaroslavl'
Ivanovo
Gor'kiy
Moscow
Vladimir
Tula
Ryazan
Stalinogorsk

0   50   100m

NON-FERROUS METALS I

COPPER, TIN, LEAD, ZINC, GOLD, SILVER
MAGNESITE AND MAGNESIUM SALTS

COPPER

Copper, a soft metal with very high conductivity of electricity, is essential to the development of an electrical industry and the supply of power. It is also used in a number of alloys. The main sources are sulphides of the metal, sometimes in combination with iron.

Copper ores have been mined in the U.S.S.R. for many decades, and were first found in the Urals and Transcaucasia. The big deposits in the Kazakh S.S.R. were discovered in the mid-1930s, and are now probably the principal source. The deposits at Dzhezkazgan are said to be the second largest in the world.

Until after the Second World War the Soviet Union imported copper, since native supplies were not fully exploited. To-day, annual production of well over 300,000 tons of metal is adequate for all ordinary needs, though there is some doubt as to whether the supply of copper will keep pace with the development of the electrical industry and the great expansion of power now in progress. Total known reserves in 1948 were estimated at about 20 million tons, which should be sufficient for probable needs for thirty to forty years. Some of the less accessible deposits in the European north, in Central Asia and in Siberia may prove to be larger than is at present established.

In Eastern Europe, only Yugoslavia produces significant quantities. Much of this production is exported to western countries. Small quantities are produced in Albania, Bulgaria and Romania, and there is some output in Slovakia and Polish Silesia. In 1953 about 20 tons of metal were also produced in East Germany.

In 1938, Soviet output of copper accounted for about 5% of the world total; by 1952 the proportion had risen to 12% The Yugoslav share in world output declined from 2·4% to 1·3%.

Production† of Copper in the U.S.S.R. and Yugoslavia

(in terms of metal)

	1938		1946		1950		1953	
	(tons)	%*	(tons)	%*	(tons)	%*	(tons)	%*
U.S.S.R.	100,000	5·0	147,000	8·2	255,000	10·0	345,400	11·6
Yugo-slavia	49,000	2·4	24,500	1·3	43,400	1·7	35,300	1·3

† Estimated.   * percentage of world total.

TIN

The largest single use of tin is in the manufacture of tinplate for containers, though in fact well over half the world consumption is in the form of alloys. These range from babbitt bearings to " copper " coinage, and include gun metal, bell metal, bronzes and many others. Some salts of tin are important in the chemicals industry. Almost all tin comes from the mineral cassiterite (SnO_2), which is found in alluvial deposits and in certain quartzes.

The first tin-mine in the U.S.S.R. was opened in 1926 at Olovyannaya (*olovo* is the Russian word for tin), and by 1949 this group of mines was producing over 4,000 tons of metal a year. The Leninogorsk group in Kazakhstan came into operation in the late 1930's, and in 1949 was the second largest source of supply, producing perhaps 1,500 tons of metal.

These supplies fall far short of Soviet needs, and imports reached a pre-war peak of 25,530 tons in 1937. Wartime

imports were higher still, though exact figures are not available. No reliable figures of current production are published, but output is perhaps about 12,000 tons a year, and imports are now much reduced. The gap between supplies and normal needs is being reduced partly by drawing on stocks built up during the war, but chiefly by severe restrictions on consumption; for instance, great energy has been devoted to developing new alloys without tin, or with smaller proportions than formerly.

The long-term prospect is less discouraging, however. Large deposits are now known to exist at Mikoyanovsk (where there has been some production since 1949) and at Pyrkakai. These deposits are, however, remote from communications and centres of consumption, and climatic conditions will make them very difficult to operate.

The only tin producing centre in Eastern Europe is near Altenberg in East Germany.

No detailed figures of Soviet production are available. If the estimate of 12,000 tons a year is correct, this would represent about 6·7% of the world total.

Production of Tin in East Germany

(in terms of metal)

	(tons)	
1947	112	Annual output in 1947–49 amounted to about 0·1% of the world total, rising to 0·3% in 1953
1948	158	
1949	159	
1950	193	
1951	260	
1952	401	
1953	572	

LEAD

Nearly a third of world output of lead is consumed in the manufacture of storage batteries and accumulators. Another sixth is used as a cover for cables, since lead is resistant to corrosion, and about 10% each for paint making (white lead), in plumbing and roofing, and for ammunition. Much the most important lead ore is galena (PbS), which is frequently found in association with zinc ores, and often with silver as well.

In 1926, the U.S.S.R. smelted only 1,342 tons of lead, and large imports were needed until the end of the Second World War; by 1937, however, home production was covering more than half of current needs, which appear to have included an allowance for stockpiling. During the war the two principal mines, at Achisay and Kolyvan', were so heavily worked that they were almost completely exhausted. The main source of supply is now the Leninogorsk group, with Sadon and Tetyukhe next in order of importance. Annual output is now probably enough to cover needs, and represents some 13% of the world total.

Reserves were estimated in 1936 at 4,123,600 tons, or adequate for nearly seventy years at the rate of consumption then current. It is not known how far new discoveries have balanced the exhaustion of Achisay and Kolyvan', and consumption has increased greatly. Even so, supplies should be adequate for several decades. The wide dispersion of the reserves in small deposits will, however, make extraction increasingly costly.

In Eastern Europe, lead ores are mined in Poland and Yugoslavia, and in smaller quantities in Czechoslovakia and Romania. Yugoslav production has been expanding rapidly for several years, and large reserves are known still to exist.

Smelter Production of Lead in the U.S.S.R. and Eastern Europe

(in terms of metal)

	1938		1946		1950		1953	
	(tons)	%*	(tons)	%*	(tons)	%*	(tons)	%*
U.S.S.R.	65,000	4·0	83,000	7·6	127,000	7·7	244,000	13·0
Poland	20,100	1·2	10,900	1·3	20,000	1·2	21,000	1·1
E. Germany	...	...	...	...	...	...	22,000	1·2
Czecho-slovakia	5,000	0·3	3,038	0·3	...	...	8,000	0·4
Romania	5,655	0·3	3,224	0·3	4,200	0·2	6,300	0·3
Yugo-slavia	8,620	0·5	32,590	2·9	57,202	3·4	70,793	3·8

There is also some very small production in Bulgaria; no figures are available. The estimates for Romania for 1950 and 1953 are highly conjectural.

* percentage of world total

ZINC

Zinc is used chiefly for galvanizing iron and steel to prevent rusting when exposed to the weather. It is also one of the constituent metals of brass, and zinc oxide is extensively used as a pigment. The chief ore is sphalerite or zinc blende (Zn S), but there are several other forms derived from it. Zinc ores are often found in association with lead.

Before 1932 little zinc was produced in the U.S.S.R. (1,888 tons in 1926, the first year for which records exist), and large quantities were imported yearly till the end of the Second World War. The first mine worked was at Sadon, but by 1932 new mines were in operation in the Salair-Zmeinogorsk area and at Tetyukhe. Eventually the Leninogorsk group became the chief source. Zinc ores are associated with lead in all areas except the Urals, where they are found in conjunction with copper and silver.

The 1953 estimated output of 228,000 tons is more than adequate for normal consumption, and represents nearly 10% of world output. Reserves were estimated in 1937 at 9·9 million tons (of which 2·43 million tons were from copper ores), but are now believed to be much higher. It would seem that known reserves will supply normal requirements for at least several decades.

Poland is also a significant producer. Production in Yugoslavia is considerable, and there is some output in Bulgaria, Czechoslovakia and Romania, as well as at Freiberg in East Germany where zinc is recovered as a by-product of other types of mining.

Smelter Production of Zinc in the U.S.S.R., Poland and Yugoslavia

	1938		1946		1950		1953	
	(tons)	%*	(tons)	%*	(tons)	%*	(tons)	%*
U.S.S.R.	80,000	4·3	90,000	5·8	142,000	7·3	228,000	9·8
Poland	70,000	3·7	56,500	3·4	...	...	142,000	6·1
Yugo-slavia	42,000	2·2	23,000	1·4	38,500	2·0	59,970	2·6

Production in Czechoslovakia was about 2,000 tons in 1938

* percentage of world total

GOLD AND SILVER

GOLD

The Soviet Union is now almost certainly the second largest producer of gold in the world; some estimates would put it even above the Union of South Africa. Unfortunately no statistics have been published during the last thirty years, and any attempt to establish actual output is a matter of guesswork.

Of the fields at present exploited, much the oldest is in the Urals, where gold has been mined at least since about 1745. The first placers to be exploited in the big Lena field were discovered around 1830, about the same time as production was beginning in the Kiya Valley. The Amur field came into operation about 1867, and the Aldan field in the closing years of the Empire. Most of these deposits are in remote parts of the country, and in Tsarist days were worked largely by political prisoners and exiles. From 1911 to 1915, total output from all fields averaged about 1,300,000 fine troy ounces a year.

Output declined sharply after the Revolution, and the average for the years 1916–1920 was only 626,000 ounces. The 1911–1915 level was not passed till 1930. Thereafter the rate of increase was rapid, to somewhere around 5½ million ounces in 1937. The expansion was brought about partly by improved techniques and a shift from placer production to mining, partly by the opening up of new fields. Production in the Aldan field surpassed that of the Lena Valley before 1930, and remained at the head of the list till well after the second World War. The Allakh-Yun field came into production in 1942, and it is now probably the largest producer of all. Output from all fields in the post-war period is variously estimated from 2 to 18 million ounces a year.

No information is available on reserves, but they are likely to be large. There is every prospect that the Soviet Union will have adequate supplies of gold for many years.

In Eastern Europe, there is considerable production in Romania (roughly 112,500 ounces in 1949) and small quantities are mined in Czechoslovakia, East Germany, Bulgaria and Yugoslavia.

SILVER

However little information there is on the subject of gold, there is less still about silver. One estimate sometimes quoted puts Soviet output of silver in 1951 at 25 million ounces, or about 12% of the world total; this figure is some 3½ times as high as the usual pre-war estimate of about 7 million ounces. Data on reserves of silver are completely lacking.

Silver is mined in Czechoslovakia, Romania and Yugoslavia. Production figures since the war are as follows:

Production of Silver Ore in Eastern Europe

(in terms of metal)

('000 f.t. oz.)

	1946	1950	1953
Czechoslovakia	600	...	...
Romania	481*	...	...
Yugoslavia (refined silver)	...	2,387	3,048

* 1947

MAGNESITE AND MAGNESIUM SALTS

Magnesium is strictly speaking a non-ferrous metal, but magnesite is often classed as a non-metallic mineral since only a small proportion of world output goes to provide metallic magnesium. The pure metal is used in alloys where lightness and rigidity are required, as a catalyst in a number of chemical processes, and as an explosive. Magnesite ($MgCO_3$) is a refractory used for lining open hearth steel furnaces, and as a source for a wide range of chemicals. Dolomite is an alternative form of the mineral, but has the disadvantage of a lower melting point. Carnallite is also a valuable source of magnesium salts and of the metal. (See pp. 76/77).

Production of magnesite in the Soviet Union passed the 100,000 ton mark in 1926, and rose to 850,000 tons by 1937; in the latter year, over a million tons of dolomite were produced. Both contributed significantly to the growth of the iron and steel industry between these years. Little in the way of magnesium salts was produced before 1926, but demand was also low; since then, the big carnallite deposit at Solikamsk has been exploited, and supplies all current needs. Reserves of all types of magnesium mineral seem to be almost unlimited. No production figures since 1937 are available.

Czechoslovakia produced up to 90,000 tons of magnesite a year before the war, but post-war production is not known. There are deposits of both magnesite and carnallite in East Germany, but no production figures are published, and information is insufficient for a useful estimate to be made.

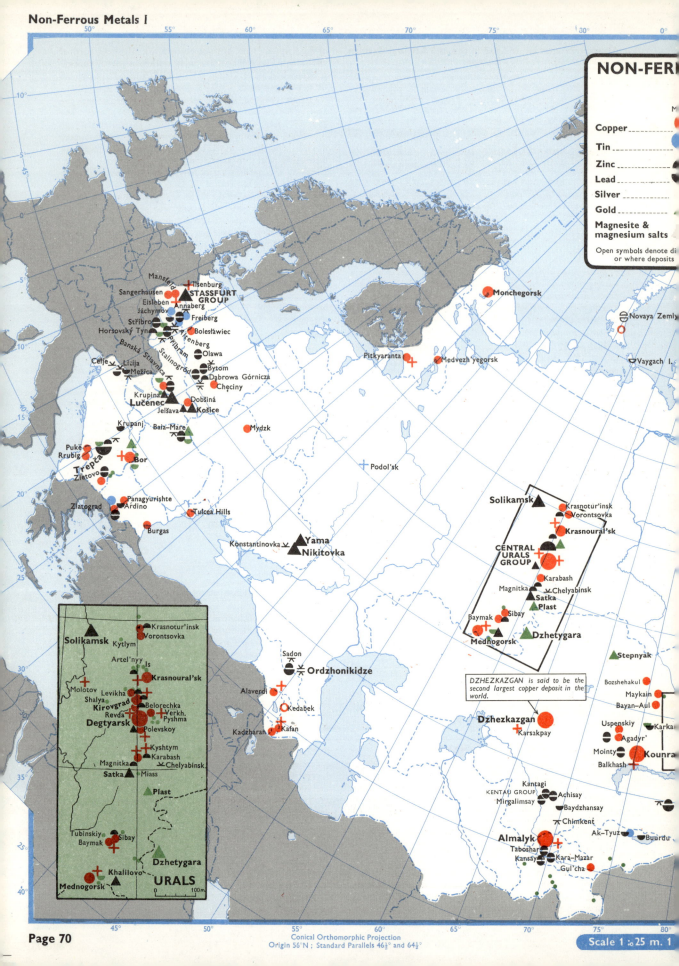

NON-FER[...]

Copper _____
Tin _____
Zinc _____
Lead _____
Silver _____
Gold _____
Magnesite &
magnesium salts

Open symbols denote di[...]
or where deposits [...]

Mansfeld
Ilsenburg
STASSFURT
GROUP
Sangerhausen
Eisleben
Jáchymov
Annaberg
Freiberg
Stříbro
Bolesławiec
Horšovský Týn
Altenberg
Banská Štiavnica
Olawa
Pribram
Stalingród
Celje
Bytom
Líčija
Dąbrowa Górnicza
Mežica
Chęciny
Krupina
Lučenec
Dobšiná
Jelšava
Košice
Krupanj
Baia-Mare
Puké
Rrubíg
Trepča
Bor
Zletovo
Zlatograd
Panagyurishte
Ardino
Burgas
Tulcea Hills

Monchegorsk
Novaya Zemly[a]
Vaygach I.
Pitkyaranta
Medvezh'yegorsk

Mydzk

Podol'sk

Konstantinovka
Yama
Nikitovka

Solikamsk
Krasnotur'insk
Vorontsovka
Krasnoural'sk
CENTRAL
URALS
GROUP
Karabash
Magnitka
Chelyabinsk
Satka
Plast
Baymak
Sibay
Dzhetygara
Mednogorsk

Stepnyak

Sadon
Ordzhonikidze

Alaverdi
Kedabek
Kadzharan
Kafan

Bozshehakul
Maykain
Bayan-Aul
Uspenskiy
Karka[...]
Agadyr'
Mointy
Kounra[...]
Balkhash

DZHEZKAZGAN is said to be the
second largest copper deposit in the
world.

Dzhezkazgan
Karsakpay

Kantagi
KENTAU GROUP
Achisay
Mirgalimsay
Baydzhansay
Chimkent
Almalyk
Ak-Tyuz
Buurdu
Taboshar
Kansay
Kara-Mazar
Gul'cha

Urals inset

Solikamsk
Kytlym
Krasnotur'insk
Vorontsovka
Artel'nyy
Is
Krasnoural'sk
Molotov
Levikha
Belorechka
Verkh.
Pyshma
Shalya
Kirovgrad
Revda
Degtyarsk
Polevskoy
Kyshtym
Karabash
Magnitka
Chelyabinsk
Satka
Miass
Plast
Tubinskiy
Sibay
Baymak
Dzhetygara
Mednogorsk
Khalilovo
URALS
0   100 m.

Conical Orthomorphic Projection
Origin 56°N ; Standard Parallels 46½° and 64½°

Scale 1 : 25 m. 1[...]

METALS

stricts   Smelting

exploitation is uncertain
t to be exploited.

Serdtse-Kamen

Pyrkakai

Seymchan

KOLYMA VALLEY   Orotukan

Ege-Khaya

Imtonzha   Endybal'sk

Noril'sk

**ALLAKH-YUN
FIELDS**   El'dikan

**VILYUY
VALLEY
FIELDS**

The ALLAKH-YUN goldfield was developed during 1942.   It may now be
one of the major producing districts.

Ayan

**ALDAN
FIELD**

**LENA-VITIM
FIELD**   Nagornyy   Lukachek   Mayskiy
Bomnak   Sofiysk
Severo-Yeniseyskiy   Bodaybo   Stoyba
**YENISEY
FIELDS**   Razdolinsk   Tyndinskiy   **ZEYA
VALLEY
FIELDS**
Never
Skovorodino

Mogocha   Mikoyanovsk   Sinancha
Bagdarin   **Tetyukhe**
Ust'-Karsk   Lifudzin
KIYA VALLEY
FIELDS   **Darasun** **Baley**   Tetyukhe-Pristan'
Salair   Belovo   Artemovsk   **Olovyannaya**   Sherlovaya Gora
Orlik   **Khapcheranga**
Kolyvan'
**LENINOGORSK
GROUP**   Chadan   Gutay
**IRTYSH
FIELDS**

Deposits in the Mongolian People's Republic
are approximately located as exact information is not available.   None are believed
to be worked.

Inset map:
Zmeinogorsk   Kolyvan'
Gornyak
Glubokoye   **Leninogorsk**
Ubaredmet   **Zyryanovsk**
Belousovka   Abaketka
**AKZHAL
GROUP**   Ust'-
Kamenogorsk
Zharma   Targyn   Cherdoyak
Alekseyevka

ALTAY KRAY &
E. KAZAKH S.S.R.
0   100 m.

0 miles approx.

NON-FERROUS METALS II

ALUMINIUM & BAUXITE, ANTIMONY, BARIUM, BERYLLIUM, COLUMBIUM & TANTALUM, MERCURY, PLATINUM, ZIRCONIUM, ARSENIC

ALUMINIUM AND BAUXITE

A light tough metal relatively resistant to corrosion and with a high conductivity of electricity, aluminium is used for a great variety of purposes, mainly in the alloy duralumin, which contains up to 95 % of aluminium, 4 % or more of copper, and small quantities of magnesium or manganese. It is especially valuable where strength and lightness must be combined, as in aircraft manufacture and in railway rolling stock and motor vehicles. Household metalware is an important consumer. Pure aluminium is also sometimes used for long-distance electrical transmission cables. By far the largest source of the metal is bauxite ($Al_2O_3.2H_2O$), but in the U.S.S.R. some is also obtained from other minerals. Most of the cryolite used in the refining of aluminium in the Soviet Union is produced synthetically.

Bauxite was first mined in the U.S.S.R. at Boksitogorsk in 1926, but it was not till 1932 that metal production began at Volkhov. In the previous year, 20,400 tons of metal had been imported. By 1937, when exploitation of the Urals deposits was beginning, output of metal reached 37,700 tons, but about 2,500 tons a year were still being imported. In 1952 the Soviet Union was the world's sixth largest producer of bauxite, and third only to the U.S.A. and Canada as a producer of aluminium metal. Production now exceeds demand, though partly from bauxite supplied from Hungary. Reserves in the U.S.S.R. appear to be adequate for at least forty years at the present rate of consumption.

In Hungary and Yugoslavia bauxite has been mined for many years, and large reserves are known to exist. Large proved reserves in Romania are not yet being exploited, but there is a little output in Albania.

Production of Aluminium Metal in the U.S.S.R., Hungary and Yugoslavia

	1938		1946		1951		1953	
	(tons)	%*	(tons)	%*	(tons)	%*	(tons)	%*
U.S.S.R.	50,000	8·7	80,000	10·6	210,000	11·8	290,000	11·6
Hungary	1,510	0·3	1,960	0·3	30,000	1·7	32,800¶	...
Yugo-slavia	1,205	0·2	565	0·1	2,820	0·2	2,792	0·1

Production† of Bauxite in the U.S.S.R., Hungary and Yugoslavia

	1938		1953	
	(tons)	%*	(tons)	%*
U.S.S.R.	250,000	6·2	1,000,000	7·5
Hungary	540,830	13·6	1,200,000	9·0
Yugoslavia	404,662	10·1	462,280	3·5

* percentage of world total   † estimated   ¶ 1954

ANTIMONY

Metallic antimony is not normally used by itself, but is combined with lead in battery plates or with other metals in bearings. Printing metal consists largely of antimony. Some of the salts are important in the chemical industry and in making paints and enamels, and are also used in treating textiles to make them flameproof. The metal is derived chiefly from stibnite (Sb_2S_3), which is commonly found in association with the ores of other metals, especially mercury.

No antimony was produced in the U.S.S.R. before 1929, and for some years after that there was only limited output at Frunze (Osh Oblast, not the capital of the Kirgiz S.S.R.) and Nikitovka. Almost all needs were supplied by imports. During the Second World War production expanded rapidly, and by 1948 some antimony was even being exported. Reserves already proved or known to exist should supply the current rate of consumption for many decades.

Czechoslovakia was a major producer before the war, mainly from mines in the Košice area, but post-war production has mostly been lower. In Yugoslavia there are two mines.

No detailed figures are available for production in the Soviet Union, but one estimate puts output in 1947 at between 3,700 and 4,500 tons of metal (9·2 % to 11·2 % of world output).

Production of Antimony in Czechoslovakia and Yugoslavia
(in terms of metal)

	1938		1946		1950		1953	
	(tons)	%*	(tons)	%*	(tons)	%*	(tons)	%*
Czecho-slovakia	Small†		2,250	7·8	2,000	4·1	...	...
Yugo-slavia	3,425	...	1,891	6·3	3,204	7·2	1,839	5·6

* percentage of world total   † 1935 production 2,429 tons

BARIUM

Metallic barium is little used, but its compounds are important in the paint, paper and chemicals industries, and the ore (barite or barytes) is used in preparing drilling-mud for oil-well boring. Barytes ($BaSO_4$) and witherite ($BaCO_3$) are the main ores, the second being comparatively rare.

The Soviet Union is badly supplied with barytes, though some of the known deposits are either under-exploited or not yet exploited at all. Production in 1940 was about 100,000 tons, which was perhaps 10 % of the world total, but this was below needs, and during the war the lack of barium minerals hindered the development of the petroleum industry. To-day there is no shortage, since large quantities are imported from East Germany, but if this source of supply were lost the position would again become serious. Reserves are probably adequate for several decades, but production is hampered by the inaccessibility of some of the major deposits in the Kazakhstan desert.

The Mühlhausen deposit in East Germany is one of the largest in the world; before the war Germany was the world's biggest producer, but no figures are available to show whether this is still so. In 1938 just over 400,000 tons were produced by Mühlhausen and its associated fields; by the end of the war output was probably down to 250,000 tons, but may since have expanded again.

Yugoslavia is also a significant producer. Exports in 1950 were 12,073 tons; in 1951, 9,850 tons; in 1952, 27,836 tons; and in 1953 jumped to 62,027 tons, or about 3½ % of world output.

BERYLLIUM

The electrical industry is the chief consumer of beryllium, which is added to copper and steel when extra resilience is needed in conjunction with a high conductivity of electricity. Beryllium oxide is also used as a high temperature refractory, having a melting point of 2400° Centigrade. The metal is found chiefly in association with granitic pegmatites, and the main deposit in the Soviet Union is at the emerald mining centre of Izumrud.

No production figures are available, but it appears that supplies are adequate for a considerable time. Metallic beryllium was first produced in 1934.

COLUMBIUM AND TANTALUM

Columbium (commonly called niobium in the U.S.A.) and tantalum are found in the same complex ore, known as columbite or tantalite according to which metal predominates. Both have the formula $(Fe, Mn) (Nb, Ta)_2 O_6$. There are also numerous variants such as loparite (which contains calcium), ilmenite, nephelite and murmanite, all of which are found in Murmansk Oblast, while ilmeno-rutile is the chief source in Chelyabinsk Oblast.

Columbium is used mainly as an anti-corrosive additive in steels likely to be subjected to high pressures or temperatures. Tantalum is employed by itself in electronic tubes; tantalum carbide is a very hard abrasive used in cutting tools; tantalum fluoride is valuable as a catalyst in synthetic rubber manufacture (p. 78/79).

The U.S.S.R. is apparently well supplied with both metals, and the loparite deposit at Lovozero is claimed as one of the largest in the world. Production began in 1936, but it was not till 1940 that it reached useful proportions. Output figures are not available.

MERCURY

Cinnabar (HgS) is almost the sole source of mercury, which is extensively used in the electrical industry, especially for thermostats and arc-rectifiers, and in the columns of thermometers and barometers. Long before the Revolution the Ukraine was among the world's chief producers, from a single mine at Nikitovka. During the Second World War Nikitovka was temporarily lost to the Germans, and the newly opened mine at Khaydarkan only partly supplied Soviet needs; but apart from that short period the Soviet Union has been self-sufficient in mercury ever since production at Nikitovka recovered from the disruption of the revolutionary period. Total reserves at Nikitovka, Khaydarkan and a few other smaller deposits are put at well over 20,000 tons (44 million lb.), and though current consumption is not known these reserves are clearly enough for many decades.

In 1938 output was of the order of 660,000 lb. (about 300 metric tons), or 5·8% of the world total. Reliable figures for later years are not available, but estimates put production in 1953 at 880,000 lb. (7·4% of the world total) and in 1954 at 935,000 lb.

Before the war Czechoslovakia was also a significant producer, but post-war output has been markedly smaller. In 1938 some 220,000 lb. of metal (1·9% of the world total) were produced, but in 1946 (the last year for which detailed data are available) output was only 64,651 lb. (29·3 metric tons). According to later estimates output in 1948 and 1949 was about 60,000 lb. a year. Yugoslavia produced about 490 tons in 1953.

PLATINUM

Very little information is available on platinum production in the U.S.S.R. Output of crude platinum is believed to be about 100,000 ounces a year (15% of the world total). This is a sharp drop from the estimated output of 175,000 ounces in 1946. Platinum reserves seem plentiful, though the older Urals deposits are probably nearing exhaustion; new sources are being exploited further north and in Siberia, especially at the new town of Noril'sk, where complex polymetallic ores are mined. At one time Tsarist Russia was the world's largest exporter of crude platinum, and it may be that the Soviet Union will again become a significant source of supply to the world market.

The Eastern European countries produce no platinum.

ZIRCONIUM

Zirconium is another metal chiefly used in the electrical industry, especially in electronic tubes and lamp filaments. It is also used in porcelain manufacture, as an explosive (in the powdered form of the pure metal), and as a deoxidizing agent in steel making. Zircon ($ZrO_2.SiO_2$), which is one of the principal ores, is a valuable refractory, used extensively in aluminium furnaces. The other chief ore is baddeleyite (ZrO_2).

There was little or no production in the Soviet Union before the Second World War, but the Zhdanov mine was in operation by 1946. The Zeya valley deposit is believed to be very large, but it is not known how fully it is being exploited; production figures are not available.

ARSENIC

The commonest ore of arsenic is arsenopyrite (FeAsS). The metal is used as a toughening agent in copper and some other alloys. Many of its salts are highly toxic, so that their principal uses are in the chemicals industry in the preparation of insecticides, rodent killers and weed killers; some are also used in glass manufacture.

The reserves available in the U.S.S.R. appear to be adequate for many years' consumption. There are no detailed figures of production available, but it is believed that output (in metal terms) was about 5,000 tons in 1947. Much of this came from deposits where arsenic is extracted in conjunction with other metallic minerals.

The only important deposit in Eastern Europe is in Polish Silesia, where about 1,500 tons a year of metal have been extracted for several decades.

Because arsenic is used principally in the chemical industry it has been plotted on the inset of the Chemicals Map, p. 79.

Mineral Production in the U.S.S.R., Eastern Europe and other major countries, 1953

(% of world production)

	U.S.S.R.	E. Europe	U.S.A.	U.K.	World's largest producers	
Aluminium (bauxite)	8	12	12	—	Dutch Guiana British Guiana	23 17
*Antimony	11†	10†	1	—	China Bolivia	25† 17
*Chromium	17	3	2	—	South Africa Turkey	21 18
Coal & lignite	17	19	23	12	U.S.A.	23
*Copper	12	1	30	—	U.S.A. N. Rhodesia	30 13
Iron ore	18	2	35	5	U.S.A.	35
*Lead	13	6	16	0·4	U.S.A. Australia	16 13
Manganese ore	55	3†	4	—	**U.S.S.R.** India	55 21
Mercury	8	10	9	—	Italy Spain	34 27
Nickel	19	—	—	—	Canada	64
Petroleum	8	2	49	0·2	U.S.A. Venezuela	49 14
Phosphates (crude)	11	—	51	—	U.S.A. French Morocco	51 17
Potash (K_2O content)	13†	29†	26	—	**E. Germany** W. Germany France	28 24 15
*Tin	7†	0·4	—	0·5	Malaya Bolivia Indonesia	32 20 19
*Zinc	10	9	19	—	U.S.A. Canada	19 14

* Ores in terms of metal   † Figures doubtful
Information for other minerals is not available

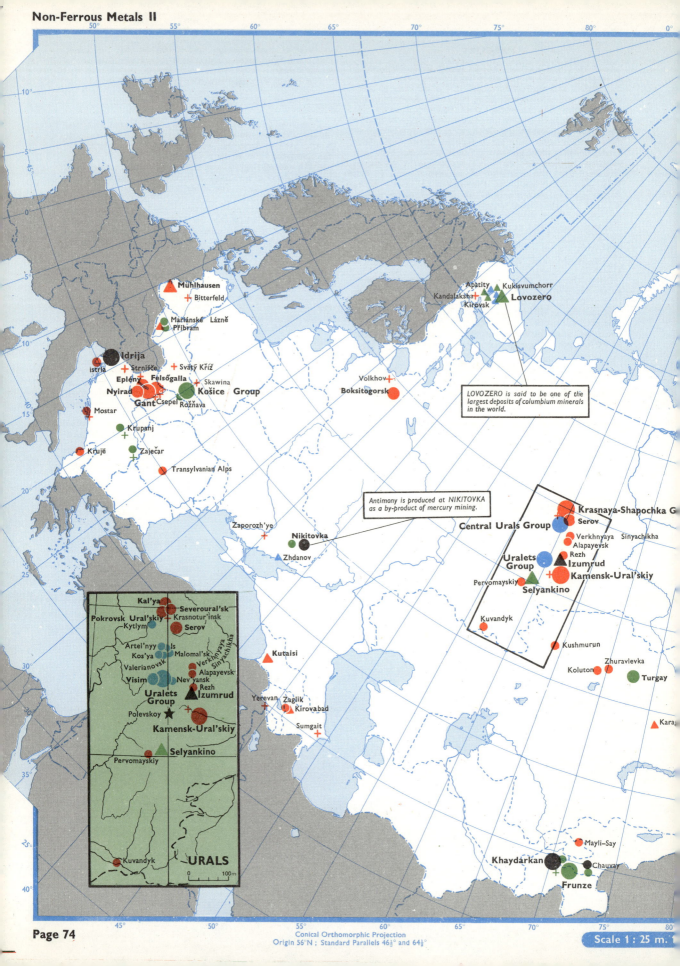

▲ Mühlhausen
+ Bitterfeld

● Mariánské Lázně
▲ Příbram

● **Idrija**
▲ istria
+ Strnišče   + Svätý Kříž
Eplény   Felsőgalla
Nyirad   + Skawina
Gant Csepel   **Košice** Group
+ Rožňava

● Mostar

● Krupanj
+

● Zaječar
+

● Transylvanian Alps

Apatity   Kukisvumchorr
Kandalaksha +
▲ **Lovozero**
Kirovsk

Volkhov +
Boksitogorsk

*LOVOZERO is said to be one of the
largest deposits of columbium minerals
in the world.*

● Zaporozh'ye

● **Nikitovka**

▲ Zhdanov

*Antimony is produced at NIKITOVKA
as a by-product of mercury mining.*

Krasnaya-Shapochka G
● Serov
Verkhnyaya   Sinyachikha
Central Urals Group   Alapayevsk
Rezh
Uralets   ▲ Izumrud
Group   **Kamensk-Ural'skiy**
Pervomayskiy
▲ **Selyankino**

Kuvandyk

Kushmurun
Zhuravlevka

Koluton   ● **Turgay**

▲ **Kutaisi**

Yerevan +   Zaglik
● Kirovabad

Sumgait +

● Kara

URALS (inset)

Kal'ya
Pokrovsk Ural'skiy
Kytlym
Severoural'sk
Krasnotur'insk
Serov

Artel'nyy   Is
Kos'ya   Malomal'sk
Valerianovsk   Verkhnyaya
Visim   Sinyachikha
Nev'yansk
Alapayevsk
Rezh
Uralets   ▲ Izumrud
Group
Polevskoy ★ +
Kamensk-Ural'skiy

Pervomayskiy   ▲ **Selyankino**

● Kuvandyk

URALS
0   100m

Mayli-Say
Khaydarkan   ● Chauvay
+
Frunze

Conical Orthomorphic Projection
Origin 56°N ; Standard Parallels 46½° and 64½°

Scale 1 : 25 m.

NON-FERROUS METALS

	Mining districts			Smelting
	Major	Secondary	Minor	
Antimony	●	●	●	+
Barium	▲	▲	▲	
Bauxite (Aluminium)	●	●	●	+
Beryllium	▲			
Columbium (incl. tantalum)	▲		▲	
Mercury	●	●	●	
Platinum	●	●	●	
Zirconium			▲	

Artificial Cryolite ★

Open circles denote districts where exploitation is uncertain
or where deposits are known not to be exploited.

Noril'sk

Vilyuy River

Tatarka Razdolinsk

Zeya Valley △

Birakan ○ ○ + Leninskoye

Salair
+ Stalinsk

Askiz

Zmeinogorsk

Chagan–Uzun

Sherlovaya Gora

NON-METALLIC MINERALS
AND THE CHEMICALS INDUSTRY

NON-METALLIC MINERALS IN THE U.S.S.R. AND EASTERN EUROPE

The eight minerals shown on the map, pp. 78/79, are all specialized in their uses, which are mainly outside the metallurgical industries. Sulphur, potash and phosphorus together form the basis of many of the most important fertilizers, and so are essential to modern agriculture, while the unique properties of mica give it a special place in the electrical and many other industries.

SULPHUR

It is practically impossible to develop a significant chemicals industry without abundant supplies of sulphur. The pure mineral occurs in several forms, and is used for explosives, insecticides and fungicides, for vulcanizing rubber and numerous other purposes. Sulphuric acid (H_2SO_4) is the most important of the heavy chemicals, and is essential to the cellulose group of industries (rayon, paper, etc.), in fertilizers (superphosphates), in many dyes, coal derivatives and petrochemicals, and in a great many lesser processes.

In some places in the world, " native " sulphur is found (i.e. sulphur uncompounded with other elements, though inevitably carrying impurities), and extraction is a simple mining or excavation process; in the U.S.S.R. there are such deposits at Gaurdak and at Shorsu. These deposits are still, however, only secondary sources for Soviet sulphur consumption; much more important are the extensive deposits of pyrites and other sulphur-bearing metallic minerals show on pp. 78/79, and the copper tailings and coal brasses in areas where sulphur is a valuable by-product of processes originally designed for other purposes. The anhydrite process for extracting sulphur from gypsum is little used, being more costly than extracting from other minerals.

Before the war, it appears that about 10% of the sulphur consumed was obtained from native deposits, about 5% was recovered (in the form of sulphuric acid) from smelter flue-gases, about 45% from pyrites, and the remaining 40% from copper and coal by-products. Total consumption was then about 600,000 tons a year (sulphur content). There is little evidence to show what changes have taken place in these proportions since then. Native sulphur is now perhaps rather more important, the Central Asian deposits having been greatly developed during the war. Techniques of extraction have also greatly improved, so that a larger production of sulphur and sulphuric acid has been achieved without a corresponding increase in output of the sulphur-bearing minerals.

An estimate for 1950 gives an output of 900,000 tons of sulphur-bearing minerals and 2,200,000 tons of sulphuric acid. The last figure is up to the level of requirements at that date. More could be consumed, especially in the production of superphosphates, but expansion is limited by lack of capital equipment. Reserves of sulphur in all its forms are likely to be adequate for many years to come.

In Eastern Europe there is little or no native sulphur, and only Poland, Romania and Yugoslavia have significant deposits of pyrites. Over half the pre-war consumption in Poland was supplied from the by-products of lead-zinc smelters. Coal brasses are becoming an increasingly important source of supply, especially in Poland, and there are reports that the anhydrite process of recovery from gypsum is also being developed.

The rather scanty information available suggests that just before the war Poland's output of pyrites was of the order of 80,000 to 90,000 tons a year and Romania's about 80,000 tons. In 1947–49 Poland is said to have produced about 55,000 tons of pyrites (roughly 23,000 tons sulphur content) and about 275,000 tons of sulphuric acid each year. Czechoslovakia developed a small post-war output of pyrites (8,000 tons a year), and produced about 215,000 tons of sulphuric acid. No post-war figures are available for Romania.

Pre-war output of pyrites in Yugoslavia ranged from 130,000 to 150,000 tons a year. In 1953, 172,995 tons were produced.

PHOSPHORUS MINERALS

Pure phosphorus is little used, though it has some importance in the chemicals industry, in making matches, insecticides, weed-killers and various other materials. It is also employed in metallurgy as a deoxidiser and in preparing phosphor bronze. In most types of ferrous metallurgy, however, the problem is more often how to remove phosphorus rather than how to add it.

By far the largest use of phosphorus compounds is in fertilizers. Phosphate rock (phosphorite) and apatite are sometimes used direct after grinding to a coarse powder, and phosphatic residues from iron works are also employed in the form of basic slag. Guano is still another type of natural phosphate fertilizer. Modern practice tends to favour, in preference to the natural forms, the more complex superphosphates, which are made by treating natural phosphates with sulphuric acid, and which have the advantage of being readily soluble.

The Soviet Union is well supplied with phosphorite and apatite, and its metallurgical industries also provide it with adequate quantities of basic slag. Phosphates were imported until 1933, but production had been greatly increased under the first Five-Year Plan. In 1937 over 500,000 tons of apatite were exported. Superphosphate production was also greatly increased under the 1928–32 Plan, but fell very sharply during the war, largely because the sulphur available was needed for military purposes. Large-scale systematic production of basic slag seems to have started only after 1945.

Production of Phosphate and Superphosphates in the U.S.S.R.
(in tons)

	Apatite	Phosphate Rock	Superphosphates
1926	26,900	73,516	80,618
1937	1,158,000	840,000	1,454,000
1946		1,600,000	1,600,000
1950		2,500,000	2,500,000
1952		2,750,000	{ 2,720,000*
			{ 3,000,000†

* Planned † Actual

Note: For years 1946, 1950 and 1952, these figures are based on estimates of the International Superphosphate Manufacturers' Association.

In 1952 the U.S.S.R. produced approximately $11\frac{1}{2}$% of the world output of phosphate rock and apatite, and about 12% of its superphosphates. The limit on production of superphosphates is imposed by the quantity of sulphur available, not by any lack of raw phosphates.

POTASSIUM MINERALS

Potassium is another element which has few uses in the pure state, though the metal and its simpler salts play a significant role in the chemicals industry. Economically the various salts known as potash are vastly more important, providing the great bulk of the alkaline fertilizers used throughout the world; potash is thus hardly less essential than the phosphate fertilizers for the maintenance and development of agriculture.

Strictly speaking potash is potassium oxide (K_2O), and conventionally quantities are quoted in terms of K_2O content. The simple oxide, however, never occurs in nature. The main sources are sylvinite (potassium chloride, KCl) and carnallite (KCl MgCl$_2$. 6H$_2$O); the two forms are often found together, as in the great Solikamsk deposit, where a thick carnallite bed lies on top of a thinner sylvinite layer. There are several other sources of potash as well, notably the complex minerals kainite and langbeinite, which form the second largest deposit in the U.S.S.R. in the Stanislav area.

Solikamsk is possibly one of the largest potash deposits in the world, and the U.S.S.R. is abundantly supplied. Detailed figures of production are not available. In 1937 total output was about 2,400,000 tons of carnallite and sylvinite from Solikamsk alone, representing a K_2O content of perhaps 300,000 tons, or about 12% of the world total. In 1950 output of potash fertilizer was about 850,000 tons, the increase being largely due to the acquisition of Stanislav from Poland after 1945.

Polish production in 1938 was over 560,000 tons of crude mineral, representing 107,000 tons K_2O equivalent. Most of the producing areas became Soviet territory in 1945. In East Germany output in 1946 was over 6 million tons crude, or about 750,000 tons K_2O equivalent.

In East Germany output in 1946 was over 6 million tons crude, or about 750,000 tons K_2O equivalent. rising to 1,900,000 tons in 1953.

MICA

All forms of mica are basically silicates of potassium and aluminium. The simplest is muscovite (H_2 $KAl_3(Si$ $O_4)$); phlogopite ((H, K, MgF).$MgAl(Si O_4)_3$) also contains magnesium and fluorine; while vermiculite is a hydrated variant on the other two and exists in several forms.

The peculiar value of mica arises from the fact that it makes an excellent insulator for heat and electricity, and can be split into thin sheets of greater or lesser translucency, which are usually flexible. Muscovite and phlogopite are easier to work, especially in the block or sheet, and are extensively employed as electrical insulators. Vermiculite does not occur in the sheet form, and is usually ground or broken down; its main uses are in sound-proofing and in non-conducting lagging for hot water pipes.

In the U.S.S.R., the Slyudyanka phlogopite deposit used to produce 4,000–5,000 tons a year, but is now largely exhausted. Vermiculite is mined at Buldymskoye, near Chelyabinsk. All the remaining deposits are believed to be muscovite. Recent output figures are not available, but reserves are probably large.

There was a small output of mica in Romania at one time, but no details are known.

ASBESTOS

Asbestos is the only known mineral fibre, and is non-inflammable and a non-conductor of electricity; it is used for insulation, the manufacture of heat-resistant fabrics and packing, and many other purposes. The commonest form is chrysotile ($3 MgO. 2SiO_2. 2 H_2O$).

Supplies in the Soviet Union are abundant. Reserves were estimated in 1937 at 19.3 million tons, and are now known to be much larger. Production is in the neighbourhood of 150,000 tons a year, of which some is exported; at this rate of extraction reserves should be adequate for at least 130 years.

In Eastern Europe some asbestos is produced in Czechoslovakia and Romania. Yugoslavia produced about 3,700 tons in 1953.

THE CHEMICALS INDUSTRY
U.S.S.R.

There is a conventional division between heavy and light chemicals, though in fact the boundaries are ill-defined. Chemicals sold by the ton may be roughly classed as heavy, while those sold by the pound (or smaller unit) may be called light. Thus sulphuric acid, fertilizers and the like are classed as heavy chemicals, while pharmaceutical products are definitely in the light category; but certain substances may be used, for instance, in different concentrations or degrees of purity for different purposes, and so may legitimately be put into either group.

Nor is it easy to define the chemicals industry as a whole. Apart from substances which can strictly be called chemicals (such as dyes, fertilizers, acids, synthetic resins and so forth), and apart also from the pharmaceuticals group, there are a number of other closely related industries which include coke-oven by-products, explosives, paint and varnish, glycerine, petrochemicals, vegetable and animal oil products, wood distillation, cellulose and industrial alcohol. It is hard to draw the boundaries between these and other industries. Petro-chemicals, for instance, are closely linked with petroleum refining, wood distillation with sawmilling and the manufacture

of synthetic fibres and paper, and (especially in the Soviet Union) industrial alcohol with the production of synthetic rubber and so with the other branches of the rubber industry, including the treatment of natural rubber.

At the time of the Revolution, there was practically no chemicals industry in Russia except for a small output of fertilizers, explosives and soap. Most products were imported. After the Revolution Leningrad became the first considerable manufacturing centre, especially for fertilizers, and it was also there that a synthetic rubber pilot plant was opened in 1928, to be followed four years later by the first big production plant. Later the Moscow area began to be developed as a centre of the industry, and the Urals later still, under the 1928–32 Plan. Since then new types of manufacture have been created in widely scattered areas. By 1941 the Soviet Union was adequately supplied with most types of chemicals, and was able to keep the industry going in the face of the exceptional demands of the war period, with relatively few imports from the U.S.A. and elsewhere.

Gross Output of Chemicals Industry in the U.S.S.R.

	Million roubles (1927 prices)	% of 1913	% of 1928
1913	457	100	71
1928	645	141	100
1933	2,301	504	354
1938	6,809	1,488	1,056

There is little information available about the progress of the industry since the war. It would seem, however, that the Soviet Union is self-sufficient in all the more important chemical products.

EASTERN EUROPE
POLAND

The Polish chemicals industry is now of considerable size, especially after absorbing much of the former German industry in Silesia. A high proportion of the production of sulphuric acid is from the by-products of lead-zinc smelters, but the anhydrite process was planned to achieve an output of 180,000 tons a year by 1955, or about a third of normal consumption, chiefly from a big works near Bolesławiec which was set up in 1951. Elsewhere the chemicals industry is based mainly on local supplies of rock salt, limestone and potash, but most of the phosphates used are imported.

In 1952 Poland produced 5,700 tons of synthetic dyes and 76,000 tons of soap and detergents. Synthetic rubber was first produced in 1949, and output reached 13,000 tons in 1951. According to the 1949–55 Plan, the chemicals industry in general was to account for 17% of industrial production (by value) in the last year of the Plan, compared with 9% in 1949.

EAST GERMANY

The East German chemical industry has been rebuilt since the war, mainly with a view to developing the production of heavy chemicals such as soda, potash and the products of coal hydrogenation. East Germany is still the main supplier of dyes, solvents and chemicals for the textile industry, and of products of coal distillation. East Germany remains the chief source of pharmaceutical products in Eastern Europe.

Output of Heavy Chemicals in Eastern Europe 1954
('000 tons)

	Poland	E. Germany	Czecho-slovakia	Hungary	Romania
Sulphuric acid	484	438	343	...	86.4*
Caustic soda	87	231*	...	...	23.8*
Nitrogen fertilizers	76	272*	42	80§	...
Phosphorous fertilizers	81	70*	90	...	3

* 1953   § Gross weight

CZECHOSLOVAKIA AND ROMANIA

The Czech chemical industry is being expanded. Special attention is being given to the production of synthetic fuel, using equipment procured from East Germany; the production of plastics (used as substitutes for non-ferrous metals); and the output of coal-tar dyes, with a view to export.

The Romanian chemical industry, of fairly recent growth, concentrates chiefly on the production of petro-chemical products.

Non-Metallic Minerals and Chemicals

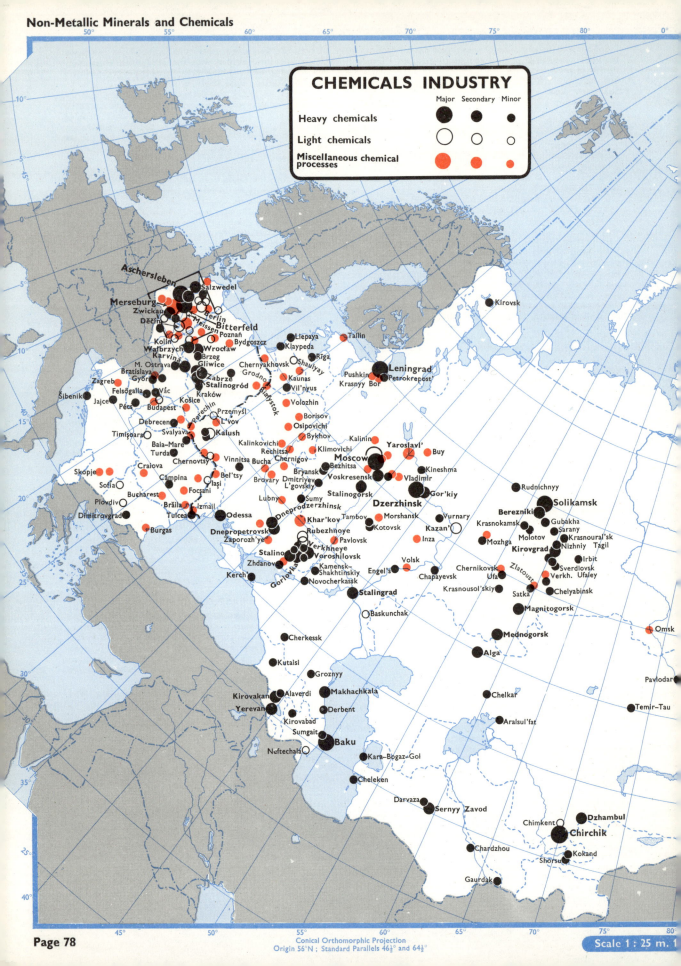

CHEMICALS INDUSTRY

Major Secondary Minor

Heavy chemicals

Light chemicals

Miscellaneous chemical processes

Aschersleben
Salzwedel
Merseburg
Zwickau Berlin Bitterfeld
Děčín Meissen
Kolín Poznań
Wałbrzych Brzeg Wrocław
Karviná Gliwice
M. Ostrava Zabrze
Bratislava Stalinogród
Zagreb Győr Kraków
Felsőgalla Vác
Šibenik Jajce Košice
Pécs Budapest Perechin
Debrecen Przemyśl
Timişoara Svalyava L'vov
Baia–Mare Kalush
Turda Vinnitsa
Chernovtsy
Skopje Craiova Câmpina
Sofia Bel'tsy
Plovdiv Bucharest Iaşi
Dimitrovgrad Focşani
Brăila Izmaïl
Burgas Tulcea

Liepaya
Klaypeda
Bydgoszcz Riga
Chernyakhovsk Shaulyay
Grodno Kaunas
Vil'nyus
Białystok
Volozhin
Borisov
Osipovichi
Bykhov
Kalinkovichi Klimovichi
Rechitsa
Chernigov
Bryansk Dmitriyev
Brovary L'govskiy
Bucha
Lubny Sumy
Dneprodzerzhinsk
Odessa
Khar'kov Tambov
Rubezhnoye
Dnepropetrovsk Pavlovsk
Zaporozh'ye
Verkhneye
Stalino Voroshilovsk
Zhdanov Kamensk-Shakhtinskiy
Gorlovka Novocherkassk
Kerch'
Cherkessk
Kutaisi Groznyy
Kirovakan Alaverdi Makhachkala
Yerevan Derbent
Kirovabad
Sumgait
Neftechala Baku
Kara–Bogaz–Gol
Cheleken

Tallin
Leningrad
Petrokrepost'
Pushkin
Krasnyy Bor
Yaroslavl' Buy
Kalinin
Moscow Kineshma
Bezhitsa
Voskresensk Vladimir
Stalinogorsk Gor'kiy
Dzerzhinsk
Morshansk
Kotovsk Kazan'
Inza
Volsk
Engel's Chapayevsk
Vurnary
Chernikovsk
Krasnosol'skiy Ufa
Stalingrad
Baskunchak
Alga
Chelkar
Aralsul'fat
Sernyy Zavod
Darvaza

Kirovsk
Rudnichnyy
Solikamsk
Berezniki
Krasnokamsk Gubakha
Sarany
Molotov Krasnoural'sk
Mozhga Nizhniy Tagil
Kirovgrad Irbit
Zlatoust Sverdlovsk
Verkh. Ufaley
Satka Chelyabinsk
Magnitogorsk
Mednogorsk
Omsk
Pavlodar
Temir–Tau
Chimkent Dzhambul
Chirchik
Chardzhou Kokand
Shorsu
Gaurdak

Page 78

Conical Orthomorphic Projection
Origin 56°N ; Standard Parallels 46½° and 64½°

Scale 1 : 25 m. 1

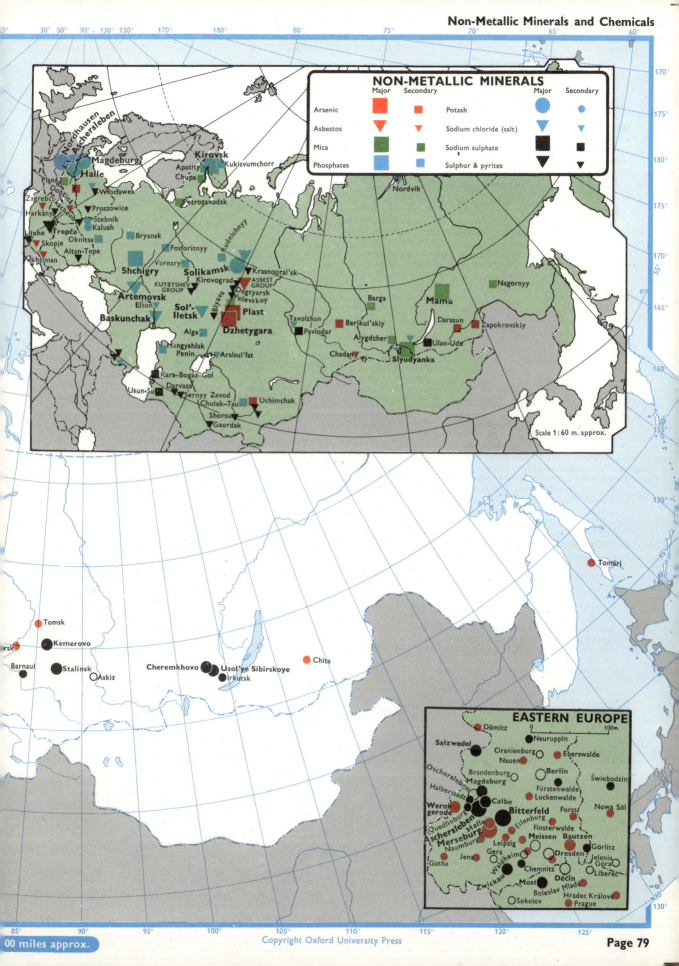

NON-METALLIC MINERALS

	Major	Secondary		Major	Secondary
Arsenic			Potash		
Asbestos			Sodium chloride (salt)		
Mica			Sodium sulphate		
Phosphates			Sulphur & pyrites		

Nordhausen
Aschersleben
Magdeburg
Halle
Planá
Zagreb
Dobšiná
Harkány
Liixhe
Trepča
Skopje
Altan-Tepe
Ikhtiman
Oknitsa
Kirovsk
Apatity
Chupa
Kukisvumchorr
Petrozavodsk
Włocławek
Proszowice
Stebnik
Kalush
Bryansk
Fosforitnyy
Rudnichnyy
Nordvik
Vurnary
Shchigry
Solikamsk
Kirovgrad
Krasnoural'sk
ASBEST GROUP
Degtyarsk
Polevskoy
KUYBYSHEV GROUP
Artemovsk
Elton
Sol'-Iletsk
Blyava
Plast
Baskunchak
Alga
Dzhetygara
Tavolzhan
Pavlodar
Berikul'skiy
Barga
Nagornyy
Mama
Darasun
Zapokrovskiy
Alygdzher
Ulan-Ude
Slyudyanka
Mangyshlak Penin.
Aralsul'fat
Chadan
Kara-Bogaz-Gol
Usun-Su
Darvaza
Sernyy Zavod
Chulak-Tau
Uchimchak
Shorsu
Gaurdak

Scale 1 : 60 m. approx.

Tomsk
Kemerovo
rsk
Barnaul
Stalinsk
Askiz
Cheremkhovo
Usol'ye Sibirskoye
Irkutsk
Chita
Tomari

EASTERN EUROPE

Dömitz
Salzwedel
Neuruppin
Oranienburg
Nauen
Eberswalde
Oschersleben
Brandenburg
Magdeburg
Berlin
Świebodzin
Halberstadt
Fürstenwalde
Luckenwalde
Werni-gerode
Calbe
Bitterfeld
Eilenburg
Forst
Finsterwalde
Nowa Sól
Quedlinburg
Halle
Ascherslebeen
Merseburg
Naumburg
Leipzig
Meissen
Bautzen
Gera
Dresden
Görlitz
Gotha
Jena
Waldheim
Chemnitz
Jelenia Góra
Zwickau
Most
Děčín
Liberec
Sokolov
Boleslav Mladá
Hradec Králové
Prague

100 m.

TRANSPORT

RAILWAYS, INLAND WATERWAYS, ROADS AND AIRWAYS IN THE U.S.S.R.

The Soviet Union has about 75,000 miles of railway and perhaps a similar length of navigable inland waterways. In Central Asia, northern Russia and in Siberia, roads connect towns to which no railway has yet been built. Other remote places can be reached only by air, and the country is covered by an extensive network of air routes, designed for freight as well as passenger carrying. The importance of an adequate transport system for the development of so vast an area is obvious, but at the same time the wide range of climatic conditions and physical features have raised numerous technical problems of construction which are still not completely solved. Thus the existence of large " permafrost " zones has greatly increased the difficulty of building roads and railways, while in other areas special precautions have to be taken against sand, floods and earthquakes. River transport is hampered by the ice which persists, for instance, on the lower Lena for six months in the year, and coastwise shipping is restricted by the fact that comparatively little of the sea coast is free from ice for more than part of each year.

RAILWAYS IN THE U.S.S.R.

(pp. 82/83)

The first railway in Imperial Russia was built in 1837 from St. Petersburg to Tsarskoye Selo, a distance of 16 miles, with a gauge of 6 feet. In 1851 work was finished on the first international link, from Warsaw (then in the Empire) to Vienna, this time with the standard European gauge of 4 feet 8½ inches. Moscow and St. Petersburg were linked in the same year by a line using still a third gauge of 5 feet; this was then established as the standard gauge for the whole Russian railway system, and lines of other gauges were later converted to it. To-day there are some narrow gauge lines used especially for goods traffic, but all the lines regularly in use for passenger carrying are of the 5 foot gauge, and so cannot be directly connected to any part of the railway system in the rest of Europe except Finland. This greatly increased the difficulties of the Germans during their operations in Russia between 1941 and 1944.

All the developments carried out during the first period of intensive building (1848–1875) were in European Russia, and by the latter year the total mileage in use was over 14,000. The Poles'ye Railway from Bryansk to Brest, completed in 1884, was a further link with Western Europe, and the first important east-west line in the Ukraine came into operation in the same year. Otherwise there was little construction after 1875 until Count Witte, the real father of the Russian railway network, became Minister of Communications in 1892 and put in hand the laying of the world's longest continuous railway, the Trans-Siberian line from Chelyabinsk to Vladivostok. The distance by the original route through Manchuria (the Chinese Eastern Railway) was over 4,000 miles. Single-line traffic along the whole of this route was initiated in 1904, and the line was of immense strategic importance during the Russo–Japanese war of the following year. The section round the southern end of Lake Baykal was not completed till 1905, and originally a ferry carried trains across the 50-mile wide lake. The northern loop through the Amur valley, designed so that the whole line should pass through Russian territory, was finished only in 1908. To-day the length of the entire line from Leningrad to Vladivostok is 5,973 miles; double-tracking was completed shortly before the Second World War.

The other principal development during Count Witte's period of office was the line, completed in 1905, from Orenburg (Chkalov) to Tashkent, which connected with an older strategic railway across the Turkmen desert to Krasnovodsk, connected in its turn by train ferry across the Caspian Sea with Baku and the Caucasian railway to Rostov. The economic significance of this and the Trans-Siberian line was enormous. Both Siberia and Central Asia were now opened up in a way which would have previously been impossible. The laying of these lines went together with the expansion of industry and agriculture southwards and eastwards, and with the spread of the Great Russian and Ukrainian settlers into areas previously occupied mainly by Asiatic tribes. Count Witte made one further contribution to

development before he died, by bringing all new railway building and many of the existing lines under the central control of the Government. By 1910 there were nearly 48,000 miles of railway in Russia, of which 10,800 were in Asia, and on the eve of the Revolution the total mileage, within the territories which then formed part of the Tsarist Empire, was somewhat over 51,000.

The loss of Poland, Bessarabia and the Baltic territories in 1919–1922 reduced the total mileage to 44,087, of which only about 39,000 could be operated for some time after the civil war. Under the Soviets, the first big piece of new construction was from Leningrad to Murmansk, and a link was built across the Urals from Kazan' to Sverdlovsk to shorten the distance between western Russia and the Trans-Siberian railway. By 1926 about 47,500 miles of track were in use. Work started in 1927 on another important route into Central Asia, the Turksib Railway from Tashkent through Alma-Ata to Semipalatinsk, which was already connected with the Trans-Siberian railway at Novosibirsk; this link was completed in 1930. During the next ten years communications between Central Asia and the rest of the U.S.S.R. were improved, especially by completing the connections of Tashkent to Orsk through Kandagach, and to Saratov by the Sol'-Iletsk-Ural'sk link. Another line was built which was later to have great economic importance from Petropavlovsk on the Trans-Siberian railway to Karaganda and on to Lake Balkhash. During the same period the main north-south link along the Urals was completed from Sverdlovsk to Orsk, and elsewhere a large amount of capital was devoted to the doubling of existing single-track lines. As a result of this new construction and the incorporation of the Baltic States railway systems into the U.S.S.R. the total mileage in use in 1940 had risen to 65,800, of which nearly a third was double track.

During the war the emphasis was largely on the strategic value of the railways. In some cases new lines had to be built to make up for lines held by the enemy; the two most important in this category were perhaps those from Obozerskaya to Belomorsk (which brought traffic from Murmansk across to the Arkhangel'sk line), and from Astrakhan' to Kizlyar, which provided an alternative route to the Caucasus and Baku after the line through Rostov had been cut. The line constructed along the Volga from Kazan' to Stalingrad was of primarily strategic importance, and the special needs of the war-time economy were to be served by the route built from Kotlas to Vorkuta to bring coal to the industrial areas of central Russia. Karaganda was linked to the Urals by a new route from Akmolinsk to Kartaly, which was also intended as the western section of the South Siberian line designed to connect the Urals with the Kuzbass at Stalinsk by a shorter route than the existing line through Novosibirsk and Omsk.

The last-mentioned project is not intended to stop at Stalinsk, but is to be carried eastwards through Abakan to join the Trans-Siberian line at Tayshet. The section from Stalinsk to Abakan was due for completion by 1950, but construction of the eastern section to Tayshet was postponed till later. Under the fourth Five-Year Plan the other long new lines completed were from Mointy to Chu round the western end of Lake Balkhash, and a line following the course of the Amu-Darya from Chardzhou to Kungrad; plans are believed to exist to carry this latter line on across the Ust'Urt Plateau to Makat and from there across the northern part of the Caspian depression to Aleksandrov Gay, which is already linked with the Volga valley line at Saratov.

The fourth Five-Year Plan provided in all for the construction of 4,493 miles of new line and the doubling of 7,800 miles of existing track (including some restoration of war-damaged lines where single-line working only had been resumed). Actual construction was only about 2,200 miles, and not all the lines planned were even begun. It was planned to convert to electric power 3,309 miles of line, which would bring the total mileage electrified to 4,575; but in the event only some 760 miles were electrified, and at the end of 1950 the total length of electric track was no more than 2,100 miles.

VOLUME OF TRAFFIC

The 1946–50 Plan aimed at a total goods traffic in the last year of the Plan of about 330,000 million ton miles. These figures were substantially exceeded. The Soviet Union now shows much the greatest ton-mileage in the world—due, of course, to the fact that hauls for heavy freight are long by the standards of most other countries.

Volume of Goods Traffic in the U.S.S.R.

	Total Tonnage (in millions)	Average Haul (in miles)	Ton-Miles (in millions)
1948	617	456	281,200
1950 Plan	771	429	c.330,000
1950 actual	830	451	374,800
1953	1,062	467	496,600
1955 Plan	...	...	525,000

No estimates are available for passenger traffic. The 1946–50 Plan envisaged a total of 1,350 million passengers in the final year, travelling an average distance of 44½ miles, representing a total traffic of about 60,000 million passenger miles. Results have not been published.

RAILWAYS IN EASTERN EUROPE

Eastern Europe is covered by a network of railways (gauge 4 feet 8½ inches) hardly less dense in parts than in Western Europe, and considerably denser than in the Soviet Union. The building of railways has had a close historical connection with frontier changes, since numerous lines have been laid to replace communications severed by newly created international boundaries, and in Poland the creation of the Corridor in 1919 made it necessary to construct a line to connect the industries of Upper Silesia with the new port of Gdynia.

In East Germany, a number of lines were destroyed at the end of the war, and others were reduced from double to single track, and some dismantled completely. Later policy reversed this process, however, and to-day East Germany probably has the greatest mileage, in relation to population and area, of any part of Eastern Europe.

INTERNATIONAL COMPARISONS

In the following table, the U.S.A. and the United Kingdom are included to compare the mileage of railways in relation to population and area.

International Railway Comparisons

	Total Mileage	Miles per 1,000 population	Miles per 1,000 square miles
U.S.S.R. (1954)*	75,000	0·357	8·4
Poland (1951)	13,375	0·535	110·5
Czech. (1949)	8,383	0·672	169·9
E. Germany (1946)	8,125	0·517	196·2
Romania (1950)	6,100*	0·360*	65·0*
Hungary (1949)	4,773	0·502	132·9
Bulgaria (1951)	2,211	0·303	51·7
Albania (1952)	63	0·056	5·9
Yugoslavia (1949)	7,265	0·461	71·2
U.S.A. (1951)	227,244	1·453	76·3
U.K. (1952)	19,471	0·386	207·2

* estimate

INLAND WATERWAYS IN THE U.S.S.R.

(pp. 86/87)

Although the mileage of navigable waterways is as great as that of the railways, and although parts of the Soviet Union cannot be reached by heavy traffic by any other way, the volume of goods carried is very much less. In recent years, no more than 8% to 10%* of all goods traffic has been carried by water. Ice is the chief obstacle to navigation, since few of the main rivers remain open throughout the year; in some other cases, a further difficulty is that wide seasonal variations in rainfall make the rivers too shallow for navigation during parts of the ice-free season.

The Volga has been in use as an artery of communication for many centuries, and it was the opening up of traffic on the Northern Dvina under Ivan the Terrible which made possible the first commercial exchanges with Great Britain. Under Peter the Great, the first of the major canals was built, linking the Volga with St. Petersburg (Leningrad). During the 18th and 19th centuries more connecting canals were built, rivers were dredged, widened and straightened, and the river fleet was greatly increased, till in the year 1913 some 48 million tons of freight and 11 million passengers were carried by inland waterways.

Since the Revolution, the development of water traffic has been closely associated with the expansion of hydro-electric power generation. The classic example was the construction of the Dneproges plant (pp. 54/55), which not only provided

* Based on tonnage originated

the Soviet Union with the world's biggest generating station but also eliminated the Dnieper rapids and made the river navigable for the greater part of its length. Other main canals constructed between the Revolution and the second World War were the White Sea–Baltic canal (completed 1933), and the Moscow–Volga canal (1937). Reconstruction of the old Dnieper–Bug–Vistula canal, linking the Black Sea and the Baltic, began in 1940 but was not completed till after 1945.

During the war many of these waterways were lost or put out of action. Apart from restoration work done since the war, the further development of the canals and rivers is now planned on an ambitious scale. The Volga-Don canal is already completed and connects Stalingrad to the Sea of Azov via the Tsimlyansk reservoir. The Rybinsk reservoir on the upper Volga helps to control the flow of water lower down, and will eventually be operated in conjunction with the new Kuybyshev scheme (pp. 52/53) and others. A canal through southern Turkmenistan is being built, and is almost complete between Mary and the Amu Darya; it will later be extended to the Caspian south of Krasnovodsk.

INLAND WATERWAYS IN EASTERN EUROPE

In Eastern Europe the Danube is still the main waterway; the Oder, Elbe, Vistula, and the Bug are also important, though on a lesser scale. Through traffic on the Elbe from Hamburg to Central Europe has declined since the division of Germany and the diversion of traffic from Czechoslovakia to the port of Szczecin (Stettin) and the Oder. In Eastern Germany, the Mittelland and associated canals link the Oder and Polish waterways in the east to Berlin, the Elbe and the Weser, and via the Dortmund-Ems canal, to the Rhine at Duisburg in the west. An important canal linking the Danube and Tisza rivers is under construction, and a canal through the Moravian Corridor from the Danube near Bratislava to the Oder near Raciborz has been suggested. A plan conceived many years ago to link the Danube with the Black Sea at Constanţa (thus eliminating the big northward bend through the delta where navigation conditions are uncertain) was revived after the war; work started, but stopped again in 1953 when the emphasis in Romanian policy shifted from large capital plans to the production of consumer goods.

ROADS IN THE U.S.S.R. AND E. EUROPE

(pp. 84/85)

The road network in Eastern Europe is almost as dense as in Western Europe, although the proportion of paved roads is considerably less, except in western Poland, western Czechoslovakia, and East Germany. In East Germany there is a system of special motor-roads, the *autobahnen*, started before 1939 but never completed.

Conditions in European Russia resemble the less developed parts of Eastern Europe. The main trunk routes are paved, but otherwise there are few paved roads away from the larger towns, and transport is often very difficult in winter and early spring. In Asiatic Russia, where roads supplement river routes or link settlements to railheads, only the most important routes are open to heavy lorries the whole year round. In Siberia the present network of tracks includes some " winter roads " in the far north; current policy favours roads against railways in the " permafrost " zones, since they are cheaper both to build and to maintain.

AIRWAYS IN THE U.S.S.R. AND E. EUROPE

(pp. 88/89)

The network of airways in Eastern Europe compares favourably with Western Europe. Most countries have internal airways and operate services with international connections. Most traffic between the U.S.S.R. and Eastern European countries is carried by the Soviet Aeroflot. Centrally placed in Eastern Europe, Prague in Czechoslovakia has developed as one of the principal air traffic centres.

Under the 1946–1950 Plan, it was intended to bring the total mileage of internal air lines in the U.S.S.R. to 110,000. Many of the remoter places can be reached only by air, and where distances are so large an adequate air service is essential to rapid transport. The map (pp. 88/89) shows all the major routes on which regular passenger and freight services are maintained, including those of Glavsevmorput', which is responsible for economic development in the Arctic.

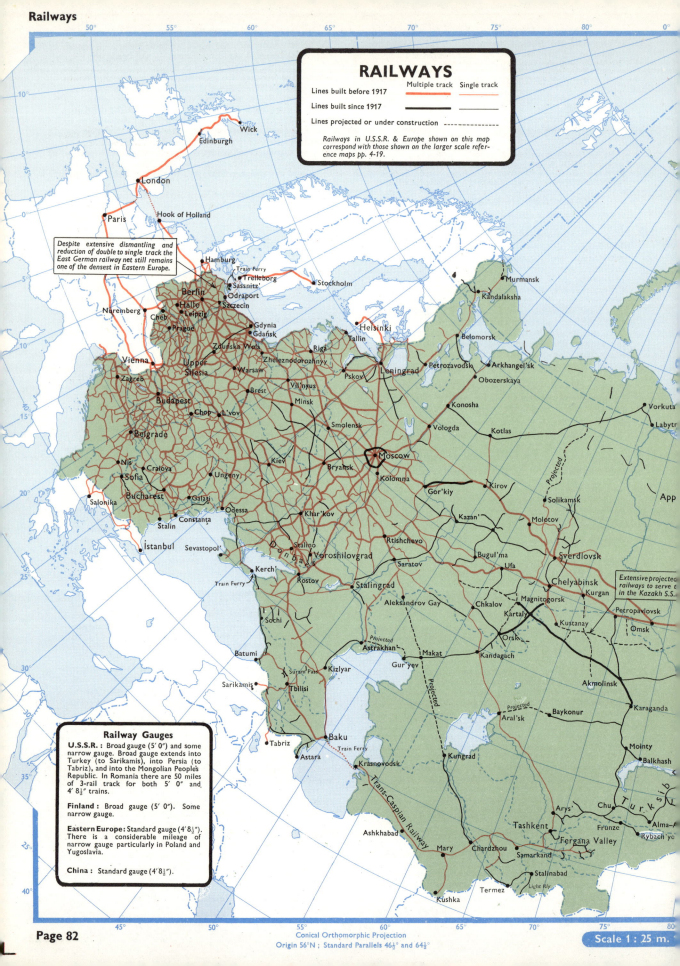

RAILWAYS

	Multiple track	Single track
Lines built before 1917		
Lines built since 1917		
Lines projected or under construction	- - - - - - - - -	

Railways in U.S.S.R. & Europe shown on this map correspond with those shown on the larger scale reference maps pp. 4-19.

Despite extensive dismantling and reduction of double to single track the East German railway net still remains one of the densest in Eastern Europe.

Railway Gauges

U.S.S.R.: Broad gauge (5′ 0″) and some narrow gauge. Broad gauge extends into Turkey (to Sarikamis), into Persia (to Tabriz), and into the Mongolian People's Republic. In Romania there are 50 miles of 3-rail track for both 5′ 0″ and 4′ 8½″ trains.

Finland: Broad gauge (5′ 0″). Some narrow gauge.

Eastern Europe: Standard gauge (4′ 8½″). There is a considerable mileage of narrow gauge particularly in Poland and Yugoslavia.

China: Standard gauge (4′ 8½″).

Wick
Edinburgh
London
Paris
Hook of Holland
Hamburg
Train Ferry
Trelleborg
Sassnitz'
Stockholm
Berlin
Odraport
Szczecin
Halle
Leipzig
Nuremberg
Cheb
Prague
Gdynia
Gdańsk
Zdunska Wola
Riga
Helsinki
Tallin
Vienna
Upper Silesia
Warsaw
Zheleznodorozhnyy
Pskov
Leningrad
Murmansk
Kandalaksha
Belomorsk
Petrozavodsk
Arkhangel'sk
Obozerskaya
Zagreb
Brest
Vil'nyus
Minsk
Smolensk
Konosha
Vologda
Kotlas
Vorkuta
Labytr
Budapest
Chop
L'vov
Kiev
Bryansk
Moscow
Kolomna
Kirov
Solikamsk
App
Belgrade
Gor'kiy
Molotov
Niš
Craiova
Ungeny
Kazan'
Sverdlovsk
Sofia
Bucharest
Galaţi
Khar'kov
Rtishchevo
Bugul'ma
Ufa
Chelyabinsk
Kurgan
Salonika
Odessa
Donbass
Stalino
Voroshilovgrad
Saratov
Chkalov
Magnitogorsk
Kartaly
Kustanay
Petropavlovsk
Stalin
Constanţa
Istanbul
Sevastopol'
Kerch'
Rostov
Stalingrad
Aleksandrov Gay
Orsk
Omsk
Train Ferry
Sochi
Astrakhan
Makat
Gur'yev
Kandagach
Akmolinsk
Batumi
Kizlyar
Karaganda
Suram Pass
Sarikamiş
Tbilisi
Aral'sk
Baykonur
Mointy
Balkhash
Tabriz
Baku
Train Ferry
Astara
Krasnovodsk
Kungrad
Trans-Caspian Railway
Turksib
Arys'
Chu
Alma-Ata
Ashkhabad
Tashkent
Frunze
Rybach'ye
Fergana Valley
Mary
Chardzhou
Samarkand
Stalinabad
Termez
Kushka
Light Rly

Comparative Railway Journey Times

Journey	Distance in miles	Overall time	
		days	hrs
Leningrad-Murmansk	901	1	14*
Leningrad-Moscow	404	–	10
Leningrad-Kiev	783	1	13
Moscow-Sochi	1226	1	23
Moscow-Baku	1580	2	15
Moscow-Minsk	466	–	14
Moscow-Tashkent	2092	3	12
Moscow-Irkutsk	3226	4	19
Moscow-Vladivostok	5800	9	5
Moscow-Peking	5609	8	12
Moscow-Paris	1815	2	13
London-Edinburgh	393	–	7½
London-Wick	729	–	22

Based on 1954-1955 timetables
* 1949 timetable

The railway from Ulan-Bator to Tsining (Chinese National Railways) reported complete in early 1955.

The projected railway from Lanchow to Alma-Ata is reported under construction from the Chinese end and is due to reach Yumen in 1957. It is believed that work has started from the Russian end but there are no progress reports.

100 miles approx.

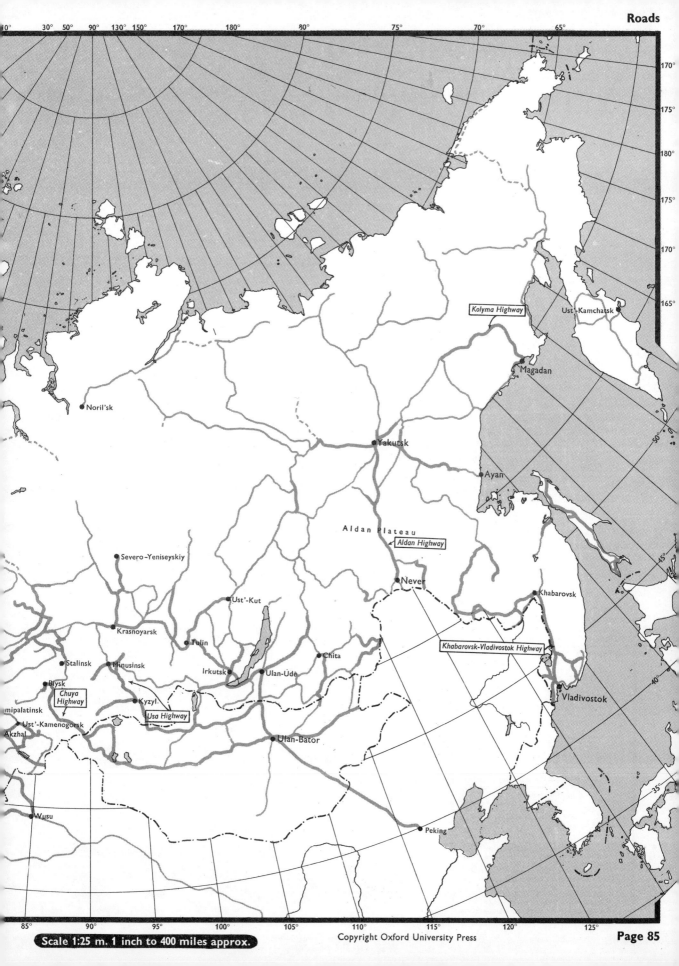

170°
175°
180°
175°
170°
165°

Kolyma Highway

Ust.-Kamchatsk

Magadan

50°

Noril'sk

Yakutsk

Ayan

Aldan Plateau
Aldan Highway

45°

Severo-Yeniseyskiy

Never

Khabarovsk

Ust'-Kut

Khabarovsk-Vladivostok Highway

Krasnoyarsk

Tulin

Chita

Stalinsk

Minusinsk

Irkutsk

Ulan-Ude

Vladivostok

Biysk

Chuya Highway

Kyzyl

Usa Highway

40°

mipalatinsk

Ust'-Kamenogorsk

Akzhal

35°

Ulan-Bator

Wusu

Peking

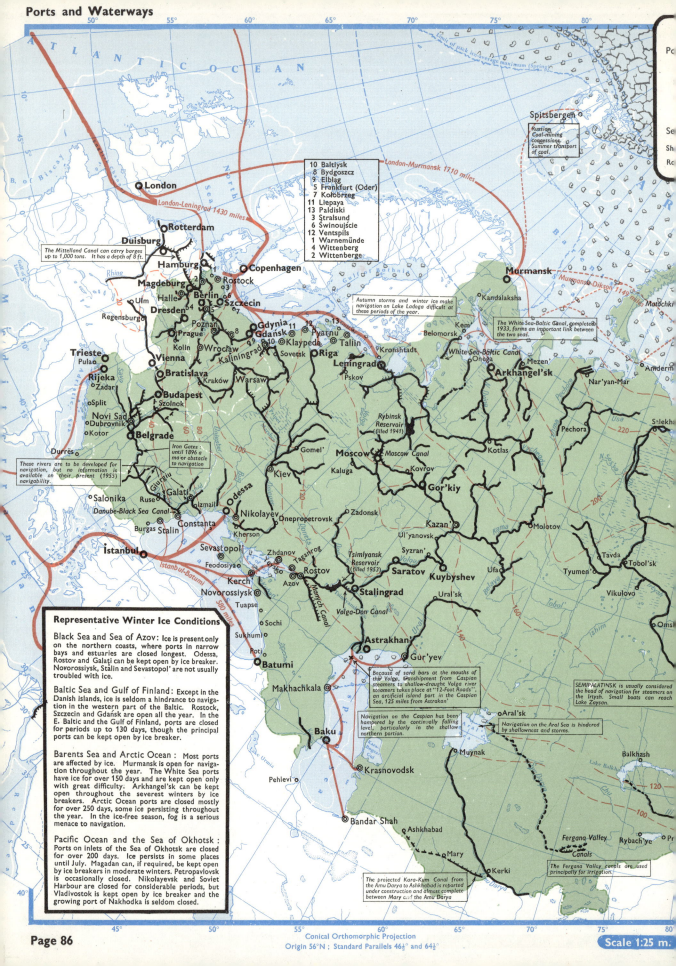

Ports and Waterways

10 Baltiysk
8 Bydgoszcz
9 Elbląg
5 Frankfurt (Oder)
7 Kołobrzeg
11 Liepaya
13 Paldiski
3 Stralsund
6 Świnoujście
12 Ventspils
1 Warnemünde
4 Wittenberg
2 Wittenberge

The Mittelland Canal can carry barges up to 1,000 tons. It has a depth of 8 ft.

Autumn storms and winter ice make navigation on Lake Ladoga difficult at these periods of the year.

The White Sea-Baltic Canal, completed 1933, forms an important link between the two seas.

Russian Coal-mining concessions. Summer transport of coal.

Iron Gates: until 1896 a major obstacle to navigation

These rivers are to be developed for navigation, but no information is available on their present (1955) navigability.

Because of sand bars at the mouths of the Volga, transhipment from Caspian steamers to shallow-draught Volga river steamers takes place at "12-Foot Roads", an artificial island port in the Caspian Sea, 125 miles from Astrakan'

Navigation on the Caspian has been hampered by the continually falling level, particularly in the shallow northern portion.

SEMIPALATINSK is usually considered the head of navigation for steamers on the Irtysh. Small boats can reach Lake Zaysan.

Navigation on the Aral Sea is hindered by shallowness and storms.

The Fergana Valley canals are used principally for irrigation.

The projected Kara-Kum Canal from the Amu Darya to Ashkhabad is reported under construction and almost complete between Mary and the Amu Darya

Representative Winter Ice Conditions

Black Sea and Sea of Azov: Ice is present only on the northern coasts, where ports in narrow bays and estuaries are closed longest. Odessa, Rostov and Galaţi can be kept open by ice breaker. Novorossiysk, Stalin and Sevastopol' are not usually troubled with ice.

Baltic Sea and Gulf of Finland: Except in the Danish islands, ice is seldom a hindrance to navigation in the western part of the Baltic. Rostock, Szczecin and Gdańsk are open all the year. In the E. Baltic and the Gulf of Finland, ports are closed for periods up to 130 days, though the principal ports can be kept open by ice breaker.

Barents Sea and Arctic Ocean: Most ports are affected by ice. Murmansk is open for navigation throughout the year. The White Sea ports have ice for over 150 days and are kept open only with great difficulty. Arkhangel'sk can be kept open throughout the severest winters by ice breakers. Arctic Ocean ports are closed mostly for over 250 days, some ice persisting throughout the year. In the ice-free season, fog is a serious menace to navigation.

Pacific Ocean and the Sea of Okhotsk: Ports on inlets of the Sea of Okhotsk are closed for over 200 days. Ice persists in some places until July. Magadan can, if required, be kept open by ice breakers in moderate winters. Petropavlovsk is occasionally closed. Nikolayevsk and Soviet Harbour are closed for considerable periods, but Vladivostok is kept open by ice breaker and the growing port of Nakhodka is seldom closed.

London-Murmansk 1710 miles
London-Leningrad 1430 miles
Murmansk-Dikson 1149 miles
Istanbul-Batumi
White Sea-Baltic Canal

Conical Orthomorphic Projection
Origin 56°N ; Standard Parallels 46½° and 64½°

Scale 1:25 m.

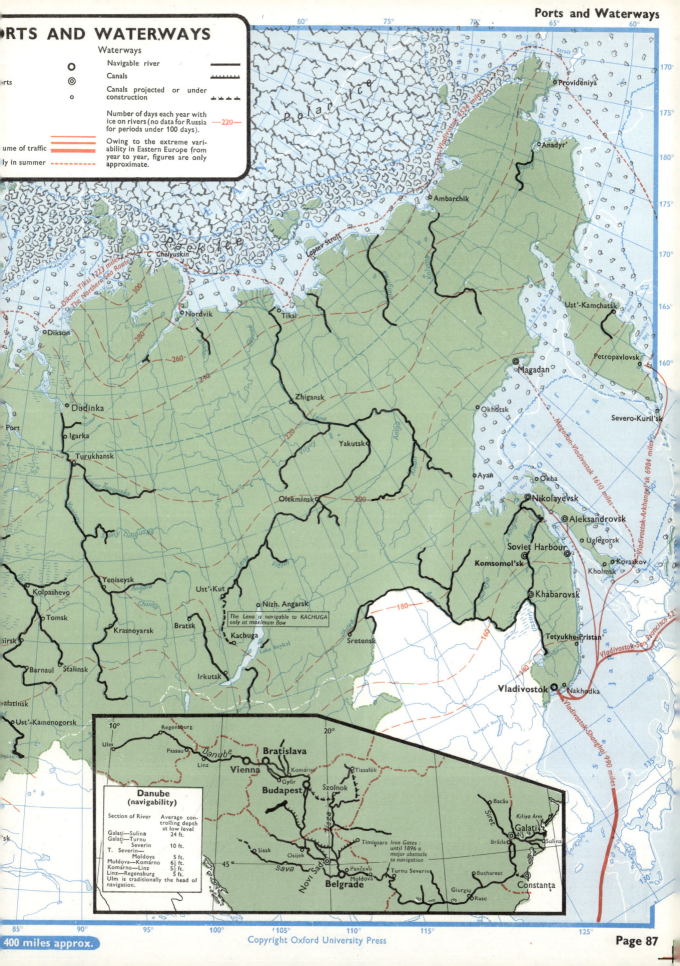

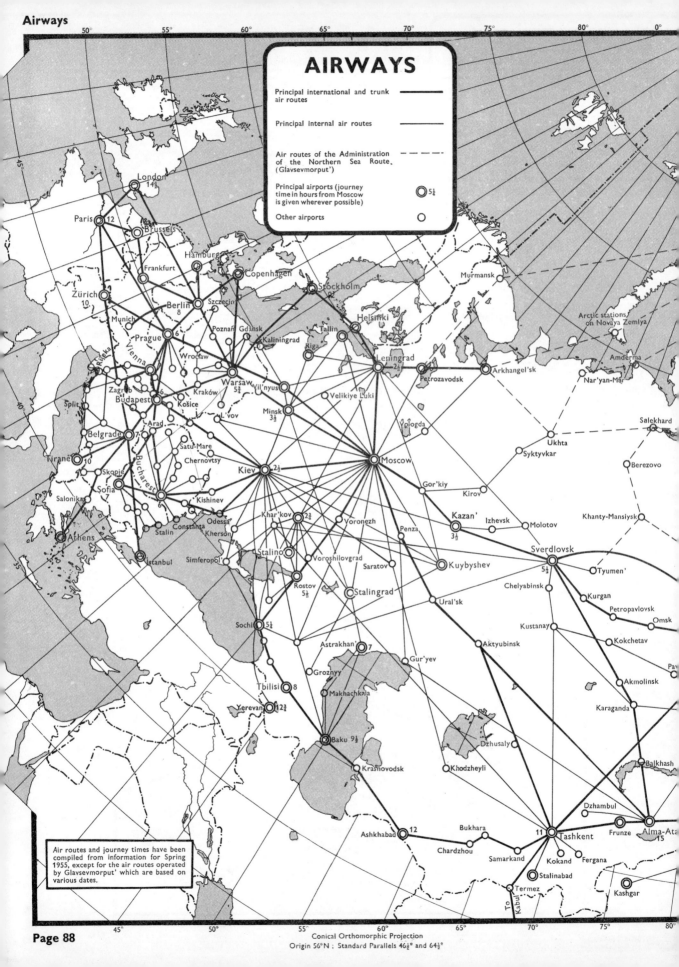

AIRWAYS

Principal international and trunk air routes	━━━━━━
Principal internal air routes	────────
Air routes of the Administration of the Northern Sea Route. ('Glavsevmorput')	─ ─ ─ ─
Principal airports (journey time in hours from Moscow is given wherever possible)	⊚ 5¼
Other airports	○

Air routes and journey times have been compiled from information for Spring 1955, except for the air routes operated by Glavsevmorput' which are based on various dates.

Conical Orthomorphic Projection
Origin 56°N ; Standard Parallels 46½° and 64½°

Uelen

Arctic stations
on Wrangel I.

Provideniya

Pil'khyn

Anadyr'

Markovo

Ambarchik

Arctic stations
on Severnaya Zemlya

Arctic stations

Zyryanka

Nordvik

Tiksi

Dikson

Magadan

Petropavlovsk

Dudinka

Noril'sk

Zhigansk

Okhotsk

Igarka

Yakutsk

Ayan

Okha

Nikolayevsk

Chulkovo

Aleksandrovsk

Aldan

Vitim

Ekimchan

Komsomol'sk

Yuzh.-Sakhalinsk
42

Bodaybo

Kolpashevo

Tsipikan

Skovorodino

Tygda

Mogocha

Svobodnyy

Birobidzhan

Khabarovsk
27½

Tomsk

Krasnoyarsk
15¾

Bikin

Novosibirsk
10½

Chita
23

Nerchinsk

Iman

Hailar

Voroshilov

Irkutsk
19

Ulan-Ude

Harbin

40½

Vladivostok

Kyakhta

imipalatinsk

Ulan-Bator
21

Ust'-Kamenogorsk

uz

Sayn Shanda

Mukden

Pyongyang

Urumchi
19½

Peking
27

Dairen

To
Lanchow

To
Lanchow

Scale 1:25 m. 1 inch to 400 miles approx.

POPULATION

Cities

50,000	⊙
100,000	⊙
500,000	○
1 million	○
5 million	○

One dot represents 20,000 people

Settled areas
(over 2·5 persons per sq. mile) in green

Unsettled or sparsely inhabited
(under 2·5 persons per sq. mile) in white

City populations, shown by black circles, are
supplementary to population shown by red dots

U.S.S.R. Populatio

Date of Foundation (A.D.)		1926	1939	1959	Date of Foundation (A.D.)	
1147	Moscow	2,029,425	4,147,018	5,032,000	14th cent.	Kazan
1703	Leningrad	1,690,065	3,191,304	3,300,000	17th cent.	Perm
8th cent.	Kiev	513,637	896,293	1,102,000	1201	Riga
885	Baku	453,333	809,347	968,000	1761	Rosto
1221	Gor'kiy	222,356	644,116	942,000	1589	Stalin
17th cent.	Khar'kov	417,342	833,432	930,000	1590	Sarate
7th cent.	Tashkent	323,613	585,005	911,000	1716	Omsk
1893	Novosibirsk	120,128	405,589	887,000	1574	Ufa-C
1586	Kuybyshev	175,636	390,267	806,000	1066	Minsk
1722	Sverdlovsk	140,300	425,444	777,000	7th cent.	Yerev
1870	Stalino	174,230	462,395	701,000	1854	Alma
4th cent.	Tbilisi	294,044	519,175	694,000	1586	Voron
1688	Chelyabinsk	59,307	273,127	688,000	1770	Zapo
14th cent.	Odessa	420,862	604,223	667,000	1250	L'vov
1787	Dnepropetrovsk	236,717	500,662	658,000	1628	Krasn

See Back Endpaper for popula

Eastern Europe: Towns with over 300,000 Inhabitants

Berlin (West)	(1958)	2,224,000
Budapest	(1956)	1,850,000
Bucharest	(1958)	1,278,814
Berlin (East)	(1958)	1,110,000
Warsaw	(1958)	1,081,000
Prague	(1958)	987,865
Sofia	(1957)	725,000
Lódz	(1958)	691,100
Leipzig	(1958)	593,908
Belgrade	(1957)	506,000
Dresden	(1958)	491,646
Kraków	(1958)	455,400
Zagreb	(1957)	441,000
Wroclaw	(1958)	401,400
Poznan	(1958)	388,500
Brno	(1959)	312,330

Map labels: Murmansk, Berlin, Prague, Brno, Bratislava, Zagreb, Riga, Tallin, Leningrad, Arkhangel'sk, Łodz, Warsaw, Vil'nyus, Budapest, Minsk, L'vov, Belgrade, Skopje, Sofia, Kishinev, Kiev, Kalinin, Moscow, Yaroslavl', Ivanovo, Kirov, Bucharest, Tula, Gor'kiy, Odessa, Dnepropetrovsk, Zaporozh'ye, Khar'kov, Voronezh, Izhevsk, Molotov, Nizh. Tagil, Stalino, Penza, Kazan', Ufa-Chernikovsk, Sverdlovsk, Rostov, Saratov, Chelyabinsk, Krasnodar, Stalingrad, Kuybyshev, Magnitogorsk, Omsk, Astrakhan', Groznyy, Tbilisi, Yerevan, Baku, Karaganda, Kzyl-Orda, Ashkhabad, Tashkent, Frunze, Alma, Samarkand, Stalinabad

Conical Orthomorphic Projection
Origin 56°N ; Standard Parallels 46½° and 64½°

Scale 1 : 25 m.

Towns over 300,000, 1959

1926	1939	1959	Date of Foundation (A.D.)		1926	1939	1959
179,023	401,665	643,000	1024	Yaroslavl'	114,277	298,065	406,000
119,776	255,196	628,000	1928	Karaganda	nil	165,937	398,000
	383,699*	605,000	17th cent.	Krivoy Rog	38,000	189,000	386,000
308,103	510,253	597,000	1617	Stalinsk	3,894	169,538	377,000
151,490	445,476	591,000	1652	Irkutsk	108,129	243,380	365,000
219,547	375,860	581,000	1899	Makeyevka	79,000	242,000	358,000
161,684	280,716	579,000	12th cent.	Tula	155,005	272,403	345,000
98,537	245,863	546,000	1725	Nizhniy Tagil	38,849	159,864	338,000
131,803	238,772	509,000	1328	Ivanovo	111,460	285,069	332,000
64,613	200,031	509,000	1858	Khabarovsk	52,045	199,364	322,000
45,395	230,528	455,000	1738	Barnaul	74,000	148,000	320,000
121,612	326,836	454,000	1794	Krasnodar	162,000	193,000	312,000
55,744	289,188	435,000	1929	Magnitogorsk	30,000	145,870	311,000
	316,177†	410,000	8th cent.	Astrakhan'	184,301§	253,655	294,000
72,261	189,999	409,000					

ns over 50,000.   * 1935.   † 1931.   § 1929.

**Total Population
U.S.S.R. 209 million**
Census, January 1959
Eastern Europe 115 million
(1958 estimate)
(incl. Yugoslavia, but
excl. West Berlin)

400 miles approx.

POPULATION

The Population of the U.S.S.R.

The first census in Tsarist Russia was held in 1724 by order of Peter the Great; it was incomplete in that it covered fully only the European part of the country, including the Ukraine and Byelorussia, while east of the Urals only the Russian population was enumerated. From time to time the 1724 figures were revised on the basis of partial censuses, and allowances were made for territories newly acquired, but it was not till 1897 that the first really comprehensive census was held. This set out to enumerate and analyse the whole population of the Empire, including the non-Russians, who were classified on an ethnic basis. From 1897 to 1926 only estimates are available.

In 1926 the Soviet authorities conducted a census even more comprehensive and thorough than that of 1897, and its results provide a social document of great interest. Not only was the ethnic analysis very fully carried out, but the figures also indicate the scale and direction of internal migration, the degree of urbanization and numerous other matters of demographic interest. A similar census was held in 1939, but the war prevented the publication of full reports. The first post-World War II census was held on January 15th, 1959; only preliminary results are so far available. In the text and tables which follow, figures for 1897, 1926, 1939 and 1959 may be taken as accurate. All others are estimates.

In January, 1959, the population of the Soviet Union was 209 million, an increase of 9·5 per cent compared with 1939. A very high birth rate, at present 25 per thousand, combined with an exceptionally low death rate, only 7·5 per thousand, is causing the population to increase at a current rate of more than 3·5 million per annum, or nearly 1·7 per cent.

In the following table figures for Great Britain and the United States are included for purposes of comparison of rates of increase.

Population Growth, Russian Empire and U.S.S.R., Great Britain and U.S.A.†

Russian Empire and U.S.S.R.			Great Britain			U.S.A.		
Date	Total pop. ('000)	Annual inc. %	Date	Total pop. ('000)	Annual inc. %	Date	Total pop. ('000)	Annual inc. %
1724	20,300		1801	10,648		1790	3,929	
1897	111,916	0·8	1901	37,000	2·5	1900	75,995	•
1914*	145,000	1·5	1911	40,831	1·0	1910	91,972	2·1
1926	147,028	•	1921	42,769	0·5	1920	105,711	1·5
1939	170,467	1·3	1931	44,795	0·4	1930	122,775	1·6
1940*	192,900	‡	1938**	46,208	0·4	1940	131,669	0·7
1950*	200,000	1·1	1951	48,854	0·5	1950	150,697	1·4
1959	208,800	1·7	1959**	50,578	0·4	1959	178,153	1·7

* End of year estimates. ** Mid-year estimates. All others are census figures.

† Frontiers as in each year. The Russian figures for 1897 and 1914 exclude Finland, Poland, Khiva and Bukhara. The U.S. figure for 1959 excludes Hawaii and Alaska.

• Increases cannot be accurately calculated owing to frontier changes, war losses, etc.

‡ Increase due to territorial changes.

Wartime population losses were extremely heavy, and have resulted in the present population being some 30 million less than would normally have been expected, as a result of direct war losses and a reduced birth rate during and immediately following the war. The effects appear in a 20 million discrepancy between the number of males and females adult at the end of hostilities, and in a relatively small number of persons currently entering the 15–20 age-group.

Regional Redistribution

The trends shown by the figures of total population conceal very wide regional differences. Parallel with the absolute growth, there has been a continuous process of redistribution both from area to area and from the countryside to the towns.

The expansion of Russian power into Siberia, described on pp. 100/101, was not at first accompanied by a pronounced eastward movement of population. This movement began on a large scale in the early part of the 19th century, and reached its pre-revolutionary peak between 1891 and 1911. During the earlier period (up to about 1880) political prisoners and exiles were the chief source of manpower for the new areas. After 1880 the movement changed its character and peasants from Old Russia were transferred in increasing numbers—sometimes voluntarily, but more usually not.

Between 1801 and 1859 about 570,000 people of Russian stock were settled in Siberia. Changes between 1859 and 1897 are shown in the following table.

Increase of Russian Population by Regions

	1859 ('000)	1897 ('000)	1897 as % of 1859
European Russia	55,205	87,384	158
Ukraine—Black Sea	10,501	19,564	186
Don-Caucasus	1,540	5,328	346
Lower Volga	2,400	3,609	150
Asiatic Russia	3,424	6,947	203
Siberia	2,288	4,659	203
Turkestan	nil	204	
Transcaucasia	nil	249	
Other regions	1,136	1,835	162

By 1911 the Russian population of Siberia had risen further to 7,996,000, an increase of 71 per cent over 1897. That of Turkestan had risen to 407,000, or just about double that of 1897.

The 1926 census included an analysis of the population according to region of birth as compared with region of residence. The whole of Asia, Ural Oblast and the Bashkir A.S.S.R., Leningrad–Karelia, Crimea and the mining districts of the Ukraine all showed a net gain in population, and all other areas a loss.

U.S.S.R.: Gains by Migration. 1926 Census

Area	Total Population ('000)	Net Gain*	Percentage
Siberian Kray	8,600	1,975	22·7
Far East	1,881	335	17·8
Turkestan	13,770	934	6·8
North Caucasian Kray	8,363	766	9·2

* Excess of those born in other regions but resident in region named over those born in the region named but resident elsewhere.

The areas which recorded the highest loss by migration were the Central Black Earth region (—11%), Byelorussia (—9%), Vyatka (—9%), Central Volga, West European and parts of the Ukraine (—8% each). The inference is that the new areas were being peopled largely by immigrants of Russian stock, and especially from the agricultural regions of European Russia.

The extent to which Russians have penetrated areas outside the Russian Republic (R.S.F.S.R.) is clearly shown in the results of the latest census. Outside the R.S.F.S.R., their numbers are highest in the Ukraine, Kazakhstan (where they form 53·1 per cent of the population) and Uzbekistan; and in the Kirgiz Republic they make up over 30 per cent of the population of 2·1 million. This move eastwards and southwards is closely connected with the spread of economic development. Voluntary recruitment is widely used to find settlers for the wheat areas in the " virgin " lands and other districts where development is being carried out, and where conditions of life and work are particularly arduous compulsory means have been employed.

During the second World War seven nationalities totalling over 1 million people—the Volga Germans, the Crimean Tartars, the Kalmyks and four Caucasian groups of lesser importance—were deported from their homes and compulsorily settled in Siberia and Central Asia. However, after Stalin's death it was admitted that such action was not justified by security considerations; the Kalmyks and many of the other smaller groups have been, or are being, resettled on their old lands, but the Volga Germans and Crimean Tartars have not yet (1960) been rehabilitated or re-instated in their original homelands.

Soviet Population by Republics
(million)

Russian Republic		117·5	Byelorussian Republic		8·1
of which: Russians	97·8		of which: Byelorussians	6·4	
Tartars	4·1		Russians	0·7	
Ukrainian Republic		41·9	Uzbek Republic		8·1
of which: Ukrainians	31·9		of which: Uzbeks	5·0	
Russians	7·4		Russians	1·1	
Kazakh Republic		9·3	Georgian Republic		4·0
of which: Russians	4·0		of which: Georgians	2·6	
Kazakhs	2·8		Russians	0·4	
Ukrainians	0·8		Armenians	0·4	

Soviet Population by Republics (continued)

Azerbaydzhan Republic		3·7	Tadzhik Republic	2·0
of which: Azerbaydzhani	2·5		of which: Tadzhiks	1·1
Russians	0·5		Armenian Republic	1·8
Moldavian Republic		2·9	of which: Armenians	1·6
of which: Moldavians	1·9		Turkmen Republic	1·5
Ukrainians	0·4		of which: Turkmen	0·9
Lithuanian Republic		2·7	Russians	0·3
of which: Lithuanians	2·2		Estonian Republic	1·2
Latvian Republic		2·1	of which: Estonians	0·9
of which: Latvians	1·3		Russians	0·3
Russians	0·6			
Kirgiz Republic		2·1		
of which: Kirgiz	0·8		Grand Total	208·8
Russians	0·6			

Soviet Population by Nationalities
(million)

Russians	114·6	Georgians	2·7
Ukrainians	37·0	Lithuanians	2·3
Byelorussians	7·8	Jews	2·3
Uzbeks	6·0	Moldavians	2·2
Tartars	5·0	Other (over 100)	19·7
Kazakhs	3·6		
Azerbaydhani	2·9	Total	208·8
Armenians	2·8		

URBANIZATION

The development of industry has meant, over the last several decades, an increase in the proportion of the population living in towns.

U.S.S.R.: Urban and Rural Population
('000)

	1939			1959		
	Urban	Rural	Urban as % of total	Urban	Rural	Urban as % of total
R.S.F.S.R.	36,296	72,083	33·5	61,447	56,017	52·3
Caucasus Republics	2,589	5,306	32·8	4,377	5,172	45·8
Central Asian Republics*	4,095	12,529	24·6	8,824	14,155	38·4
Western Republics†	17,429	40,219	30·2	25,156	35,700	41·3

* Excluding Mongolian People's Republic.

† Excluding Karelo-Finnish A.S.S.R.

Sizes of towns

The next table analyses the urban population by size of towns over 50,000 in 1926, 1939 and 1959. While in the thirteen years 1926 to 1939 the population of such towns increased by over a quarter, in the twenty years 1939–59 the increase was only 16·6 per cent. In both periods by far the greatest increase was in towns of 500,000 and over, but while the population of these towns nearly tripled in the earlier period, it rose by only 88 per cent between 1939 and 1959. It is worth noting that Moscow, Leningrad and Kiev are still the only cities with over 1,000,000 people, though Baku, Gor'kiy, Khar'kov and Tashkent all have more than 900,000 inhabitants.

U.S.S.R. Towns over 50,000 Population 1926, 1939, 1959

	1926		1939		1959	
	No. of towns	Pop. million	No. of towns	Pop. million	No. of towns	Pop. million
50,000–100,000	58	3·9	99	7·1	151	10·6
100,000–500,000	30	5·8	78	15·6	123	24·4
500,000 and over	3	4·2	11	12·8	25	24·1
Total over 50,000	91	13·9	188	35·5	299	59·1
% of total population		9·5		18·6		28·3

U.S.S.R.: Towns of over 200,000 by Region

Region	Population ('000)		Increase 1939–1959
	1939	1959	%
R.S.F.S.R.	17,643	25,774	46·1
Caucasus Republics	2,331	3,101	33·0
Central Asian Republics*§	1,231	2,375	92·9
Western Republics†	4,832	6,874	42·3

* Excluding Mongolian People's Republic.

† Excluding Karelo-Finnish A.S.S.R.

§ Including Ashkhabad, which though possessing only 170,000 inhabitants in 1959 is a Union Republic capital.

From the dates of foundation shown in the table (on the map), it will be noticed that nearly all the largest towns are old settlements, though some remained small till recent years. Karaganda is the most striking example of a town founded since the Revolution which has grown to great size. Others include Magnitogorsk (1959 population 311,000), Komsomol'sk (177,000) and Stalinogorsk (107,000). Stalinsk, which had under 4,000 people in 1926, had 377,000 in 1959; Stalinabad (1926: 5,600) had 244,000; Murmansk rose in the same period from 8,800 to 226,000; Dzerzhinsk from 8,900 to 163,000; and Prokopyevsk from 10,700 to 282,000.

THE POPULATION OF EASTERN EUROPE

Frontier changes make historical comparisons both difficult and of little value. The table below therefore shows in general only post-war census figures and the latest available estimates, together with the area covered and the density of population.

The Population of Eastern Europe

	Area sq. km.	Population	Density per sq. km.	Anural Rate of Natural increase 1953–57 %
Poland	311,730			
1950 Census		25,008,179	80·1	1·9
1958 (est.)		28,997,000	93·0	
Czechoslovakia	127,859			
1950 Census		12,338,450	96·4	1·0
1959 (est.)		13,564,000	106·0	
East Germany	107,431			
1950 Census		17,199,098	160·0	—0·9
1958 (est.)		16,255,000	151·3	
Berlin (West)	481			
1956 Census		2,223,777	4,623·7	0·4
1958 (est.)		2,224,000	4,623·7	
Berlin (East)	403			
1950 Census		1,189,074	2,950·3	—1·3
1958 (est.)		1,100,000	2,729·5	
Hungary	93,030			
1949 Census		9,204,799	98·9	0·6
1959 (est.)		9,917,000	106·6	
Romania	237,500			
1956 Census		17,489,450	73·6	1·4
1958 (est.)		18,059,000	76·0	
Bulgaria	111,493			
1956 Census		7,600,525	68·1	0·9
1959 (est.)		7,793,000	69·8	
Albania	28,748			
1955 Census		1,391,499	48·3	2·9
1958 (est.)		1,507,000	52·4	
Yugoslavia	255,804			
1953 Census		16,990,617	66·4	1·4
1959 (est.)		18,448,000	72·1	

Eastern Europe: Urban and Rural Population

	Urban (%)	Rural (%)
Poland (1958 est.)	46·4	53·6
Czechoslovakia (1947)	48·8	51·2
East Germany (1958)	71·5	28·5
Hungary (1949)	34·5	65·5
Romania (1958 est.)	31·7	68·3
Bulgaria (1956)	33·5	66·5

Albania, for which no figures are available, has a very largely rural population.

The size of towns

Berlin, both East and West, Budapest, Bucharest and Warsaw, all have populations of over 1 million and Prague is nearing the million mark. There are 54 towns with between 100,000 and a million inhabitants; 19 are in Poland, 9 in East Germany, 5 in Czechoslovakia, 7 in Romania, 6 in Yugoslavia, 5 in Hungary and 3 in Bulgaria. The largest town in Albania is Tirane, the capital, which had 108,183 inhabitants in 1955.

The names of towns of over 300,000 in the U.S.S.R. and in Eastern Europe are listed on the map on pp. 90/91.

Ethnic Composition of the U.S.S.R., 193[

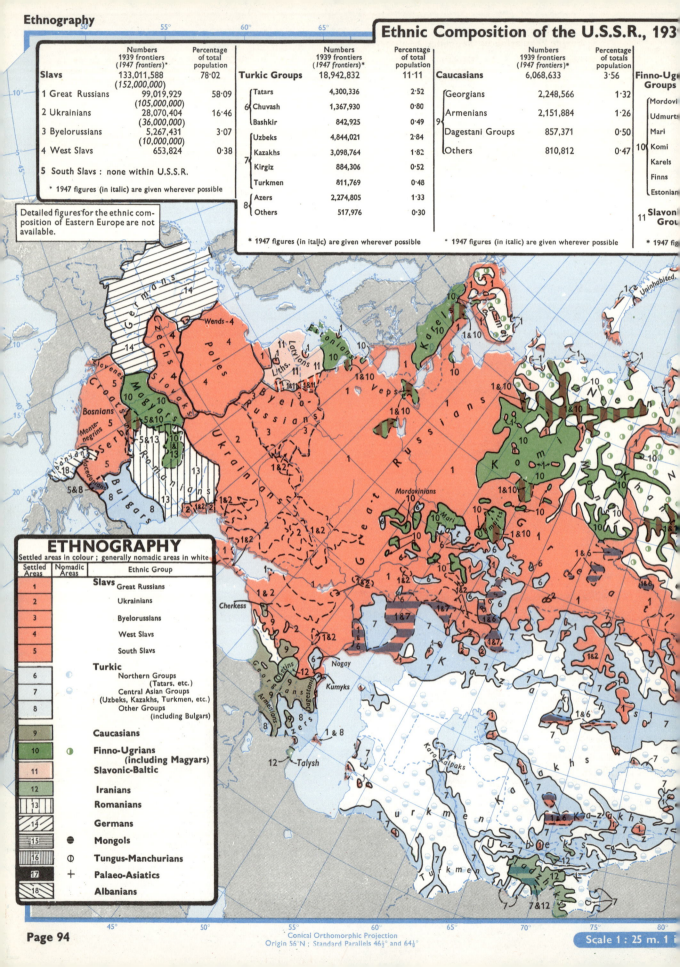

Slavs	Numbers 1939 frontiers (1947 frontiers)*	Percentage of total population
	133,011,588 (152,000,000)	78·02
1 Great Russians	99,019,929 (105,000,000)	58·09
2 Ukrainians	28,070,404 (36,000,000)	16·46
3 Byelorussians	5,267,431 (10,000,000)	3·07
4 West Slavs	653,824	0·38
5 South Slavs : none within U.S.S.R.		

* 1947 figures (in italic) are given wherever possible

Turkic Groups	Numbers 1939 frontiers (1947 frontiers)*	Percentage of total population
	18,942,832	11·11
6 Tatars	4,300,336	2·52
Chuvash	1,367,930	0·80
Bashkir	842,925	0·49
7 Uzbeks	4,844,021	2·84
Kazakhs	3,098,764	1·82
Kirgiz	884,306	0·52
Turkmen	811,769	0·48
8 Azers	2,274,805	1·33
Others	517,976	0·30

* 1947 figures (in italic) are given wherever possible

Caucasians	Numbers 1939 frontiers (1947 frontiers)*	Percentage of totals population
	6,068,633	3·56
9 Georgians	2,248,566	1·32
Armenians	2,151,884	1·26
Dagestani Groups	857,371	0·50
Others	810,812	0·47

* 1947 figures (in italic) are given wherever possible

Finno-Ug[
Groups

Mordovi[
Udmurts
Mari
10 Komi
Karels
Finns
Estonian[

11 Slavoni[
Grou[

* 1947 fig[

Detailed figures for the ethnic composition of Eastern Europe are not available.

ETHNOGRAPHY

Settled areas in colour ; generally nomadic areas in white

Settled Areas	Nomadic Areas	Ethnic Group
		Slavs
1		Great Russians
2		Ukrainians
3		Byelorussians
4		West Slavs
5		South Slavs
6	○	**Turkic** Northern Groups (Tatars, etc.)
7	◔	Central Asian Groups (Uzbeks, Kazakhs, Turkmen, etc.)
8		Other Groups (including Bulgars)
9		**Caucasians**
10	◖	**Finno-Ugrians** (including Magyars)
11		**Slavonic-Baltic**
12		**Iranians**
13		**Romanians**
14		**Germans**
15	⬤	**Mongols**
16	⊕	**Tungus-Manchurians**
17	+	**Palaeo-Asiatics**
18		**Albanians**

Conical Orthomorphic Projection
Origin 56°N ; Standard Parallels 46½° and 64½°

Scale 1 : 25 m. 1 i[

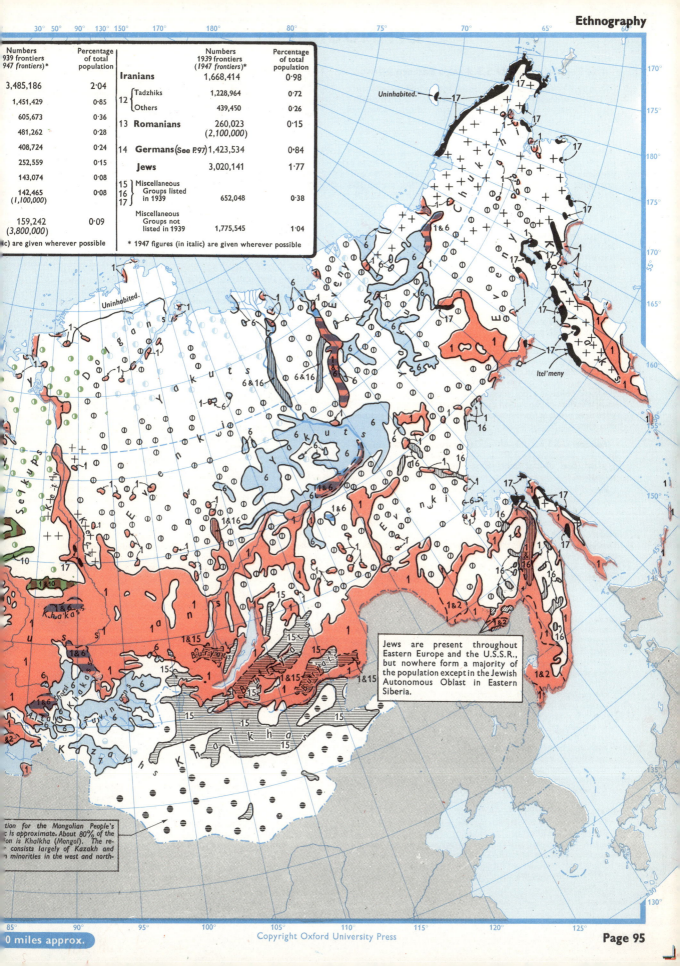

Numbers 939 frontiers (947 frontiers)*	Percentage of total population		Numbers 1939 frontiers (1947 frontiers)*	Percentage of total population
3,485,186	2·04	**Iranians**	1,668,414	0·98
1,451,429	0·85	12 { Tadzhiks	1,228,964	0·72
605,673	0·36	{ Others	439,450	0·26
481,262	0·28	13 **Romanians**	260,023 (2,100,000)	0·15
408,724	0·24	14 **Germans** (See P.97)	1,423,534	0·84
252,559	0·15	**Jews**	3,020,141	1·77
143,074	0·08	15 16 17 } Miscellaneous Groups listed in 1939	652,048	0·38
142,465 (1,100,000)	0·08	Miscellaneous Groups not listed in 1939	1,775,545	1·04
159,242 (3,800,000)	0·09			

ic) are given wherever possible

* 1947 figures (in italic) are given wherever possible

Jews are present throughout Eastern Europe and the U.S.S.R., but nowhere form a majority of the population except in the Jewish Autonomous Oblast in Eastern Siberia.

...tion for the Mongolian People's
...c is approximate. About 80% of the
...ion is Khalkha (Mongol). The re-
...consists largely of Kazakh and
...n minorities in the west and north-

Uninhabited.

Uninhabited.

ETHNOGRAPHY

The Peoples of the U.S.S.R. and Eastern Europe

The 1926 census of the Soviet Union recognized the existence of 169 " nationalities ". These were classified on a composite basis of ancestry, physical type, language, culture, sometimes religion, and sometimes on no other standard than that of having formed a distinct community for a reasonably long time. Many were simply sub-groups of much larger communities—for instance, the Jews were divided into five " nationalities "—and the list was reduced to twenty-five main headings. By further grouping, without offence to national characteristics or to ethnographic practice, it is possible to reduce the list to still smaller proportions, giving a total of twelve for the U.S.S.R.; only one more (the Albanians) need be added to make up the tale for Eastern Europe as well.

Since the Revolution, the administrative framework of the Soviet Union has been based largely on nationalities. The sixteen Republics are named after the national groups which form the majority of their populations, and in the R.S.F.S.R. and a few of the others special administrative status is given to areas where ethnic minorities predominate; hence the Autonomous Soviet Socialist Republic (A.S.S.R.), the Autonomous Oblast and the National Okrug, the last of which is reserved for peoples of a relatively low level of cultural and economic development.

In Eastern Europe an attempt was made after 1918 to redraw the map on a " national " basis in accordance with the principle of self-determination, but the considerable minorities who often had to be left on the wrong side of the frontier prevented the new political order from acquiring full stability. Large-scale transfers of population during and after the Second World War have brought frontiers and ethnic boundaries more closely into line.

For the U.S.S.R., only the 1926 figures are really comprehensive; although the 1939 census worked on a similar basis, a great deal of the information obtained has never been published.

THE SLAVS

The dominant ethnic group throughout the Soviet Union and Eastern Europe is the Slavs. Apart from their common Indo-European origin and a strong linguistic connection, there is little to identify some sections of the Slavs with others, and different historical experience, as well as the admixture of non-Slav ethnic elements, has caused a number of quite distinct characteristics to make their appearance. Five main groups can be distinguished.

Great Russians. Settled in south and central Russia since prehistoric times, the Great Russians outlasted invasions from the east and from the north, but became partly mixed with Turkic (Tatar) elements, some Scandinavian blood, and also with indigenous races in the areas into which they expanded. They adopted the Greek Orthodox form of Christianity, modified slightly into the Russian form, under the authority of a Russian Patriarch. In 1947 they numbered rather over 100 million.

we are not russians!

Ukrainians or Little Russians. These are a branch of the Great Russians distinguished chiefly by linguistic variations, and claim to be of purer Slav stock; they are also Orthodox by religion, but include some Uniates. In 1947 there were about 36 million Ukrainians in the U.S.S.R., and a small number in Poland. The small group of Sub-Carpathian Russians, or Ruthenes, are really Ukrainians.

Byelorussians (White Russians). A western branch of the Great Russians, influenced by contact with their western Polish and Lithuanian neighbours, and including some Scandinavian blood, the White Russians in 1939 were divided about equally between the U.S.S.R. and Poland; since 1945 there have been about 10 million in the U.S.S.R. East of the 1939 frontier they are mainly Orthodox, to the west largely Roman Catholic.

The West Slavs include the Poles, Czechs, Slovaks, Wends and a few smaller groups; all bear marked traces of their contacts with the Germans; there are also Scandinavian influences in Poland and Magyar influences among the Czechs and Slovaks. The Wends are a small Slavic group found mainly in Lusatia. The Poles and Slovaks are predominantly Roman Catholic; a majority of the Czechs are Protestants.

The South Slavs comprise the Slovenes, Croats, Bosnians, Serbs, Montenegrins and Macedonians, and their distinctive characteristics derive mainly from Turkish occupation; the Croats and Slovenes show marked Germanic influence and there is an admixture of Magyar elements in the Danube and Tisza valleys. The Slovenes and Croats are mainly Roman Catholics, the Bosnians Moslem, and the remainder Greek Orthodox.

TURKIC GROUPS

In the Soviet Union, nationalities of Turkic or Turanian origin form the second largest ethnic group, though amounting to only 11 % of the total population as compared with 78 % Slavs (1939; the proportion of Slavs is now somewhat higher since the incorporation of eastern Poland and Ruthenia). The Turkic groups are of Central Asian origin, and are overwhelmingly Moslem by religion.

Tatars. The main concentration of Tatars is in the Volga region, but other smaller groups are found widely scattered; all are descendants of the invaders from the east during the Middle Ages, when the Tatars formed a screen to the advancing Mongols. In 1926 the Volga Tatars and associated groups numbered about 3,350,000, and in 1939 about 4,300,000.

Bashkirs. Distinct from the Tatars, the Bashkirs are a nomadic steppe people who settled west of the Urals in the time of Genghis Khan. In 1926 they numbered, with related groups, about 843,000. They are Sunni Moslem by religion.

Kazakhs. These were formerly the largest Turanian group in the U.S.S.R. (about 4 million in 1926), but their numbers fell by a quarter between 1926 and 1939. They are a nomadic people with considerable Mongol influences, and are mainly Sunni Moslems.

Uzbeks. Also about 4 million in 1926, rising to nearly 5 million in 1939, the Uzbeks now greatly outnumber the Kazakhs. They are a settled people with strong Persian influences in their language and culture, and are also Sunni Moslems.

Other Central Asian Turanians include the Kirgiz, Turkmen, Kara-Kalpaks, Uygurs, Dungans and a few minor groups. The Kirgiz and Dungans are nomadic and show Mongol influences; the rest are mainly sedentary.

Siberian Turks. The largest group is the Yakuts (241,000 in 1926), who show marked Mongol characteristics but are Orthodox by religion. Other groups are the Khakass and the Tuvinians in southern Siberia, and the Dolgans in the extreme north.

Caucasian Turkic Groups. The Azers or Azerbaydzhanis (2,274,805 in 1939) are the largest Turkic group of the Caucasus region, and migrated there from Central Asia round the southern end of the Caspian Sea, thus absorbing many Persian influences into their culture. They are Shi'ite Moslems by religion. The other groups (Kumyks, Karachay, Balkar and Karapapakh) are all Sunni Moslems. The Karachay and Balkars were accused of collaborating with the Germans during the war, and were deprived of their national status and dispersed.

Bulgars. The original Bulgar invaders of south-east Europe came from Central Asia, and were the southern branch of a larger movement west. Others settled in the Volga region, several centuries before the Tatars. They adopted Slav speech and culture and the Orthodox religion, and are now more Slav than Turkic in character. While most of the Volga Bulgars were dispersed or absorbed, the related Chuvash have retained a distinct identity (1,367,930 in 1939).

FINNO-UGRIANS

The original home of the Finno-Ugrians is believed to be in the Altay mountains of southern Siberia, from which they were driven north-westwards by pressure from other nomadic peoples of Central Asia The main concentrations are in the extreme north-west of the U.S.S.R. and in Finland, and in the area between the Volga and the Urals. The largest sub-groups in the U.S.S.R. are the Mordovinians (1,451,429 in 1939), the Udmurt (605,673) and the Karels in the north-west (252,559) Apart from the Mordovinians and Udmurts, the Volga–Urals group includes the Mari and the Komi (further north). The north-western group consists of the Karels and Finns, the Estonians (Protestants, with strong German influence), and the Veps, Vod', Izhora and Saami (otherwise known as Lopars, and closely related to the Lapps). Two groups in northern Siberia, the Khanty (Ostyaks) and the Nentsy (Samoyeds) are probably Finno-Ugrian. Apart from the Estonians, most of the Finno-Ugrian peoples are Orthodox by religion.

The Magyars of Hungary and Central Europe are probably distantly related to the Finno-Ugrians, though strongly differentiated by historical experience. The languages have certain similarities.

SLAVONIC—BALTIC GROUPS

The Latvians, Latgals and Lithuanians are the descendants of Slavonic groups who settled in the Baltic area before the arrival of the Scandinavians or Russians. Their languages and cultures form a distinct group, though both the Roman Catholic Lithuanians and the Protestant Latvians have been greatly influenced by their ties with the cultures of Germany and Poland.

GERMANS

The strictly German groups include about 47 million in Western Germany, perhaps 17 million in Eastern Germany (including Berlin), and the great majority of the 7 million people of Austria. Other German settlements have existed at various times throughout central and eastern Europe. People of German language and culture are thus the second largest ethnic group in Europe, totalling over 70 million in all. Along the fringes of the main area they occupy, they are heavily mixed with other stocks—Slav in eastern districts of Prussia, Saxony and the Czech borderlands, Magyar and Latin in Austria.

The former Volga German Republic, which owed its origins to political and religious minorities who settled in the Saratov area in the 18th century, was abolished in 1941 after its inhabitants were accused of collaborating with the invading armies; the people were dispersed elsewhere. Most of the Germans in Eastern Europe, as in the Sudetenland or the Volksdeutsche settlement in the Yugoslav Banat, were compulsorily transferred to Germany after the Second World War; there are now few left to the east of the Oder-Neisse line or in the Balkans.

ROMANIANS

By tradition the Romanians are the descendants of Roman legionaries stationed on the Black Sea coast, but numerous other elements have combined to make up the present inhabitants of Romania, Moldavia and Bessarabia. The language is, however, of the Latin group, though the Moldavians east of the Prut use the Cyrillic script. Most Romanians are of Orthodox religion.

CAUCASIAN GROUPS

The term " Caucasian " is usually applied to the Georgians (Gruzians), Armenians and the complex population of the mountain valleys, among whom nearly 40 different sub-groups were listed in the 1926 census. All are Japhetic. The Georgians are Orthodox, though where the Turkish influence has been strong, as in the area of Batumi, there are many Moslems; the Armenians follow a Gregorian rite of their own; the Dagestanis, Chechen-Ingush, Cherkess, Kabardinians and other associated groups are Sunni Moslem, though the Abkhaz are partly Orthodox.

IRANIANS

Also in the Caucasus are found a number of groups of Iranian origin, of which the Osetins are the most important. This group (under 300,000 in 1926, 354,000 in 1939) is mainly Orthodox in the North Osetian Autonomous Soviet Socialist Republic and Sunni Moslem in the South Osetian Autonomous Oblast. The other Caucasian–Iranian groups (Talysh, Tats and Yezids) are Shi'ite Moslem; the 60,000 Kurds are Zoroastrian.

In Central Asia the very important Tadzhik group (980,000 in 1926, about 1,250,000 in 1939) is of old Persian or Sogdian origin, out-dating the surrounding Turkic peoples by many centuries. The Tadzhiks are Sunni Moslems, though a small mountain group in the Pamirs known as the Yagnobts follows the Ismail sect of which the Aga Khan is head.

MONGOLS

The original home of the Mongols was in the lower Amur valley; they spread into what is now called Mongolia in the 12th and 13th centuries, before their great advance to the west under Genghis Khan. In the Mongolian People's Republic the dominant group is the Khalkhas, who form about 80% of the population; there are several minority groups, of whom some are Turkic.

In the U.S.S.R. the largest Mongol group is the Buryats (237,500 in 1926). The Kalmyks are a western extension of the Mongols who were driven out of European Russia in the 18th century; their main concentration by the Caspian Sea was dispersed after 1945 because they were suspected of collaboration with the Germans. Most of the Mongols are Lamaist Buddhist by religion, but there is a small group of Moslems (the Sart-Kalmyks) in the Altay area.

TUNGUS-MANCHURIAN

The Tungus (37,500 in 1926) are the largest of the pre-Turkic and pre-Russian nationalities of Siberia, and are related to the Manchurians, The main group is the Evenki, a semi-nomadic people whose chief centre is in the north-east of Krasnoyarsk Kray (the Evenki National Okrug). Smaller groups are found scattered in the Far East. A connection is believed to exist between the Tungus and the American Indians.

PALAEO-ASIATIC GROUPS

A few other small groups, mainly in north-eastern Siberia, do not fit into any of the main categories. These include the Chukchi, Koryaks and Itel'meny (Kamchadals), all of whom are related to the Eskimos. In Sakhalin is a curious aboriginal group known as the Ainos, and in the Amur region are several other aboriginal communities.

JEWS

In 1926 nearly 2,700,000 Jews were listed, some 2,600,000 of them being scattered about European Russia, the Ukraine and Byelorussia. Distinct Jewish communities together numbering about 70,000 were settled in the Crimea, Dagestan, Georgia and Central Asia; there were also 8,300 Karaim, who are believed to descend from one of the offshoots of the first dispersion in the 7th century B.C. After the acquisition of new territories in Europe in 1939, the total number of Jews in the Soviet Union increased to nearly 5 million, but was reduced to perhaps 2 million during the war in consequence of German attempts at extermination.

The Jewish Autonomous Oblast was set up in 1934, and in 1939 included over 50,000 Jews out of a total population of 108,000.

In Eastern Europe the number of Jews was heavily reduced by extermination and emigration during the war, but there are still appreciable numbers in Romania, Poland and elsewhere.

ALBANIANS

The 1,200,000 Albanians are probably a survival of an ancient Balkan tribe who managed to preserve their identity through many centuries of occupation by various conquerors. Some 65% are Moslems, the rest being about equally divided between Greek Orthodox and Roman Catholic.

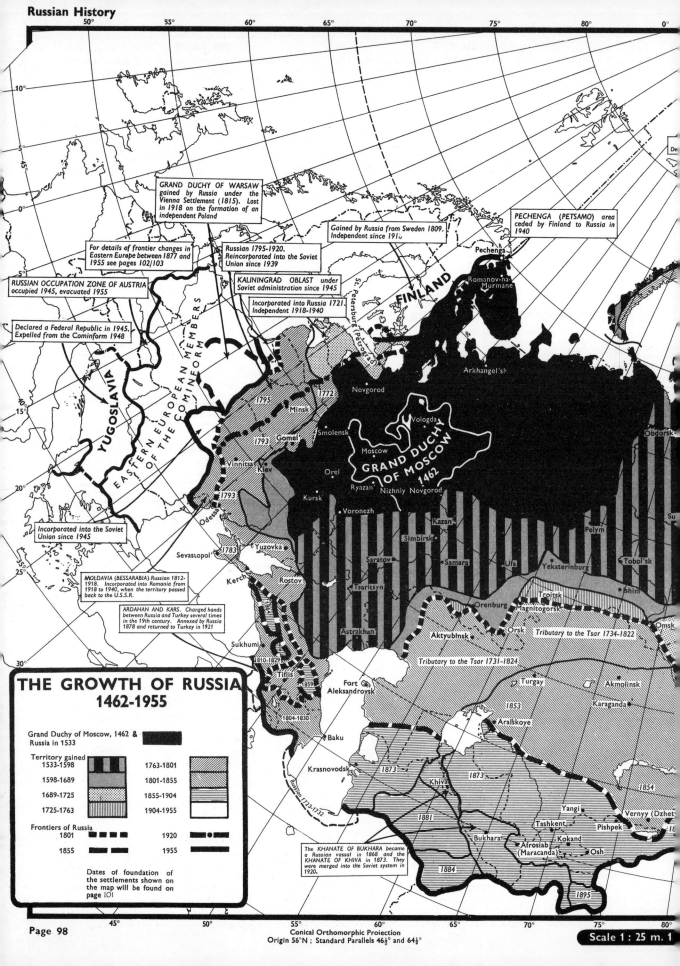

GRAND DUCHY OF WARSAW gained by Russia under the Vienna Settlement (1815). Lost in 1918 on the formation of an independent Poland

PECHENGA (PETSAMO) area ceded by Finland to Russia in 1940

Gained by Russia from Sweden 1809. Independent since 1917.

For details of frontier changes in Eastern Europe between 1877 and 1955 see pages 102/103

Russian 1795-1920. Reincorporated into the Soviet Union since 1939

Pechenga

RUSSIAN OCCUPATION ZONE OF AUSTRIA occupied 1945, evacuated 1955

KALININGRAD OBLAST under Soviet administration since 1945

Romanov-na-Murmane

Incorporated into Russia 1721. Independent 1918-1940

Declared a Federal Republic in 1945. Expelled from the Cominform 1948

St. Petersburg (Petrograd)

FINLAND

Arkhangel'sk

Incorporated into the Soviet Union since 1945

YUGOSLAVIA

EASTERN EUROPEAN MEMBERS OF THE COMINFORM

1795

1772

Novgorod

Vologda

Obdorsk

Minsk

1793

Gomel

Smolensk

Moscow

GRAND DUCHY OF MOSCOW 1462

Vinnitsa

Kiev

Orel

Ryazan

Nizhniy Novgorod

1793

Odessa

Kursk

Voronezh

Kazan

Pelym

Incorporated into the Soviet Union since 1945

Simbirsk

Tobol'sk

Sevastopol'

1783

Yuzovka

Saratov

Samara

Ufa

Yekaterinburg

Su

MOLDAVIA (BESSARABIA) Russian 1812-1918. Incorporated into Romania from 1918 to 1940, when the territory passed back to the U.S.S.R.

Kerch

Rostov

Tsaritsyn

Troitsk

shim

Magnitogorsk

Orenburg

Tributary to the Tsar 1734-1822

Omsk

ARDAHAN AND KARS. Changed hands between Russia and Turkey several times in the 19th century. Annexed by Russia 1878 and returned to Turkey in 1921

Sukhumi

Astrakhan

Aktyubinsk

Orsk

Tributary to the Tsar 1731-1824

1810-1829

Tiflis

Fort Aleksandrovsk

Turgay

Akmolinsk

THE GROWTH OF RUSSIA
1462-1955

1859

1804-1830

Baku

1853

Karaganda

Aral'skoye

Grand Duchy of Moscow, 1462 & Russia in 1533

Krasnovodsk

1873

1854

Territory gained
1533-1598	1763-1801
1598-1689	1801-1855
1689-1725	1855-1904
1725-1763	1904-1955

Russian 1723-1732

Khiva

1873

Yangi

Vernyy (Dzhet

Frontiers of Russia
| 1801 | 1920 |
| 1855 | 1955 |

1881

Tashkent

Pishpek

18

Dates of foundation of the settlements shown on the map will be found on page 101

Bukhara

Afrosiab (Maracanda)

Kokand

Osh

The KHANATE OF BUKHARA became a Russian vassal in 1868 and the KHANATE OF KHIVA in 1873. They were merged into the Soviet system in 1920.

1884

1895

Conical Orthomorphic Projection
Origin 56°N ; Standard Parallels 46½° and 64½°

Scale 1 : 25 m. 1

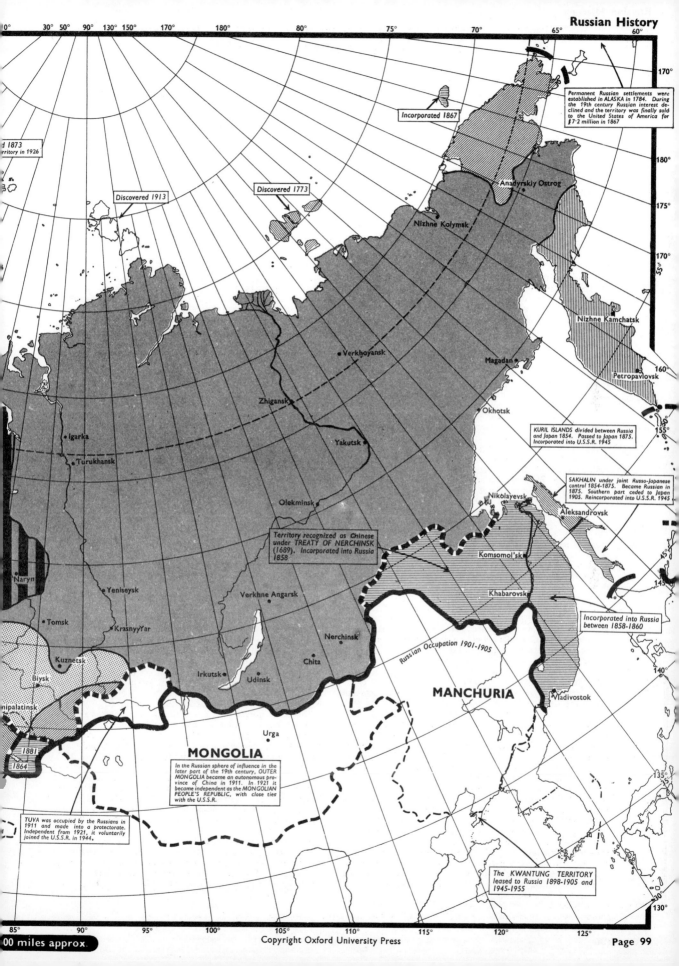

d 1873
rritory in 1926

Discovered 1913

Discovered 1773

Incorporated 1867

Permanent Russian settlements were
established in ALASKA in 1784. During
the 19th century Russian interest declined and the territory was finally sold
to the United States of America for
$7·2 million in 1867.

Anadyrskiy Ostrog

Nizhne Kolymsk

Nizhne Kamchatsk

Verkhoyansk

Magadan

Petropavlovsk

Zhigansk

Okhotsk

Igarka

KURIL ISLANDS divided between Russia
and Japan 1854. Passed to Japan 1875.
Incorporated into U.S.S.R. 1945

Turukhansk

Yakutsk

SAKHALIN under joint Russo-Japanese
control 1854-1875. Became Russian in
1875. Southern part ceded to Japan
1905. Reincorporated into U.S.S.R. 1945

Nikolayevsk

Olekminsk

Aleksandrovsk

Territory recognized as Chinese
under TREATY OF NERCHINSK
(1689). Incorporated into Russia
1858

Komsomol'sk

Naryn

Yeniseysk

Khabarovsk

Verkhne Angarsk

Incorporated into Russia
between 1858-1860

Tomsk

KrasnyyYar

Nerchinsk

Russian Occupation 1901-1905

Kuznetsk

Chita

MANCHURIA

Biysk

Irkutsk

Udinsk

nipalatinsk

Vladivostok

1881

1864

Urga

MONGOLIA

In the Russian sphere of influence in the
later part of the 19th century, OUTER
MONGOLIA became an autonomous province of China in 1911. In 1921 it
became independent as the MONGOLIAN
PEOPLE'S REPUBLIC, with close ties
with the U.S.S.R.

TUVA was occupied by the Russians in
1911 and made into a protectorate.
Independent from 1921, it voluntarily
joined the U.S.S.R. in 1944.

The KWANTUNG TERRITORY
leased to Russia 1898-1905 and
1945-1955.

00 miles approx.

THE GROWTH OF RUSSIA

Long before the name Russia appeared on the historical scene, there had been three areas of advanced civilization on the territory of the present-day Soviet Union. One was Central Asia, where the oases of Zeravshan (ancient Sogdiana), Khorezm and Murgab (ancient Margiana) had been among the oldest centres of civilization in the world. The capital of Sogdiana, Samarkand (ancient Afrosiab) was conquered by Alexander the Great in 329 B.C. The second area was Transcaucasia, parts of which belonged to the Urartu Kingdom (for some time the rival of Assyria), and which later was a bone of contention between Rome and the Parthians. Thirdly, there was the Black Sea littoral, where Greek colonies sprang up in the 6th and 5th centuries B.C., including the present cities of Sevastopol' (ancient Chersonesos) and Kerch' (ancient Panti-capeion); the latter in Roman times became the capital of the Bosporan Kingdom.

The origins of Russia proper are, however, to be found further north. The first Russian state was Kievan Russia, which in the 9th–13th centuries embraced much of present-day European Russia. The majority of its population consisted of Eastern Slavs but the ruling House of Rurikids was of Varangian (Norse) origin. According to tradition, the Varangians established themselves in Novgorod in 862, but soon moved the capital to Kiev, further south along the great trade route between Scandinavia and Byzantium. From the 11th century the Kievan state gradually broke up into a number of feudal principalities which gravitated to one or other of the three main centres—the kingdom of Galicia and Volhynia in the south-west, the Grand Principality of Vladimir in the north-east, and Novgorod in the north, which soon became a republic and played an important role in the trade of northern Europe.

From the earliest times wave after wave of Asiatic invaders had crossed the southern steppes of Russia, culminating in the great Mongol conquests which in the 13th century brought under single control, for a time, a huge area including European Russia, Central Asia, Siberia and China. Soon after Genghis Khan's death in 1227 his huge empire broke up into separate khanates, one of which comprised Central Asia and another (later to be known as the Golden Horde) included Russia, which was conquered in 1237–40 by the great Khan's grandson Batu. The Mongol or, as it came to be called in Russia, Tatar domination lasted for over 200 years, during which period the originally insignificant principality of Moscow gradually absorbed most of its neighbours and by 1380 felt strong enough to challenge the Tatars. However, only a hundred years later did the Grand Prince of Muscovy, Ivan III (1462–1505), finally stop paying tribute to the Tatars. He also subdued the Novgorod Republic (which had earlier successfully withstood the Livonian Knights) in 1478, and annexed the Grand Principality of Tver' (1485) and the Republic of Vyatka (1489). His son acquired Pskov in 1510 and Ryazan' in 1521. In the west, the Muscovite policy of " gathering the Russian lands " brought Moscow into conflict with the Grand Duchy of Lithuania, which had absorbed the western and southern parts of the former Kievan state. It was in Ivan III's reign that, after the conquest of Constantinople by the Turks, the doctrine of Moscow as a successor to Byzantium, " Moscow the third Rome ", gained popularity.

His grandson Ivan IV (1533–84), called the Terrible, assumed the title of Tsar (i.e. Emperor) in 1547, and through the conquest of the khanates of Kazan', Astrakhan', and Siberia (into which the former Golden Horde had split up) transformed Muscovy into a multi-national state. After Ivan IV's death small groups of Cossacks advanced rapidly through Siberia, using the rivers as their main routes, building fortified towns in key positions (Tyumen' 1586, Tobol'sk 1587, Tomsk 1604, Krasnoyarsk 1628, Yakutsk 1632). From Yakutsk the Russian advance took three directions—to the north-east, the south-west and the south. In the north-east they founded Verk-hoyansk (the coldest place in the northern hemisphere) in 1638, circumnavigated the Chukchec Peninsula, thus discovering the Bering Straits in 1648, founded Okhotsk—the first Russian settlement on the Pacific coast—in 1649 and annexed Kamchatka in 1697. In the south-west the Baykal area was annexed in the middle of the 17th century, and Irkutsk founded in 1652. In the south, on the Amur river, a fort was established in 1652 on the site of the modern Khabarovsk. The

Russian pioneers came into conflict with the northward-moving Chinese and in 1689 the Far Eastern territories were partitioned between the two countries by the Treaty of Nerchinsk.

Meanwhile in the west an independent Ukrainian Cossack state was formed in 1649 on the territory traversed by the middle course of the Dnieper, previously subject to the Polish-Lithuanian Commonwealth. By the Act of Pereyaslav in 1656 the Cossack state joined Muscovy as an autonomous unit. This precipitated a war with Poland which ended in 1667 with a treaty that gave the right-bank Ukraine except the city of Kiev back to Poland, though Muscovy gained Smolensk further up the river.

The territorial acquisitions of Peter the Great (1682–1725) were comparatively small but important. In the Northern War against Sweden he gained for Russia the territory east and south of the Gulf of Finland—western Karelia with Vyborg, the area on the banks of the Neva river, Estland and Livland. The new port of St. Petersburg was built at the mouth of the Neva in 1703, which superseded Arkhangel'sk as the country's principal sea port, and ten years later became the new capital of Russia. After the victorious conclusion of the Northern War Peter assumed the Western title of Emperor. In the south, however, Peter was less successful and had in the end to return to the Turks the fortress of Azov at the mouth of the Don which he had earlier conquered. In a war against Persia he annexed the western and southern shores of the Caspian Sea, including Baku, but these were also subsequently lost. The gradual annexation of southern Siberia continued during Peter's reign and later in the 18th century. Alaska was discovered by a Russian expedition in 1741, was annexed to Russia and its fur riches exploited by the Russian-American Company until 1867 when it was sold to the U.S.A.

The main advance southwards in Europe was not renewed until the reign of Catherine II (1762–96) when the condition of the Ottoman Empire was beginning to cause concern to Europe. Her two wars against the Porte brought Russian control to the Crimea, which was occupied in 1783. The area west of the lower Dnieper was finally annexed in 1789, and the chief Black Sea port of Odessa founded. The south-west frontier was established along the river Dniester, where it remained until 1812, when Bessarabia was ceded to Russia by the Turks.

The advance to the Caucasus was taking place simultaneously. Rostov-on-Don was founded in 1761, Yekaterinodar (now Krasnodar) on the Kuban' in 1794, and Vladikavkaz (now Ordzhonikidze) in central Caucasus in 1784. Decisive for the future development of Russian policy in the Caucasus was the establishment of a Russian protectorate over the East Georgian kingdom in 1783. The last king of Eastern Georgia, threatened by Persia, ceded his country to Russia in 1800, and most princes of Western Georgia followed in 1803–4. The khanates of northern Azerbaydzhan, semi-independent from Persia, were occupied by Russia in the following years (Baku 1806), and after a war with Persia Eastern Armenia with Yerevan was annexed in 1828. After the capture of Yerevan Russia began to meet the growing British influence in Persia. With almost the whole of Transcaucasia in Russian possession, a protracted war began against the tribesmen of the mountainous North Caucasus, who were only finally subdued in 1864.

The other outstanding territorial changes under Catherine II were brought about by the three partitions of Poland in 1772, 1793 and 1795, as a result of which Russia acquired the right-bank Ukraine, the whole of Byelorussia, Lithuania and Courland. Central Poland passed under Russian control in 1815, when by a decision of the Congress of Vienna the Duchy of Warsaw (re-named the Kingdom of Poland) was incorporated into the Russian Empire as an autonomous unit. In the north, Finland had been gained from Sweden in 1809; unlike the Ukraine and Poland, Finland retained internal autonomy throughout the period of its association with the Russian Empire.

Two more great advances were carried out before the First World War. The first of these was the expansion of Russian power into Central Asia after the middle of the 19th century. The steppes of western and northern Kazakhstan had been finally annexed to Russia in the early 19th century (their Khans having for the previous century been vassals of the Russian Emperor). The last independent Kazakh state, caught between the Russians, the Chinese and the Uzbek khanate of Kokand, submitted to Russia in 1854. In a series of campaigns during the 1860s–70s, the Russians conquered all three Uzbek states of Turkestan (taking Tashkent in 1865, Samarkand and Bukhara in 1868, Khiva in 1873 and Kokand in 1876); the Kokand Khanate was abolished and annexed, while the central part of the Khiva khanate and the larger part of the Bukhara emirate became protectorates. The only serious resistance to Russian advance in Central Asia was put up by a part of the Turkmen people (the Akhal-Teke tribe), and when this was overcome the whole of Turkmenia was annexed in the early 1880s. In Central Asia, as in the Caucasus, British influence was met, and in 1888 an agreed frontier was delimited between Russia and Afghanistan.

At the end of the century expansionist tendencies shifted to the Far East, where the weakness of China was to the advantage of the Great Powers and of the newly-expansionist Japan, who became Russia's chief antagonist in the area. The Amur region, left to China in 1689, was ceded to Russia in 1858, and two years later the Pacific coast as far south as the Korean frontier was also annexed. A treaty with Japan in 1854 divided the Kuril Islands between the two countries, but left Sakhalin under joint control; in 1875 Russia exchanged her share of the Kurils for exclusive control of Sakhalin. Strong Russian pressure was exerted on China, especially during the 1895 Sino-Japanese war. A Russian sphere of influence was established in northern Manchuria and Port Arthur was leased from the Chinese. The construction of the Chinese Eastern Railway across Manchuria to Vladivostok and its branch line south to Port Arthur

extended Russian influence throughout Manchuria, the northern part being occupied after the Boxer rebellion. The defeat in the Russo-Japanese war of 1904–5 cost Russia the southern half of Sakhalin and her position in Manchuria. Finally, on the eve of the First World War, the Chinese province of Tuva became a Russian protectorate.

The First World War, revolution, and the ensuing civil war and foreign intervention lost Russia (reconstituted as the Union of Soviet Socialist Republics in 1922) Finland, the Baltic States, Poland—all of which became independent—western Byelorussia, western Volhynia—both ceded to Poland by the Treaty of Riga in 1920—Bessarabia—whose annexation by Romania in 1918 was never recognized by the Soviets—and Kars, which was returned to Turkey. Against this, the former Central Asian protectorates, Bukhara and Khiva, were incorporated in 1920.

The 1922 frontiers remained unchanged until 1939, when, by agreement with Germany, Eastern Poland (that is, western Byelorussia, western Volhynia and eastern Galicia) was annexed. In 1940, as a result of the Soviet-Finnish war, Finland was forced to cede western Karelia; the Baltic States were occupied by the Red Army and annexed; and Romania, under ultimatum, gave up Bessarabia and northern Bukovina. In 1944 the Soviet Union quietly annexed Tuva, nominally an independent " people's republic " since 1921. After the war all these gains were retained (except for a revision of the frontier with Poland which gave back to Poland territories with a Polish majority, including the towns of Bialystok and Przemysl) and new ones made—the northern part of East Prussia, including Konigsberg (re-named Kaliningrad), Transcarpathia (hitherto an autonomous region within Czechoslovakia), the southern part of Sakhalin and the Kuril Islands.

DATES OF FOUNDATION (or first mentioning) OF SELECTED SETTLEMENTS

Original names given with the later forms in brackets. Date of incorporation under Moscow/St. Petersburg given in brackets.

	Date of foundation		Date of foundation
European Russia (incl. the Urals)		**Siberia** (cont.)	
Arkhangel'sk	1583	Irkutsk	1652
Astrakhan'	Founded as Itil in 8th cent. A.D.; refounded as Astrakhan' in 13th cent. A.D. (1556)	Krasnyy Yar (Krasnoyarsk)	1628
		Kuznetsk (Stalinsk)	1617
Gomel'	12th cent. A.D. (1772)	Magnitogorsk	1929
Kazan'	14th cent. A.D. (1552)	Nerchinsk	1653
Kerch'	6th cent. B.C. (1774)	Obdorsk (Salekhard)	1595
Kiev	8th cent. A.D. (1667)	Omsk	1716
Kursk	1095, destroyed 1240 (re-built 1586)	Tobol'sk	1587
Minsk	1067 (1793)	Tomsk	1604
Moscow	1147	Troitsk	1743
Nizhniy Novgorod (Gor'kiy)	1221 (1392)	Turukhansk	1607
		Verkhneudinsk	
Novgorod	9th cent. A.D. (1478)	(Ulan-Ude)	1666
Odessa	14th cent. A.D. (1791)	Verkhoyansk	1638
Orel	1564	Yakutsk	1632
Orenburg	1743	Yekaterinburg	
Orsk	1735	(Sverdlovsk)	1721
Romanov-na-Murmane		Yeniseysk	1618
(Murmansk)	1915	**Central Asia and Kazakhstan**	
Rostov-on-Don	1761	Afrosiab (Samarkand)	ca. 3000 B.C. (1868)
Ryazan'	1095 (1521)	Akmolinsk	1830
St. Petersburg, (Petrograd) (Leningrad)	1703	Bukhara	Before 9th cent. A.D. (1868)
		Karaganda	1928
Samara (Kuybyshev)	1586	Kokand	10th cent. A.D. (1876)
Saratov	1590	Krasnovodsk	1869
Sevastopol'	5th cent. A.D. (1783)	Osh	Before 8th cent. A.D. (1874)
Simbirsk (Ul'yanovsk)	1648	Pishpek (Frunze)	1846 (1873)
Smolensk	865 (1667)	Semipalatinsk	1718
Tsaritsyn (Stalingrad)	1589	Tashkent	7th cent. A.D. (1865)
Ufa	1574	Vernyy (Alma Ata)	1854
Vinnitsa	14th cent. A.D. (1793)	Yangi (Dzhambul)	5th cent. A.D. (1864)
Vologda	1147 (1397)	**Far East**	
Voronezh	1586	Aleksandrovsk	1881
Yuzovka (Stalino)	1869	Anadyrskiy Ostrog	
Transcaucasia		(Anadyr')	1649
Baku	885 (1806)	Khabarovsk	1858
Sukhumi	7th cent. B.C. (1810)	Komsomol'sk	1858
Tiflis (Tbilisi)	4th cent. A.D. (1801)	Magadan	1933
Siberia		Nikolayevsk	1850
Biysk	1709	Okhotsk	1649
Chita	1653	Petropavlovsk	1740
Igarka	1928	Vladivostok	1860

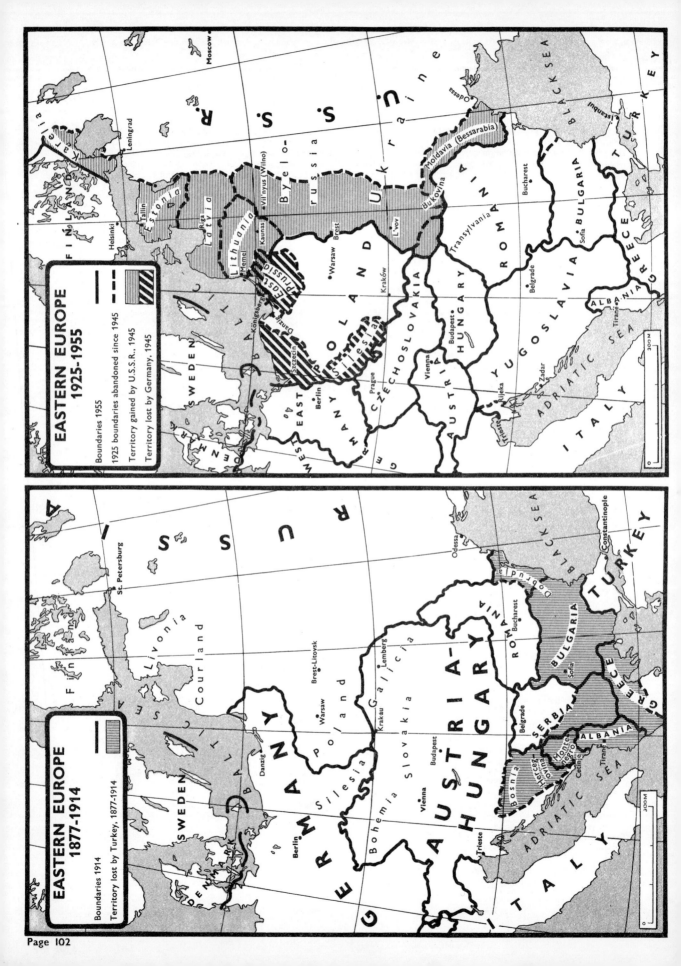

EASTERN EUROPE 1925-1955

Boundaries 1955

1925 boundaries abandoned since 1945

Territory gained by U.S.S.R., 1945

Territory lost by Germany, 1945

EASTERN EUROPE 1877-1914

Boundaries 1914

Territory lost by Turkey, 1877-1914

THE GROWTH OF EASTERN EUROPE

In ancient times the Roman, and later the Byzantine, Empire extended as far north as the Danube and the Carpathians, but all to the north was inhabited only by primitive tribes. The Asiatic invasions of the early Middle Ages brought new peoples to the Danube valley and the north European plain, and by the year 1071 the control of Byzantium was confined to the Balkan peninsula. In the 14th century most of the Balkans came under Turkish rule, which was consolidated after the fall of Byzantium in 1453. The Turkish Empire reached its greatest extent late in the 17th century.

Further north stable states made their appearance much later. During the Middle Ages Scandinavian, German, Lithuanian and Slav principalities competed for power, and a definite pattern emerged only with the growth of the Grand Duchy of Moscow, the Kingdom of Poland (united with Lithuania in 1386) and the Kingdom of Hungary (united with Poland in 1370, but later separated). Most of Hungary was occupied by the Turks from 1541 to 1686, only the western part being under Austrian control, but in 1699 all except the Banat was incorporated as a separate kingdom under the Habsburg Empire. By 1721 the Scandinavians had been driven out of most of Eastern Europe south of the Baltic by the defeat of Charles XII of Sweden in the Northern War. Meanwhile a new factor had appeared in the north-west of the area with the effective union of Brandenburg and Prussia in 1660, which became the Kingdom of Prussia in 1701.

From 1795 to 1817 the whole of Eastern Europe (except Finland, which was annexed to Russia in 1809 as a self-governing Grand Duchy) was divided between the four great empires of Russia, Prussia, Austria-Hungary and Turkey. The Serbs revolted against Turkish control in 1803, and in 1817 secured their independence. Greece followed in 1829, and Romania (the principalities of Wallachia and Moldavia), though under Turkish suzerainty till 1878, was virtually independent from 1856. Bulgaria was created by the Treaty of San Stefano (March 1878) but was split in two by the Congress of Berlin a few months later, to prevent Russia from acquiring too direct a means of access to the Mediterranean; in 1885 it was reunited. In 1912 Albania declared its independence, and in the First Balkan War the last remnants of Turkish power were driven out of Europe except for the city of Constantinople and a small area round it. The Second Balkan War of the following year was fought by her late allies and Romania against Bulgaria; Serbia and Greece took over most of Macedonia, and Romania acquired the southern Dobruja, but Bulgaria was left in control of the coastal strip on the Aegean west of the Maritsa river, while Turkey regained a European frontier on the line from Enos on the Aegean to Midia on the Black Sea.

The annexation of Bosnia and Hercegovina by Austria in 1908 gave formal effect to an arrangement which had existed de facto since 1878. Otherwise there were no frontier changes in Eastern Europe north of the Ottoman lands from 1871 (when the German Empire was finally established) to 1919.

The treaties of 1919–1920 between the Allied Powers and Germany (Versailles), Austria (St. Germain), Hungary (Trianon), Bulgaria (Neuilly) and Turkey (Sèvres) brought about a complete reorganization of much of the region. Hungary was separated from Austria, and the new states of Poland and Czechoslovakia were created; Romania was greatly enlarged by the acquisition of Transylvania from Hungary; the former Kingdoms of Serbia and Montenegro were joined with the Habsburg provinces of Slovenia, Croatia, Bosnia, Hercegovina, Vojvodina and part of the Banat to form the Kingdom of the Serbs, Croats and Slovenes which in 1929 took the name of Yugoslavia; and in the extreme south Albania was formally recognized, while Greece acquired western Thrace from Bulgaria and eastern Thrace up to Çatalca from Turkey.

The eastern frontier of Poland was originally fixed along the "Curzon line", but could not be enforced because Russia was still in a state of civil war, and the Bolshevik government refused to recognize the new frontier. After Poland had invaded Russia and Russia had carried out an unsuccessful counter-attack, the frontier finally agreed in 1921 was well to the east of the Curzon line. In 1919 the city of Vilna (now Vil'nyus) was allotted to the restored state of Lithuania, but was seized by the Poles in 1920. Poland was also given access to the Baltic through the Corridor, which separated East Prussia from the rest of Germany. Danzig (Gdańsk) was made a free city, but Poland took the precaution of building the port of Gdynia at the outlet from the Corridor.

The treaty arrangements of 1919 left a number of territories with indeterminate status, subject to plebiscite. The main areas affected were:

Marienwerder, Allenstein and Masuria. Returned to Germany (East Prussia) after a plebiscite in 1920.

Upper Silesia. Divided after plebiscite in 1921; the rest to Germany.

Teschen and adjoining areas. Divided after plebiscite in 1920 between Czechoslovakia and Poland.

The other main changes in the immediate post-war period were unconnected with the peace treaties. Finland declared its independence of Russia in 1918, and after a civil war the White régime was recognized by the U.S.S.R. in 1920. Estonia, Latvia and Lithuania were similarly recognized. In 1918 Romania annexed Bessarabia and the northern Bukovina from Russia, but the U.S.S.R. never recognized this change.

There were no more alterations of frontier from 1923 to 1938, when the Munich agreement awarded the German-speaking areas of Czechoslovakia (the Sudetenland) to Germany. Poland then annexed Teschen and a few other border areas, and Hungary annexed several areas along the southern frontier of Czechoslovakia, as well as the province of Ruthenia (Sub-Carpathian Russia). In March 1939 the Czech lands (Bohemia and Moravia) were virtually annexed by Germany, while Slovakia became a separate state subordinate to Germany. At the same time Germany also compelled Lithuania to cede the territory of Memel (Klaipeda), which had been provisionally administered by the League of Nations from 1919 to 1923, when it was forcibly annexed by Lithuania. A few months later the German demand for access to the Polish Corridor and for the incorporation of Danzig became the starting point of the Second World War. The U.S.S.R. then occupied eastern Poland, early in 1940 the Vyborg (Viipuri) district and several areas further north were acquired from Finland after a short war; and the three Baltic Republics, together with Bessarabia and Bukovina, were occupied in June and formally annexed to the U.S.S.R. in August, 1940.

After the Second World War the main territorial changes were as follows:

Finland ceded to the U.S.S.R. the Petsamo (Pechenga) district, some border areas in Karelia, and the Vyborg (Viipuri) district.

The U.S.S.R. retained possession of the Baltic Republics, which became separate republics of the Union, after minor boundary adjustments.

The eastern frontier of Poland was fixed approximately on the Curzon line of 1919.

The U.S.S.R. annexed Memel and the northern part of East Prussia (now Kaliningrad Oblast, R.S.F.S.R.).

The U.S.S.R. acquired from Czechoslovakia the province of Ruthenia (now Transcarpathian Oblast, Ukrainian S.S.R.).

Bessarabia and the northern Bukovina were incorporated in the U.S.S.R., partly as the Moldavian S.S.R., partly into the Ukrainian S.S.R.

Poland acquired (subject to confirmation by an eventual peace treaty with Germany) the southern part of East Prussia, Danzig, the whole of eastern Germany up to the line of the rivers Oder and Görlitzer Neisse (including the port of Stettin, now called Szczecin) and Lower Silesia.

The whole of Transylvania was returned to Romania, and Hungary ceded to Czechoslovakia a small area across the Danube from Bratislava.

Yugoslavia acquired from Italy the ports of Zara (Zadar) and Fiume (Rijeka), several islands, and the greater part of Venezia Giulia. The Trieste area remained unsettled till 1954, when practically all except the town itself and the area to the west was handed over to Yugoslavia.

Bulgaria retained the southern Dobruja.

The parts of eastern Germany occupied by the Red Army under the Potsdam agreement of 1945 became virtually a separate state after 1949.

Territorial Changes in E. Europe, 1938–1954
(Areas in square miles)

	1938	1954		1938	1954
Poland	148,580	120,359	Bulgaria	39,825	42,796
Czech.	54,380	49,354	Yugoslavia	96,136	99,268
Hungary	35,935	35,902	Albania	10,628	10,632
Romania	114,000	91,700	E. Ger. (excl. Berlin)	—	41,379

INDUSTRIAL PROGRESS IN THE U.S.S.R.

As soon as the October Revolution of 1917 had been brought to a successful conclusion and the Bolsheviks were installed in power, the first steps were taken towards converting the economy to a basis of public ownership and central planning. Fully elaborated plans did not, however, start till 1929, and in the twelve years after the Revolution there were numerous modifications of structure and method. A number of distinct phases can be identified:

1917–1921	The period of war communism
1921–1929	The New Economic Policy (NEP)
1929–1946	The war period and the beginnings of reconstruction
1946–1955	The fourth and fifth Five-Year Plans
1956–1958	A period of re-adjustment when the Sixth Five-Year Plan was drawn up and then abandoned
1959	The start of the Seven-Year Plan 1959–1965

1917–1921 War Communism

For the first four years, administration was extremely loose, due partly to the break-up of the Tsarist administrative system and the difficulty of finding people with the necessary training, partly to the fact that parts of the country were still the scene of fighting. Between November 1917 and April 1918 a series of decrees was passed providing a temporary basis for the re-organization of the economy. All land was nationalized. The banks were transferred to state ownership, and the state also expropriated the larger industrial enterprises and the employers' syndicates; in April 1918 the state assumed a complete monopoly of foreign trade. The system which resulted from these early measures was described as " state capitalism ", and was never intended to be more than transitional. A vigorous attempt was made to institute workers' control of industry, but this experiment was never fully carried through. The greater part of the distributive system was taken out of private hands, and consumer co-operatives were set up under the control of the local revolutionary committees or soviets. The agricultural policies of this period are described on p. 36–37.

The general oversight of economic affairs was entrusted to a Supreme National Economic Council set up in December 1917. During 1919 this body issued a series of " directives " to fifteen industries, which did no more than state what levels of production these industries were expected to reach and what amounts of state financial support they would get. In February 1920 a more positive step towards central planning was taken with the setting up of the State Electrification Commission (Goelro), which was made responsible for all electrification plans throughout the country, and with particular emphasis on the exploration and development of hyrdo-electric potentialities (p. 52–53). Finally on 21st February, 1921 the first general planning authority came into existence with the title of State Planning Commission (Gosplan). For the first few years of its life, however, the functions of Gosplan were mainly advisory; its extensive executive powers did not come till later.

1921–1929 The New Economic Policy

Lack of knowledge and administrative experience caused these early experiments in planning to fail almost completely, and peasant resistance prevented the implementation of the agrarian side of the policy. So in 1921 the process of transformation of the economy was deliberately slowed down. Private ownership of land was partially restored. A halt was called to the nationalization of industry, and private undertakings were allowed to exist side by side with the public sector, though the state kept full control of mining and of the basic manufacturing industries. Compulsory deliveries to the state of the produce of agriculture and manufacturing industry were reduced and replaced by taxation. Investment funds were entirely channelled through the state, and by far the greatest part of the investment carried out was in the public sector, especially coal, petroleum, iron ore mining, electricity and the production of iron and steel.

As a result of these policies, and also because the civil war and the war of intervention were now over and conditions throughout the country were becoming more settled, industrial output increased substantially during the NEP period. By 1928 most of the principal types of production were above the levels of 1913, and output of electricity was $2\frac{1}{2}$ times as high. Iron ore and steel were still below the 1913 level, though steel production caught up in 1929.

Meanwhile work on future plans was going ahead. Gosplan was enlarged in 1925, and was divided into three functional sections; subordinate commissions were also created for the R.S.F.S.R. and other Republics of the Union. The nature and timing of the first Five-Year Plan were not decided without considerable argument, and the form it eventually took, with its single-minded insistence on heavy industry at the expense of consumer goods, bears the marks of Stalin's own hand and personality. When eventually it was adopted in April 1929, its start was ante-dated to the beginning of the previous year.

1929–1941 The Five-Year Plans

The first Five-Year Plan ended at the end of 1932. Reporting to the Central Committee of the Communist Party in January 1933, Stalin claimed that " capitalist elements have been completely and irrevocably eliminated from industry ", and that the proportion of industrial (i.e., non-agricultural) output had been raised from 48% of the national product in 1928 to 70% in 1932.

In the Second Five-Year Plan, from 1933 to 1937, although heavy industry was still to receive the lion's share of the resources available, there was to be an expansion of the consumer goods industries as well. At the end of the five years, the Soviet authorities stated that the output of industry on a whole had more than doubled, and that in the final year 80% of all industrial output was from factories built or reconstructed during the first two Plans. The consumer goods part of the Plan was not fulfilled. The balance between the various sections was modified as time went on, and again there was extra military expenditure; " this " it was reported at the end " required a considerable quickening of the expansion of heavy industry, and this to a certain extent at the cost of slowing down the growth of light industry ". (Military expenditure increased from 1,400 million roubles in 1933 to 17,500 million roubles in 1937, or from 3·5% to 17·4% of all budgetary expenditure).

The third Five-Year Plan was intended to run from the beginning of 1938 to the end of 1942, but was interrupted by the German invasion in June 1941. To a greater extent than its predecessors, the Plan provided for industrial expansion in and east of the Urals and in Central Asia—partly on strategic grounds, partly because the potential resources in the east were known to be enormous—and after the invasion the eastward shift was greatly accelerated.

1941–1946 The War Period

Quite apart from the imperative need to concentrate on production for defence, the temporary loss of the western regions forced Soviet industry into a new pattern. The lost areas had in 1940 comprised 40% of the population and nearly 50% of the industrial potential of the country. Official Soviet figures show that the war and the action of the occupying forces destroyed 31,850 factories, 5 million kw. of electrical capacity, 4,100 railway stations, large quantities of industrial equipment and nearly 1·25 million houses. The first six months of hostilities reduced production in the U.S.S.R. as a whole by nearly 50%; thereafter an improvement began, but it was not until 1944 that total output regained the 1940 level. The eastward shift of industry is illustrated by the fact that industrial output in the eastern regions rose in volume by 130% between 1940 and 1944.

Reconstruction began as soon as the invaders had been ousted, with first emphasis put on the rebuilding and re-equipment of factories; but, although the whole country was freed of enemy forces by August 1944, it still took more than a year to decide the form and scope of the next plan, which came into effect at the beginning of 1946.

1946–1955 The Fourth and Fifth Five-Year Plans

Industrial production under the fourth Five-Year Plan (1946–1950) was intended to reach, on average, a level of 48% above that of 1940, and transport was given pride of place, followed by cement and electricity. The first years after the war were hardly less severe from the consumer's point of view than the most difficult periods of the 1920s. None of the main branches of industry passed their 1940 levels till 1947, and petroleum production and iron and steel not till 1949. This meant shortages of consumer goods, with little hope of improvement until the demands of the major industries were satisfied. In 1950 the index of production of capital goods stood at 205 (1940 = 100), the index of consumer goods at 123. Output of cotton cloth was still only 98% of the 1940 level.

A currency reform of December 1947 increased the purchasing power of the rouble, and at the same time had a generally deflationary effect. In the later years up to 1954 there was an all-round reduction in prices each spring, but this was not repeated in 1955. After the death of Stalin in March 1953 the new leaders changed the emphasis of the economy in favour of consumer goods, and increased the volume of imports; later, and especially after the resignation of Mr. Malenkov from the post of Prime Minister in February 1955, there was a partial return to the old policy of concentration on heavy industry at the expense of consumer goods.

The fifth Five-Year Plan from 1951 to 1955 aimed at an 80% increase over 1950 in capital goods and a 65% increase in consumer goods (269% and 103% over 1940). In July 1955 it was announced that the Plan had actually been fulfilled eight months ahead of schedule.

1956–1958 A period of re-adjustment

The next two years saw the launching and abandoning of the sixth Five-Year Plan, which was to have raised industrial output by a further 65% by 1960. The Plan contained ambitious targets for the coal, steel, electricity, automobile, chemical and textile industries, and aimed to initiate a new *Drang nach Osten* to exploit hitherto untapped resources of minerals, water power, etc. Labour shortages, underfulfillment of targets in

certain key sectors, advances in technology, the need to reorganize economic administration, however, led to a major revision of targets in 1957 and to the eventual abandonment of the whole Five-Year Plan.

The administrative reorganization was directed towards decentralization of control. One hundred and five economic regions, each with an economic council responsible for industrial operations within the region and for drafting regional plans within the general pattern of the Central Plan, were created and many central economic ministries were disbanded. Gosplan became the co-ordinator of regional plans and its executive powers were transferred to the Council of Ministers, who exercise control of industry through the governments of the Union Republics.

1959 The Start of the Seven-Year Plan 1959–1965

The reorganization occupied nearly two years. The current Seven-Year Plan embodies the lessons learnt by the Soviet economy in the 'fifties and is directed towards making the Soviet Union a leading world economic power and raising the standard of living of the Soviet peoples. It aims at an annual growth rate in gross industrial production of 8·6% (producers' goods 9·3%, consumers' goods 7·3%) compared with 10·5% in the abandoned sixth Five-Year Plan, and 13·1% in the fifth plan period (1951–1955). Greatest emphasis is laid on increasing output in the petroleum, chemical, plastics, artificial fibres and electricity industries. Automation is to be speeded up and a large programme of technical education is to be carried through.

The following table shows actual production of certain commodities in selected years from 1932 to 1959 and compares the targets of the now abandoned sixth Five-Year Plan with those of the current Seven-Year Plan.

Growth of Industrial Production and Targets for the Seven-Year Plan, 1959–1965

	1932	1937	1940	1950	1955	Sixth Plan 1960 Targets	Actual production 1959	Seven-Year Plan Targets 1960	Seven-Year Plan Targets 1965
Coal and lignite (million tons)	64·4	128·0	165·9	261·1	391·0	593·0	506·5	515·1	600–612
Electricity ('000 million kwh.)	13·5	36·2	48·3	91·2	170·1	320·0	264·0	291·0	500–520
Petroleum (million tons)	21·4	28·5	31·1	37·9	70·8	135·0	129·5	144·0	230–240
Pig iron (million tons)	6·2	14·5	14·9	19·2	33·3	53·0	43·0	46·9	65–70
Crude steel (million tons)	5·9	17·7	18·3	27·3	45·3	68·3	59·9	65·0	86–91
Metal cutting machine tools ('000 units)	19·7	48·5	58·4	70·6	117·8	200·0	146·0	...	190–200
Motor vehicles ('000 units)	23·9	199·9	145·4	362·9	445·3	650·0	495·0	...	750–856
Tractors ('000 units)	48·9	51·0	31·6	108·8	163·4	322·0	213·5	152·0†	...
Grain combines ('000)	10·0	43·9	12·8	46·3	48·0	140·2	58·5	54·0†	...
Mineral fertilizers ('000 tons)	921·0	3,240	3,027	5,492	9,629	19,600	12,900	13,500	37,200
Cement (million tons)	3·5	5·5	5·7	10·2	22·5	55·0	38·8	45·5	75–81
Paper ('000 tons)	471	832	812	1,193	1,862	2,722	2,300	...	3,500
Cotton fabrics (million metres)	2,694	3,448	3,954	3,899	5,904	7,270	4,600*	...	7,700–8,000
Rail freight (million ton miles)	105,117	220,355	259,715	374,800	603,300	853,762	887,938	...	1,126,544

* Million sq. metres; in 1958 production was 5,789 million metres.　　　　　　　　　　　† Deliveries to agriculture.

INDUSTRIAL PROGRESS IN EASTERN EUROPE

Throughout the post-war period Eastern Europe has been in the Soviet sphere of influence: Albania came under communist control in the late autumn of 1944, and by February 1948, when the first, broadly based, post-war Czech government was overthrown and replaced by a fully communist one, all other influences were excluded from the area.

Poland and East Germany suffered heavy war damage and East Germany, as the principal enemy country, was severely punished. Parts of her territory were put under Soviet or Polish administration and the German inhabitants expelled, thus creating a serious refugee problem in the rest of Germany. Heavy reparations were exacted in the form both of deliveries from current production and of dismantled factories and equipment, and for a time key factories were transferred to Soviet control. In 1949, however, following Soviet recognition of the newly-formed East German Communist Government, punitive measures ceased. Hungary and Romania were less heavily punished for their part in the war, though some major industries were temporarily transferred to joint companies in which the Soviet Union held a half interest. Their position was regulated by peace treaties signed in September 1947.

Initially, there was little attempt at economic co-ordination within the area, but between 1947 and 1949 all countries embarked on short-term plans approximating to the Soviet model, and when these were completed launched a series of five-year plans (six years in Poland). Two features common to all of them were extension of social ownership to the greater part of mining and manufacturing, and the enlargement or creation of heavy industries.

Since 1956 the concept of economic autarchy within each country of the Soviet bloc has been gradually abandoned in favour of co-ordinated economic planning for the area as a whole. Comprehensive machinery for intra-bloc consultation has been set up within the Council for Mutual Economic Aid, and the economic plans have been reorganized to run concurrently with the Soviet Seven-Year Plan. For certain key sectors it has been decided to develop even longer-term plans, running for ten to fifteen years. Broadly, the aim is that each country should concentrate on those branches of industry for which its resources best fit it: thus, for example, East Germany, Poland and Czechoslovakia will be able to develop their engineering, vehicles, machine-building, and chemical industries, each specializing on certain products; Czechoslovakia and Poland will become major centres of textile production; Romania will concentrate on petroleum and petro-chemicals, Bulgaria on mining and metal refining and Hungary on electrical engineering and aluminium.

The following tables give some indication of the extent of economic progress in the area in recent years and the different rates of growth achieved by the different countries. The effects of the 1959 uprising on economic activity are clearly apparent both in the index of Hungarian production and in the volume of output of selected goods.

Index of General Industrial Production

(1953 100)

	1954	1955	1956	1957	1958	1959
Poland	111	124	135	149	163	176*
E. Germany	110	119	126	136	151	170
Czechoslovakia	104	116	127	139	155	172
Hungary	98	105	95	108	122	135
Romania**		100	111	121	132	141
Bulgaria	109	117	134	150	173	210‖

* Annual rate based on Jan. Nov. ** 1955 100. Average Jan. Sept.

Volume of Output of Selected Goods in Eastern Europe†

	1954	1955	1956	1957	1958
Hard coal‡—million metric tons					
Total	118·6	122·3	124·1	123·7	126·8
of which: Poland	91·6	94·5	95·1	94·1	95·0
Czechoslovakia	21·6	22·1	23·4	24·2	25·8
Crude petroleum§—million metric tons					
Total	11·4	12·5	12·6	12·3	12·6
of which: Romania	9·7	10·6	10·9	11·2	11·3
Hungary	1·2	1·6	1·2	0·7	0·8
Electric power—'000 million kwh.					
Total	65·4	73·3	79·8	85·2	94·1
of which: East Germany	26·0	28·7	31·2	32·7	34·9
Poland	15·5	17·7	19·5	21·2	24·0
Czechoslovakia	13·6	15·0	16·6	17·7	19·6
Crude steel•—million metric tons					
Total	12·7	13·8	14·8	15·6	16·8
of which: Poland	3·9	4·4	5·0	5·3	5·7
Czechoslovakia	4·3	4·5	4·9	5·2	5·5
East Germany	2·3	2·5	2·7	2·9	3·0
Hungary	1·5	1·6	1·4	1·4	1·6
Cement—million metric tons					
Total	12·0	13·6	14·4	15·8	17·5
of which: Poland	3·4	3·8	4·0	4·5	5·1
Czechoslovakia	2·6	2·9	3·1	3·7	4·1
East Germany	2·6	3·0	3·3	3·5	3·6
Romania	1·5	1·9	2·1	2·4	2·6
Cotton yarn—'000 metric tons					
Total	354	377	372	394	423
of which: Poland	109	115	116	123	135
Czechoslovakia	67	76	82	87	95
East Germany	62	63	62	61	64
Romania	43	46	43	45	47
Passenger cars—units					
Czechoslovakia	5,400	12,480	15,080	34,560	43,440
East Germany	19,680	22,200	28,080	35,640	38,400
Poland	1,680	3,960	5,760	7,920	11,520

† Poland, East Germany, Czechoslovakia, Hungary, Romania, Bulgaria, except where otherwise stated.

‡ Excluding small quantities produced in Romania.

§ Excluding East German synthetic production and small quantities of natural crude produced in Czechoslovakia.

• Excluding small quantities produced in Bulgaria: 1957 production, 159,000 tons.

YUGOSLAVIA

The post-war economic development of Yugoslavia differs importantly from that of other East European countries. From the end of the war to the break with the Cominform in June 1948, socialization proceeded considerably faster than in the other countries, and the Five-Year Plan adopted in April 1947 was the first long-term plan in Eastern Europe. It was based on the development of the country's rich mineral resources and aimed, with the help of very high investment, to make Yugoslavia nearly self-supporting in manufactured goods by 1951. Soviet disapproval of a plan directed primarily towards the interests of one country to the exclusion of the rest of the Soviet bloc was the principal reason for the clash with the Cominform. Isolation from its eastern neighbours greatly damaged the Yugoslav economy and forced the government to divert more resources to defence. An acute economic crisis in 1950 was overcome only with help from the U.S.A., the U.K. and France.

After 1950 economic administration was decentralized, a distinction drawn between those sectors of the economy owned by the public authorities and those owned co-operatively, local authorities were given greater power in economic matters, and

workers' councils began to play a large part in factory management.

Since 1953/54 emphasis has shifted towards the development of agriculture and light and consumers' goods industries, and investment has been directed towards achieving a better balance in the economy. These aims are embodied in the 1957–1961 Plan. As the table below shows, considerable progress has been achieved; and with a National Income equal to some £110 a head, Yugoslavia has now graduated from being an under-developed country to being a semi-industrialized one.

Yugoslavia: Output of Selected Commodities

	1954	1955	1956	1957	1958
Index—All industries (1953 100)	114	132	146	170	189
Lignite (million metric tons)	12·7	14·1	15·9	16·8	17·8
Electricity ('000 million kwh.)	3·4	4·3	5·1	6·3	7·4
Aluminium ('000 metric tons)	3·5	11·5	14·6	18·1	21·7
Crude steel ('000 metric tons)	616	805	887	1,049	1,120
Cement ('000 metric tons)	1,392	1,572	1,560	1,980	1,968
Cotton fabrics (million sq. metres)	166	174	182	208	218

THE TRADE OF THE U.S.S.R. AND EASTERN EUROPE

Before the Second World War, neither the Soviet Union nor any of the East European countries played a large role in international trade. In 1937 the total imports of the U.S.S.R. were valued at U.S. $253 million and exports at U.S. $326·1 million, amounting respectively to 1% and 1·3% of the world total. This was due partly to the comparative self-sufficiency of the area covered by the Soviet Union, but even more to difficulties of political origin or in connection with payments. In the same year the imports of the East European countries (excluding the U.S.S.R. and Yugoslavia) were valued at U.S. $972·4 million (3·8% of the world total) and their exports at U.S. $1,103·8 million (4·4%). In 1958 the total turnover of world trade amounted to some U.S. $220,000 million, of which the trade of the Soviet Union and its east European satellites (including intra-bloc trade) accounted for about 10%. Since this is virtually the same proportion as in 1953, communist trade has done no more than keep pace with the growth in world trade.

The Pattern of Trade before 1945

In 1937 for every country except the Soviet Union itself and Albania, Germany was the largest single trading partner, accounting for proportions ranging from 14% in the case of Czechoslovakia to 49% in the case of Bulgaria. Albania's trade was dominated to an even greater extent by Italy.

Table I
The Pattern of Trade in 1937
(Trade with selected countries as percentages of total trade)

	U.S.S.R.	Yugo.	Other E. Europe	Ger.	U.K.	U.S.A.	Others
U.S.S.R.							
Imports	—	—	1·6	15·0	14·3	18·2	50·9
Exports	—	—	2·0	6·2	32·7	7·8	51·3
Poland							
Imports	1·2	0·3	6·1	14·5	11·9	11·9	54·1
Exports	0·4	0·5	7·0	14·5	18·3	8·4	50·9
Czechoslovakia							
Imports	1·1	3·7	8·8	15·5	6·3	8·7	55·9
Exports	0·8	5·0	9·7	13·7	8·7	9·3	52·8
Hungary							
Imports	—	4·9	17·6	25·9	5·3	4·7	41·6
Exports	—	2·2	9·1	24·0	7·2	2·9	54·6
Bulgara							
Imports	0·02	0·5	13·6	54·8	4·7	2·0	24·4
Exports	—	0·7	11·4	43·1	13·8	3·8	27·2
Romania							
Imports	0·2	0·6	21·9	28·9	9·4	3·9	36·1
Exports	0·1	1·4	14·5	19·2	8·8	1·7	54·3
Albania							
Imports	—	11·2	20·1	4·7	4·9	4·6	*54·5
Exports	—	0·7	0·9	0·1	0·7	8·5	*89·1
Yugoslavia							
Imports	—	—	16·8	32·4	7·8	6·0	37·0
Exports	—	—	15·9	21·7	7·4	4·6	50·4

* Italy took 78·6% of Albania's exports and supplied 24·0% of her imports.

This German predominance was of enormous political and economic importance. Most European countries accumulated balances in Reichsmarks and were sometimes obliged to accept imports of manufactures which they did not really want. During the occupation from 1941–1944 45, Germany's commercial stranglehold over Eastern Europe was greatly intensified and interchanges between the countries themselves were reduced to a minimum.

The position of the U.S.S.R. was very different; the Soviet authorities were concerned to export only as much as was necessary to finance large-scale imports of machinery and certain raw materials essential to the drive for industrialization. In 1937 metals and metal manufactures, mainly machinery and equipment, formed 55% of all imports into the U.S.S.R., while agricultural produce (including cotton, leather and rough timber) accounted for 48% and oil, coal and manganese for a further 8·4% of exports. For the first fifteen years after the Revolution, the U.S.S.R. showed a trade deficit with the rest of the world, and found considerable difficulty in financing it.

From 1933 onwards, however, surpluses became the rule, though trade with the U.S.A. remained in heavy deficit for political reasons. During the war the U.S.S.R. enormously increased its imports of raw materials, machinery and armaments, but the greater part of the very large deficit was financed under the Lend-Lease programme.

The Post-War Pattern of Soviet Trade

Until the mid-'fifties it is difficult to obtain any comprehensive picture of the true commercial trade of the Soviet Union. Statistical data from the Soviet side is lacking; many of the goods imported up to 1953/54 came in the form of reparations deliveries or under arrangements for specially low-priced sales of certain products to the U.S.S.R. (e.g. Polish coal or Hungarian bauxite); and, moreover, the imposition of strategic controls on exports from many western countries distorted trade flows between East and West.

However, with the publication of official Soviet trade statistics from 1955 onwards, some analysis of the post-war direction and composition of trade becomes possible. As the following tables (III and IV) show, trade remains heavily concentrated with the Eastern bloc (including China, second only to East Germany as a trading partner in 1958); but the easing of East-West tension and the continuing Soviet need for certain western raw materials and for the most modern western machinery is reflected in the growth in the proportion of imports from non-Communist countries between 1955 and 1958. Although the differences between the composition of trade in 1955 and 1958 are not very striking, the differences between these figures and pre-war figures show clearly the Soviet Union's growth as an industrial power. In 1938 machinery and equipment accounted for only 5% of the value of exports, while it made up 35% of the value of imports in that year.

According to Soviet sources the volume of total foreign trade turnover rose by some 580% between 1938 and 1958, the rise between 1955 and 1958 being in the order of 40%. In terms of current prices, trade has developed as follows:

Table II
U.S.S.R. Trade 1955–1959
(million U.S. $)

	1955	1956	1957	1958	1959
Imports	3,081	3,614	3,938	4,349	5,075
Exports	3,468	3,612	4,382	4,290	5,425
Balance	+387	−2	+444	−52	+350

Table III
Soviet Trade with Socialist* Countries
(% of total)

	1955	1956	1957	1958
Imports	91	79	72	73
Exports	80	74	76	70

* Eastern Europe, Yugoslavia, China, Mongolia, North Korea.

Table IV
Composition of Soviet Trade
(% of total)

	Imports		Exports	
	1955	1958	1955	1958
Food and drink	17	14	12	12
Fuel and raw materials	14	14	13	20
Base metals and manufactures	7	9	13	16
Machinery and equipment	26	25	17	18
Textile fibres and yarns	5	7	10	7
Others	31	31	35	27
Total	100	100	100	100

Part of the recent increase in trade is the result of the more positive attitude of the Soviet Government to foreign trade which followed Stalin's death, the slackening of the cold war and the ending of punitive trade policies in Eastern Europe.

Soviet economic aid, mainly in the form of long-term credits at low interest, has been extended to countries both within and outside the bloc and has in some cases (e.g. China) been an important factor in stimulating trade.

THE POST-WAR PATTERN OF EASTERN EUROPEAN TRADE

After the war the Soviet Union emerged as Eastern Europe's principal trading partner. During the first post-war years Russia secured a very favourable rate of exchange in trade with Eastern Europe, and imports of goods from the area were in some cases more like reparations deliveries than normal commercial transactions. Trade with the countries outside the Soviet bloc was in most cases reduced to a bare minimum and was discouraged by strategic embargoes imposed by the West on trade with the Communist bloc.

Table V

East European* Trade with the Communist Bloc†

(Expressed as percentages of total trade)

	1948	1950	1952	1954	1956	1958
Poland						
Imports	42	61	68	71	66	58
Exports	40	57	65	69	58	59
Czechoslovakia						
Imports	31	54	71	75	66	71
Exports	33	52	72	75	64	68
Hungary						
Imports	34	56	72	68	61	69
Exports	35	66	73	73	63	65
Bulgaria						
Imports	81	85	89	87	81	....
Exports	74	86	89	86	87	...
East Germany						
Imports	76	76	75	74	72	71
Exports	74	68	75	78	74	77
Romania						
Imports	60	79	78	80	78	...
Exports	80	89	93	80	78	...
Total of Six Countries						
Imports	44	65	72	74	69	68
Exports	45	63	74	76	68	70

* Poland, Czechoslovakia, Hungary, Bulgaria, East Germany and Romania. † Including China.

With the gradual lessening of political tension between Russia and the West the governments of Eastern Europe were encouraged to develop their export industries and to resume their trading relations with the West. The East German precision engineering industry, the Polish shipbuilding industry and the Czech machine tool and motor industries were rapidly developed along with the Hungarian electro-technical and electronic industries, and became major instruments in the foreign trade drive of the Soviet bloc. The mounting dependence of the six major Eastern European satellites on trade with the Communist bloc up to 1954 and its subsequent decline shows clearly in Table V.

During the early 'fifties Russia's economic relations with Eastern Europe underwent basic changes. By sponsoring the development of the export industries of the East European countries Russia had to recognize their economic shortcomings and grant them financial assistance. Long-term trade and payments agreements were concluded between the Soviet Union and the countries of Eastern Europe, not only on more favourable terms but also with a guarantee of increased Soviet supplies of raw materials and industrial equipment.

In 1957 the U.S.S.R. embarked on the policy of integrating economic planning throughout Eastern Europe. The Council for Mutual Economic Aid (Comecon), originally set up in 1949 as an advisory body, gradually developed into a central economic planning organization attempting to integrate the entire trade of the Soviet bloc. As a result of the Council's recommendations trade among the countries of Eastern Europe was increased to promote economic co-operation and industrial specialization inside each country. Several inter-bloc development schemes were launched: East Germany and Czechoslovakia are helping to develop the Polish brown coal mines; Poland, East Germany and Czechoslovakia are promoting the development of the cellulose and power industries in Romania; Russia, East Germany and Czechoslovakia are helping to equip Bulgarian

industries; and Russia and East Germany are helping to develop the Romanian petroleum-chemical industry. Thus the policy of international division of labour, which has been hovering in the background for so long, is now being put into effect.

Table VI

Growth of Eastern European Trade

	Imports			Exports		
	1953	1958	%	1953	1958	%
	million U.S. $		increase	million U.S. $		increase
Bulgaria	200	366	83	206	374	81
Czechoslovakia	879	1,357	54	993	1,567	58
Eastern Germany	983	1,680	71	968	1,890	95
Hungary	471	630	34	503	680	35
Poland	774	1,227	58	831	1,059	27
Romania	385	415	8	341	430	26
Total of six countries	3,692	5,675	54	3,842	6,000	56
Albania	...	79	...	...	29	...

THE COMMODITIES TRADED

Food. The Soviet Union is again a regular exporter of cereals (557 million tons a year), mainly to the Communist bloc countries. It is an importer of animal products, oilseeds, vegetable fats, rice and sugar, which come both from the bloc (China is an important partner in this trade) and from such non-bloc countries as Denmark, Argentina, France and Burma. In Eastern European countries agricultural exports no longer play the dominant part which they did before the war, except in the case of Bulgaria which, on the recommendation of Comecon, is to develop its horticultural industry. Poland and Czechoslovakia and Hungary export limited quantities of tinned meat, and Poland has a considerable trade in bacon with the U.K.

Fuel. Soviet exports of fuel have grown substantially in recent years, both as regards solid fuels and petroleum and petroleum products; exports of the latter rose from 1·1 million tons in 1950 to 18 million tons in 1958, and there has been some penetration of Western markets. Poland continues to export hard coal, though shipments have dropped with the growth of supplies in western Europe, and Romania exports petroleum.

Other minerals. There has been a striking increase in Soviet exports of iron and steel both to Eastern and Western markets: they rose from 7,000 tons in 1938 to 11·9 million tons in 1958. Rolled products are also exported by Czechoslovakia and Poland. Certain types of steel products are imported into the bloc from the West. Hungary is an exporter of bauxite, Bulgaria of lead and zinc, Albania of chromite; the Soviet Union exports aluminium, and its tin exports threatened to reach such proportions that the Soviet government agreed with the International Tin Council to limit them to 13,500 tons in 1959. The bloc is dependent on imports of copper from the West.

Fibres. The Soviet Union has a two-way trade in cotton, exporting to the bloc and importing mainly from Egypt. Imports of wool are considerably higher than before the war, amounting to some 50,000 tons a year. East Germany is an exporter of synthetic fibres.

Rubber and chemicals. In spite of growing domestic output of synthetic rubber, the Soviet Union continues to import natural rubber, mainly of Malayan and Indonesian origin. East Germany is an exporter of synthetic rubber. Heavy chemicals figure mainly in intra-bloc trade, pharmaceuticals and some specialized products are imported from the West.

Machinery. Table IV shows the importance of machinery and equipment in Soviet trade; in 1958 imports were valued at $1,065 million, exports at $782 million. The most important exports were complete factory installations, tractors, agricultural machinery, motor vehicles and parts. Imports were principally metal working machinery, power generating equipment, and machinery for the chemical, food and building industries. Most exports go to other Communist countries, but shipments to underdeveloped countries are increasing. Imports from the West are mainly of highly specialized machinery for the chemical, plastics, synthetic fibre and food industries.

While under the Comecon plans the Soviet Union will develop a full range of engineering products, Eastern European countries are specializing in the production of, and trade in, certain goods: East Germany on machine tools, engines, generators, precision instruments, chemical plant; Poland on ships, rolling stock, mining equipment; Czechoslovakia on generators, machine tools, tractors, vehicles, chemical plant; Hungary, electrical equipment; Romania, oil drilling equipment; Bulgaria, mining equipment and electric motors.

Consumer goods. With the shift in emphasis from heavy industry and the movement to raise standards of living, trade in consumer goods is growing. Exports from the Soviet Union are growing only slowly, but imports reached $885 million in 1958—mainly textiles, clothing, footwear, furniture, drugs and cosmetics. Hungary and Czechoslovakia are the most important exporters of manufactured consumer goods in Eastern Europe.

THE PRINCIPAL TRADING PARTNERS

Tables III and V show the extent to which foreign trade is concentrated within the Soviet bloc. The Soviet Union's most important partners in the area are East Germany, China, Czechoslovakia, Poland and Romania; in non-communist Europe trade is mainly with the highly industrialized countries, though there is a large exchange of goods with Finland which, during the days when it paid reparations, adapted some of its industries to Soviet requirements. Soviet trade with the underdeveloped countries remains comparatively small, though it is still growing. Its pattern is to some extent dictated by the U.S.S.R.'s need for raw materials (Malaya), to some extent it follows Russian aid (United Arab Republic, India), to some extent it is set by Russia's ability to take up surplus commodities (in Latin America, Soviet oil has been bartered against coffee).

Direction of Soviet Trade 1958
S million

	Imports	Exports
Total	4,350	4,298
Soviet bloc		
of which: East Germany	816	780
China	881	623
Czechoslovakia	512	447
Poland	265	377
Romania	234	251
Western Europe		
of which: Finland	137	117
U.K.	73	146
France	81	87
West Germany	77	66
Rest of World		
of which: Egypt	107	88
India	51	130
Malaya	118	—

Eastern Europe's main trading partners in the non-communist world in 1958 are shown in the following table (the figures are taken from the partners' returns):

Main Non-Communist Trading Partners of Eastern Europe, 1958
S million

	Imports from E. Europe	Exports to E. Europe
Western Germany	214	224
U.K.	120	68
France	79	70
Other O.E.E.C. countries	463	468
U.S.A.	48	105
Yugoslavia	137	77
Egypt	102	100
Finland	54	51

The main exporters from Eastern Europe to the non-Communist world are East Germany (with such things as precision instruments, potash, and chemicals); Poland (coal, zinc alloys, ships), and Czechoslovakia (machine tools, vehicles, shoes). Eastern Europe's main western imports are iron ore, pyrites, rubber, special steels, fibres and industrial equipment.

Western European Trade with the U.S.S.R. and Eastern Europe, 1958*
(million U.S. S)

	Imports from		Exports to	
	U.S.S.R.	E. Europe	U.S.S.R.	E. Europe
Food, drink and tobacco	111	293	63	134
Cereals and products	63	37	4	7
Live animals and meat products	3	114	—	—
Sugar	10	38	—	3
Materials and semi-finished products	603	416	204	379
Solid fuel	72	138	—	—
Petroleum and products	148	50	—	1
Base metals	99	61	138	207
Manufactures	48	312	285	381
Metal-working machinery	1	19	7	15
Other machinery	14	58	101	107
Electric machinery	2	20	29	44
Made up textiles, clothing	—	23	3	6
Total (inc. unspecified)	765	1,095	562	937

* Trade between East and West Germany is not included.

— Nil or negligible.

Population figures. The most reliable and latest figures available have been given for towns over 50,000, irrespective of whether these are from censuses or estimates. Comparative dates are given for places with interesting changes of population.

Use of symbols. These are explained by running footnotes. In addition some items of general interest (e.g. health resorts) have been included in the Gazetteer but not mapped.

Abbreviations of Administrative Divisions of the USSR.

AO.	Autonomous Oblast
ASSR.	Autonomous Soviet Socialist Republic
NO.	National Okrug
RSFSR.	Russian Soviet Federated Socialist Republic
SSR.	Soviet Socialist Republic

See Back Endpaper for Population of major towns (Post 1959)

ABAKAN—ARBAT BANKS

	Page	Lat.	Long.
Abakan: RSFSR. *Iron ore.*	15	54N	91E
Abakan Mts.: RSFSR.	15	52N	88E
Abaza: RSFSR. *Iron ore.*	†	52N	91E
Abdulino: RSFSR.	13	54N	54E
Abez': RSFSR.	18	67N	61E
Abinskaya: RSFSR.	9	45N	38E
Abkhaz ASSR.: Georgian SSR.	EP	43N	41E
Ablaketka: Kazakh SSR. *Tin.*	17	50N	83E
Abrud: Romania.	11	46N	23E
Abyy: RSFSR.	19	68N	145E
Achinsk: RSFSR.	15	56N	91E
Achisay: Kazakh SSR. *Lead, zinc. (Kentau Group)*	17	44N	69E
Achi-Su: RSFSR. *Petroleum* (Izberbash Group)*	12	43N	48E
Adimi: RSFSR.	19	48N	148E
Adjud: Romania.	11	46N	27E
Adler: RSFSR.	12	45N	40E
Adyge AO.: Krasnodar Kray, RSFSR.	EP	45N	40E
Adzhar ASSR.: Georgian SSR.	EP	42N	42E
Aga Buryat Mongol NO.: Chita Oblast, RSFSR.	EP	52N	115E
Agadyr': Kazakh SSR. *Copper.*	17	48N	73E
Agata: RSFSR.	19	67N	93E
Agdam: Azerbaydzhan SSR.	12	40N	47E
Aginskoye: RSFSR.	12	51N	114E
Agnita: Romania.	11	46N	25E
Agrakhanskiy Penin.: RSFSR..	12	44N	48E
Agryz: RSFSR.	13	57N	53E
Agvali: RSFSR.	12	43N	46E
Aim: RSFSR.	19	59N	134E
Aitos: Bulgaria	11	43N	27E
Aiud: Romania	11	46N	24E
Ajka: Hungary. *Coal, hydro-elec., aluminium*.*	10	47N	18E
Ak-Bulak: RSFSR.	13	51N	56E
Akchatau: Kazakh SSR. *Tungsten.*	17	48N	74E
Akhalkalaki: Georgian SSR.	12	41N	44E
Akhaltsikhe: Georgian SSR. *Lignite.*	12	42N	43E
Akhtme: Estonian SSR.	8	59N	28E
Akhtopol: Bulgaria	11	42N	28E
Akhtyrka: Ukrainian SSR.	9	50N	35E
▲Akmolinsk: Kazakh SSR. *Pop. 1954: 100,000. Engineering.*	15	51N	72E
▲Akmolinsk Oblast: Kazakh SSR.	EP	50N	70E
Aksay: RSFSR.	9	48N	44E
Aksenovo-Zilovskoye: RSFSR.	12	53N	117E
Aksha: RSFSR.	12	50N	112E
Akstafa: Georgian SSR.	12	41N	46E
Aksuat: Kazakh SSR.	14	50N	64E
Aksuat: Kazakh SSR.	17	48N	83E
Aktash: RSFSR.	13	55N	52E
Aktyubinsk: Kazakh SSR. *Pop. 1954: 80,000. Ferro-alloys, eng., lignite, elec. power.*	14	50N	57E
Aktyubinsk Oblast: Kazakh SSR..	EP	48N	60E
Ak-Tyuz: Kirgiz SSR. *Lead.*	17	43N	76E
Akzhal: Kazakh SSR. *Gold nearby.*	17	49N	82E
Alagir: RSFSR.	12	43N	44E
Alai Range: USSR.	17	39N	70E
Alakul': *lake:* USSR.	17	46N	82E
Alapayevsk: RSFSR. *Iron ore, iron & steel, bauxite, eng.*	13	57N	62E
Alatyr': RSFSR. *Textiles.*	14	55N	46E
Alaverdi: Armenian SSR. *Copper, copper smelting, chemicals.*	12	41N	45E

	Page	Lat.	Long.
Alba Iulia: Romania.	11	46N	24E
Al'bertin: Byelorussian SSR.	7	53N	26E
Aldan: RSFSR. *Gold.*	19	59N	125E
Aldan, R.: RSFSR.	23	59N	133E
Aldan Plateau: RSFSR. *Gold.*	19	57N	126E
Aleksandriya: Ukrainian SSR. *Lignite.*	9	49N	33E
Aleksandrov: RSFSR.	8	56N	39E
Aleksandrovac: Yugoslavia	11	43N	21E
Aleksandrov Gay: RSFSR.	14	50N	49E
Aleksandrovka: RSFSR..	9	47N	39E
Aleksandrovsk: RSFSR.	13	59N	58E
Aleksandrovsk-Sakhalinskiy: RSFSR. *Port. Coal, saw-milling, eng.*	19	52N	142E
Aleksandrovskiy: RSFSR.	9	51N	37E
Aleksandrovskiy Zavod: RSFSR.	12	51N	118E
Aleksandrovskoye: RSFSR.	15	60N	78E
Aleksandrów: Poland	6	53N	19E
Alekseyevka: Kazakh SSR.	15	52N	71E
Alekseyevka: Kazakh SSR. *Gold.*	15	48N	86E
Alekseyevka: RSFSR.	9	51N	39E
Alekseyevka: RSFSR.	14	52N	48E
Alekseyevo-Lozovskoye: RSFSR.	9	49N	41E
Alekseyevsk: RSFSR.	19	58N	109E
Aleksin: RSFSR.	8	54N	37E
Aleksinac: Yugoslavia	11	44N	22E
Alesd: Romania	11	47N	22E
Alevin, C.: RSFSR.	19	59N	151E
Alexandria: Romania	11	44N	25E
Aleysk: RSFSR.	15	53N	83E
Alga: Kazakh SSR. *Phosphorite, chemicals.*	14	50N	57E
Ali-Bayramly: Azer. SSR.	12	40N	49E
Alitus: Lith. SSR. *Textiles.*	7	54N	24E
Allakh-Yun': RSFSR. *Gold.*	19	61N	138E
Allaykha: RSFSR.	19	70N	147E
ALMA ATA [Vernyy]: Kazakh SSR. *Pop. 1954: 320,000. Eng., textiles, elec. power*	17	43N	77E
Alma-Ata Oblast: Kazakh SSR.	EP	43N	78E
Almásfüzito: Hungary. *Petroleum refining.*	†	48N	18E
Almalyk: Uzbek SSR. *Copper copper smelting.*	17	41N	70E
Almaznaya: Ukr. SSR. *Iron.*	†	48N	39E
Altai Range: Mongolian People's Republic	19	46N	95E
Altan Bulak: Mongolian People's Republic	12	50N	106E
Altan Tepe [Topolog]: Romania. *Sulphur & pyrites.*	†	45N	28E
Altay Kray: RSFSR.	EP	52N	80E
Altenberg: E. Germany. *Tin, tin smelting.*	6	51N	14E
Altenburg: E. Germany. *Pop. 1946: 51,800. Textiles* (Central German Group)*	6	51N	12E
Altnay: RSFSR.	13	57N	62E
Aluksne: Latvian SSR.	8	57N	27E
Alushta: Ukrainian SSR.	9	45N	34E
Alyaty: Azer. SSR. *Petroleum* (Alyaty Group).*	12	40N	49E
Alygdzher: RSFSR. *Mica.*	19	53N	98E
Amangel'dy: Kazakh SSR.	14	50N	65E
Amazar: RSFSR.	19	54N	121E
Ambarchik: RSFSR. *Port.*	19	70N	162E
Ambla: Estonian SSR.	‡5	59N	25E

	Page	Lat.	Long.
Ambrolauri: Georgian SSR. *Hydro-elec.*	12	42N	43E
Amderma: RSFSR. *Port.*	18	70N	60E
Amga: RSFSR.	19	61N	132E
Amga, R.: RSFSR.	23	62N	134E
Amgu: RSFSR.	19	46N	138E
Amgun, R.: RSFSR.	23	53N	138E
Amiradzhany: Azer. SSR.	†	40N	50E
Amo: RSFSR.	19	63N	104E
Amu Darya, R.: USSR.	22	38N	67E
Amur Oblast: RSFSR.	EP	54N	127E
Amur, R.: RSFSR.	23	52N	139E
Amurzet: RSFSR. *Iron ore*.*	19	48N	131E
Amvrosiyevka: Ukrainian SSR.	9	48N	38E
Anabar, R.: RSFSR.	23	72N	113E
Anadyr': RSFSR. *Port.*	19	65N	177E
Anadyr' Plateau: RSFSR.	19	67N	165E
Anadyr' Range: RSFSR.	19	67N	180E
Anadyr', R.: RSFSR.	23	65N	174E
Anan'yev: Ukrainian SSR.	9	48N	30E
Anapa: RSFSR.	9	45N	37E
Andizhan 1: Uzbek SSR. *Pop. 1954: 105,000. Textiles*.*	17	41N	72E
Andizhan 2 (12 miles S. of Andizhan 1): Uzbek SSR. *Petroleum.*	‡17	41N	72E
▲Andizhan Oblast: Uzbek SSR.	EP	42N	74E
Andreapol': RSFSR.	8	57N	32E
Andreyevka: Kazakh SSR.	17	46N	80E
Andreyevka: RSFSR.	14	52N	52E
Andreyevsk: RSFSR.	19	58N	114E
Andrijevica: Yugoslavia	10	43N	20E
Andrushevka: Ukrainian SSR..	9	50N	29E
Andrychów: Poland	7	50N	19E
Andryushka: RSFSR.	19	69N	155E
Angara, R.: RSFSR.	23	59N	97E
Angarsk: RSFSR.	12	53N	104E
Angermünde: E. Germany.	6	53N	14E
Angren: Uzbek SSR. *Coal.*	17	41N	70E
Anikshchyay: Lith. SSR.	‡5	56N	25E
Animovka: Ukrainian SSR.	9	47N	35E
Anipemza: Armenian SSR.	12	40N	44E
Aniva Bay: USSR.	19	47N	143E
Anna: RSFSR.	9	51N	40E
Annaberg: E. Ger. *Cobalt, tin, uranium.*	6	51N	13E
Annenskiy Most: RSFSR.	8	61N	37E
Antopol': Byelorussian SSR.	7	52N	25E
Anzhero-Sudzhensk: RSFSR. *Pop. 1954: 85,000. Coal.*	15	56N	86E
Apatin: Yugoslavia	10	46N	19E
Apatity: RSFSR. *Phosphates, zirconium.*	†	68N	33E
Apostolovo: Ukrainian SSR.	9	48N	34E
Aprel'sk: RSFSR.	19	58N	115E
Apsheron Penin.: Azer. SSR.	12	40N	50E
Apsheron-Port: Azer. SSR.	†	40N	50E
Apsheronsk: RSFSR. *Petroleum* (Kuban' Field)*	12	44N	40E
Apuseni Mts.: Romania	11	47N	23E
Arad: Romania. *Pop. 1948: 87,291. Textiles.*	11	46N	21E
Aragats, Mt.: Armenian SSR.	12	40N	44E
Aral Sea: USSR.	16	45N	60E
Aral'sk: Kazakh SSR. *Port.*	16	47N	62E
Aralsul'fat: Kazakh SSR. *Sodium, chemicals.*	16	47N	62E
Aramil': RSFSR. *Textiles.*	13	57N	61E
Arandelovac: Yugoslavia	11	44N	20E
Ararat: Armenian SSR.	12	40N	45E
Arbat Banks: Sea of Azov	9	46N	35E

* Located by symbol, but not named, on special topic map. ‡ See Inset map. **EP** Front Endpaper map.
† On relevant special topic map only. For list of topics see p. VIII. Place names in [square] brackets are former, or alternative, names.
▲ See appendix on p.134

	Page	Lat.	Long.
Arbuzinka: Ukrainian SSR.	9	48N	31E
Archangel *see* Arkhangel'sk.			
Ardatov: RSFSR.	14	55N	46E
Arded: Romania	11	48N	23E
Ardino: Bulgaria. *Lead.*	11	42N	25E
Ardon: RSFSR.	12	43N	44E
Arefino: RSFSR.	8	58N	39E
Arka: RSFSR.	19	60N	142E
Arkadak: RSFSR.	9	52N	44E
Arkhangel'sk [Archangel]: RSFSR. *Pop. 1954: 325,000. Largest sawmilling centre and timber port in USSR; eng., hydro-elec.*	18	65N	40E
Arkhangel'sk Oblast: RSFSR.	EP	63N	40E
Arkhangel'skoye: RSFSR.	9	51N	41E
Arkona, C.: E. Germany.	6	55N	14E
Armavir: RSFSR. *Pop. 1954: 100,000.*	9	45N	41E
Armenian SSR.	EP	40N	45E
Armyansk: Ukrainian SSR.	9	46N	34E
Arsen'yev: RSFSR.	†	44N	133E
Arsen'yevo: RSFSR.	8	54N	37E
Arshintsevo: Ukrainian SSR.	9	45N	36E
Arsk: RSFSR.	14	56N	50E
Artel'nyy: RSFSR. *Gold, platinum.*	13	59N	60E
Artem: RSFSR. *Pop. 1954: 75,000. Coal.*	19	43N	132E
Artem I.: Azerbaydzhan SSR. *Petroleum.*	12	40N	50E
Artemovka: Ukrainian SSR.	9	50N	35E
Artemovsk: RSFSR. *Gold.*	19	54N	93E
Artemovsk [Bakhmut]: Ukr. SSR. *Pop. 1954: 60,000. Salt.*	9	49N	38E
Artemovskiy: RSFSR. *Gold* (Lena-Vitim Field).*	19	58N	115E
Artemovskiy: RSFSR. *Anthracite.*	13	57N	62E
Artik: Armenian SSR.	12	41N	44E
Artsiz: Ukrainian SSR.	11	46N	29E
Arys': Kazakh SSR.	17	42N	69E
Aryta: RSFSR.	19	64N	126E
Arzamas: RSFSR. *Pop. 1954: 50,000. Textiles.*	8	55N	44E
⏘Arzamas Oblast: RSFSR.	EP	55N	45E
Arzgir: RSFSR.	9	45N	44E
Aš [*Ger.* Asch]: Czechoslovakia.	6	50N	12E
Asbest: RSFSR. *Asbestos (centre of group of mines).*	13	57N	62E
Asbestovskiy: RSFSR. *Asbestos* (Asbest Group).*	13	58N	61E
Aschersleben: E. Ger. *Lignite*, potash, eng., chemicals, textiles.*	6	52N	11E
Asenovgrad: Bulgaria	11	42N	25E
Asha: RSFSR.	13	55N	57E
ASHKHABAD [Poltoratsk]: Turkmen SSR. *Pop. 1954: 134,000. Textiles, eng., (cinema studios), elec. power.*	16	38N	58E
⏘Ashkhabad Oblast: Turkmen SSR.	EP	39N	58E
Asht: Tadzhik SSR.	‡17	41N	70E
Asino: RSFSR.	15	57N	86E
Askino: RSFSR.	13	56N	56E
Askiz: RSFSR. *Chemicals, barium.*	†	53N	90E
Astara: Azerbaydzhan SSR.	16	39N	49E
Astrakhan': RSFSR. *Pop.1954: 325,000. Port (outport at " 12-Foot Roads "), elec. power, eng.*	12	46N	48E
⏘Astrakhan-Bazar: Azer. SSR.	12	39N	48E
⏘Astrakhan' Oblast: RSFSR.	EP	47N	47E
Atasuskiy: Kazakh SSR. *Iron ore.*	17	49N	72E
Atbasar: Kazakh SSR.	14	52N	68E
Atka: RSFSR.	19	61N	152E
Atkarsk: RSFSR.	9	52N	45E
Augustów: Poland	7	54N	23E
Auschwitz *see* Oświęcim.			
Avchala: Georgian SSR.	12	42N	45E
Avrig: Romania	11	46N	24E
Ayaguz: Kazakh SSR. *Eng., textiles.*	17	48N	80E
Ayakhta: RSFSR.	15	59N	94E
Ayan: RSFSR. *Port. Gold.*	19	57N	138E
Aydyrlinskiy: RSFSR. *Nickel, cobalt.*	13	52N	60E
Aynazhi: Latvian SSR.	‡5	58N	24E
Ayon I.: RSFSR.	19	70N	168E
Ayzpute: Latvian SSR.	‡5	57N	21E
Azerbaydzhan SSR.	EP	40N	48E
Azov: RSFSR. *Port.*	9	47N	39E
Azov, Sea of: USSR.	9	46N	37E
Azovskoye: Ukrainian SSR.	9	45N	35E
Azovy: RSFSR.	18	65N	65E
Babadag: Romania	11	45N	29E
Babanka: Ukrainian SSR.	9	49N	30E
Babayevo: RSFSR.	8	59N	36E

	Page	Lat.	Long.
Babintsy: Ukrainian SSR.	9	51N	30E
Babushkin [Mysovsk]: RSFSR. *Iron ore.*	12	52N	106E
Babushkin [Losinoostrovskaya]: RSFSR. *Pop. 1954: 100,000. Eng.*, chemicals*, textiles*. Suburb of Moscow.*	8	56N	38E
Bacău: Romania. *River port. Eng., sawmilling, textiles, petroleum.*	11	47N	27E
Bačka Palanka: Yugoslavia	10	45N	19E
Bačka Topola: Yugoslavia	10	46N	20E
Bagdarin: RSFSR. *Gold.*	12	54N	114E
Bagrationovsk [*Ger.* Preussisch Eylau]: RSFSR.	7	54N	21E
Baia de Aramă: Romania	11	45N	23E
Baia-Mare: Romania. *Gold, lead, silver, zinc, chemicals, uranium.*	11	48N	24E
Baia-Sprie: Romania	11	48N	24E
Băileşti: Romania	11	44N	23E
Bailovo: Azerbaydzhan SSR.	†	40N	50E
Baja: Hungary. *Textiles.*	10	46N	19E
Bakal: RSFSR. *Iron ore..*	13	55N	59E
Bakaly: RSFSR.	13	55N	54E
Bakanas: Kazakh SSR.	17	45N	76E
Bakchar: RSFSR.	15	57N	82E
Bakharden: Turkmen SSR.	16	39N	57E
Bakhchisaray: Ukrainian SSR.	9	45N	34E
Bakhmach: Ukrainian SSR.	9	51N	33E
Bakony Forest: Hungary.	10	47N	18E
Bakr-Uzyak: RSFSR.	13	53N	59E
Baksan: RSFSR.	12	44N	44E
BAKU: Azerbaydzhan SSR. *Pop. 1954: 890,000. Petroleum refining, petroleum, eng., chemicals, textiles, elec. power. Extensive suburbs form city of Greater Baku, covering the Apsheron Peninsula, with a total pop. approx. twice that of the central city.*	12	40N	50E
Bakuriani: Georgian SSR.	12	42N	44E
Balabanovo: RSFSR.	8	55N	36E
Baladzhary: Azerbaydzhan SSR. *Petroleum..*	†	40N	50E
Baladzholskiy: Kazakh SSR.	17	49N	82E
Balagansk: RSFSR.	12	54N	104E
Balakhany: Azerbaydzhan SSR. *Petroleum..*	†	41N	50E
Balakhna: RSFSR..	8	56N	44E
Balakhta: RSFSR.	15	55N	92E
Balaklava: Ukrainian SSR.	9	44N	34E
Balakleya: Ukrainian SSR.	9	49N	37E
Balakovo: RSFSR.	14	52N	48E
Balanda: RSFSR.	9	51N	44E
Balashov: RSFSR. *Pop. 1954: 100,000. Eng.*	9	52N	43E
⏘Balashov Oblast: RSFSR.	EP	52N	43E
Balassagyarmat: Hungary	6	48N	19E
Balaton, L.: Hungary	10	47N	18E
Balbagar: RSFSR. *Iron ore.*	†	53N	109E
Balchik: Bulgaria.	11	43N	28E
Baley: RSFSR. *Gold, elec. power.*	12	52N	116E
Balezino: RSFSR.	13	58N	53E
Baligród: Poland	7	49N	22E
Balkan Mts.: Bulgaria	11	43N	26E
Balkany: RSFSR. *Tungsten.*	13	53N	60E
Balkhash: Kazakh SSR. *Pop. 1954: 50,000. Copper smelting and refining, elec. power*..*	17	47N	75E
Balkhash, L.: Kazakh SSR.	17	47N	75E
Ballsh: Albania	10	41N	20E
Bals: Romania	11	44N	24E
Balta: Ukrainian SSR.	9	48N	30E
Baltiysk [*Ger.* Pillau]: RSFSR. *Port (naval base).*	6	55N	20E
Balygychan: RSFSR.	19	64N	155E
Balyksa: RSFSR.	15	53N	89E
Band: Romania	11	47N	24E
Banja Luka: Yugoslavia. *Hydro-elec.**	10	45N	17E
Bánovce: Czechoslovakia	6	49N	18E
Banovo Jaruga: Yugoslavia	10	45N	17E
Banská Bystrica: Czechoslovakia. *Textiles.*	6	49N	19E
Banská Stiavnica: Czechoslovakia. *Gold, silver, lead, zinc, copper.*	6	48N	19E
Bansko: Bulgaria	11	42N	24E
Bar: Ukrainian SSR.	7	49N	28E
Bar: Yugoslavia	10	42N	19E
Barabinsk: RSFSR.	15	55N	78E
Baran': Byelorussian SSR.	8	54N	30E
Baranchinskiy: RSFSR.	13	58N	60E
Baranovichi [*Pol.* Baranowicze]: Byelorussian SSR. *Pop. 1954: 60,000. Engineering*.*	7	53N	26E
Baranovka: Ukrainian SSR.	7	50N	28E
Bârca: Romania	11	44N	24E
Barcs: Hungary. *Sawmilling.*	10	46N	18E

	Page	Lat.	Long.
Barda: Azerbaydzhan SSR.	12	40N	47E
Barda: RSFSR.	13	57N	56E
Bardejov: Czechoslovakia	7	49N	21E
Barents Sea: USSR.	18	73N	45E
Barga: RSFSR. *Mica.*	†	61N	90E
Barguzin: RSFSR.	12	53N	109E
Barguzin Range: RSFSR.	12	54N	110E
Bârlad: Romania. *Textiles.*	11	46N	28E
Barlinek [*Ger.* Berlinchen]: Poland	6	53N	15E
Barmash: Albania	10	40N	21E
Barnaul: RSFSR. *Pop. 1954: 225,000. River port. Textiles. chemicals, eng., sawmilling, elec. power.*	15	53N	84E
Barth: E. Germany.	6	54N	13E
Bartoszyce: Poland.	7	54N	21E
Barvenkovo: Ukrainian SSR.	9	49N	37E
Baryshevo: RSFSR.	8	61N	29E
Barzas: RSFSR. *Coal.*	15	56N	86E
Basargechar: Armenian SSR.	12	40N	46E
Bashkir ASSR.: RSFSR.	EP	55N	55E
Bashtanka: Ukrainian SSR.	9	47N	32E
Basil'kovka: Ukrainian SSR.	9	48N	36E
Basil'sursk: RSFSR.	8	56N	46E
Baskunchak: RSFSR. *Chemicals, salt. (See Nizhniy, & Verkhniy Baskunchak).*	†	48N	47E
Batagay-Alyta: RSFSR.	19	63N	131E
Batak: Bulgaria.	11	42N	24E
Batamshinskiy: Kazakh SSR. *Nickel.*	†	51N	58E
Bátaszék: Hungary	10	46N	19E
Bataysk: RSFSR. *Pop. 1954: 50,000. Engineering.*	9	47N	40E
Batetskiy: RSFSR.	8	59N	30E
Batumi: Georgian SSR. *Pop. 1954: 140,000. Port. Petroleum refining, engineering.*	12	42N	42E
Bauska: Latvian SSR.	‡5	56N	24E
Bautzen: E. Germany. *Chemicals, textiles.*	6	51N	14E
Bayan-Aul: Kazakh SSR. *Copper.*	15	51N	76E
Bayan-Gol: Mongolian People's Republic	19	49N	106E
Baychunas: Kazakh SSR. *Petroleum..*	16	47N	53E
Baydzhansay: Kazakh SSR. *Lead.*	17	43N	70E
Bayevo: RSFSR.	15	53N	80E
Baykal: RSFSR.	12	52N	104E
Baykal, L.: RSFSR.	13	53N	107E
Baykal Range: RSFSR.	12	54N	108E
Baykonur: Kazakh SSR. *Lignite.*	17	48N	66E
Baymak: RSFSR. *Copper, copper smelting.*	13	53N	58E
Bayram-Ali: Turkmen SSR.	16	38N	62E
Baysun: Uzbek SSR.	17	38N	67E
Bazar-Dara Range: Tadzhik SSR.	17	38N	73E
Bazar-Dyuzi, mtn.: Azerbaydzhan SSR.	12	41N	48E
Bazarny Syzgan: RSFSR.	14	54N	47E
Becej: Yugoslavia	11	46N	20E
Beclean: Romania	11	47N	24E
Bednodem'yanovsk: RSFSR.	8	54N	43E
Będzin: Poland. *Coal* (Silesian Coalfield).*	†	50N	19E
Begovat: Uzbek SSR. *Steel, hydro-elec.*	17	40N	69E
Beiuş: Romania	11	47N	22E
Bekdash: Turkmen SSR.	16	42N	53E
Békéscsaba: Hungary. *Pop. 1941: 52,404.*	11	47N	21E
Bela Crkva: Yugoslavia	11	45N	21E
Bela Palanka: Yugoslavia	11	43N	22E
Bela Slatina: Bulgaria	11	43N	24E
Belaya, R.: RSFSR.	22	54N	56E
Belaya Berezovka: RSFSR.	9	52N	34E
Belaya Glina: RSFSR.	9	46N	41E
Belaya Kalitva: RSFSR. *Coal* (Donbass).*	9	48N	41E
Belaya Krinitsa: Ukrainian SSR.	9	51N	30E
Belaya Tserkov': Ukrainian SSR.	9	50N	30E
Bełchatów: Poland	6	51N	19E
Belebey: RSFSR.	13	54N	54E
Belev: RSFSR.	8	54N	36E
Belgorod: RSFSR. *Pop. 1954: 50,000. (Produces half of all writing chalk in USSR); engineering.*	9	51N	37E
Belgorod-Dnestrovskiy [*Rom.* Cetatea Albă]: Ukrainian SSR. *Pop. 1930: 21,000; 1941: 8,000.*	9	46N	30E
Belgorod Oblast: RSFSR.	EP	51N	37E
BELGRADE [*Serb.* Beograd]: Yugoslavia. *Pop. 1953: 469,988. River port. Eng. (at Zeleznik and Rakovica), textiles, paper.*	11	45N	20E

* Located by symbol, but not named, on special topic map. ‡ See Inset map. EP Front Endpaper map.
† On relevant special topic map only. For list of topics see p. VIII. Place names in [square] brackets are former, or alternative, names.
⏘ See appendix on p.134

Page III

	Page	Lat.	Long.
Belinskiy: RSFSR.	8	53N	43E
Belišće: Yugo. *Sawmilling.*	10	46N	18E
Bel'kachi: RSFSR.	19	59N	132E
Beloglazovo: RSFSR.	15	52N	82E
Belogorsk: Ukrainian SSR.	9	45N	35E
Belogradchik: Bulgaria. *Anthracite.*	11	44N	23E
Belokany: Azerbaydzhan SSR..	12	42N	46E
Belokholunitskiy: RSFSR.	14	59N	51E
Belokorovichi: Ukrainian SSR.	7	51N	28E
Beloles'ye: Ukrainian SSR.	9	46N	30E
Belolutsk: Ukrainian SSR.	9	50N	39E
Belomorsk [*Fin.* Sorokka]: Karelo-Finnish SSR. *Port. Sawmilling.*	18	64N	35E
Beloomut: RSFSR.	8	55N	39E
Belopol'ye: Ukrainian SSR.	9	51N	34E
Belorechenskaya: RSFSR.	9	45N	40E
Belorechka: RSFSR. *Copper, zinc.*	13	57N	60E
Beloretsk: RSFSR. *Manganese, iron.*	13	54N	58E
Belousovka: Kazakh SSR. *Lead, zinc.*	15	50N	83E
Belovo: RSFSR. *Pop. 1954: 50,000. Coal, zinc processing.*	15	55N	86E
Beloye: RSFSR.	9	51N	36E
Beloye, L.: RSFSR.	8	60N	38E
Belozersk: RSFSR.	8	60N	38E
Belsh: Albania	10	41N	20E
Bel'tsy [*Rom.* Bălţi]: Moldavian SSR. *Pop. 1954: 90,000. Chemicals, eng.*	11	48N	28E
Belyayevka: Ukrainian SSR.	9	46N	30E
Belyy: RSFSR.	8	56N	33E
Belyy I: RSFSR.	18	73N	70E
Belyye Berega: RSFSR.	8	53N	34E
Belz: Ukrainian SSR.	7	50N	24E
Belzig: E. Germany	6	52N	12E
Bendery [*Rom.* Tighina]: Moldavian SSR. *Pop. 1930: 31,384; 1941: 15,075. Textiles.*	11	47N	29E
Benkovac: Yugoslavia	10	44N	16E
Beograd *see* Belgrade			
Berat: Albania	10	41N	20E
Berchogur: Kazakh SSR. *Coal.*	16	48N	59E
Berdichev: Ukrainian SSR. *Pop. 1954: 70,000.*	7	50N	28E
Berdigyastyakh: RSFSR.	19	62N	127E
Berdsk: RSFSR.	15	55N	83E
Beregovo [*Czech.* Berehovo, *Hung.* Beregszász]: Ukrainian SSR.	7	48N	23E
Berestechko: Ukrainian SSR.	7	50N	25E
Berettyóújfalu: Hungary	11	47N	22E
Bereza-Kartuzskaya: Byelorussian SSR.	7	52N	25E
Berezhany: Ukrainian SSR.	7	49N	25E
Berezino: Byelorussian SSR.	8	54N	29E
Berezniki: RSFSR. *Pop. 1954: 75,000. Salt*, chemicals, paper.*	13	59N	57E
Berezno: Ukrainian SSR.	7	51N	27E
Berezovka: RSFSR.	13	60N	54E
Berezovka: Ukrainian SSR.	9	47N	31E
Berezovo: RSFSR.	18	64N	65E
Berezovskaya: RSFSR.	9	50N	44E
Berezovskiy: RSFSR. *Tungsten.*	13	57N	61E
Berezovskoye: RSFSR. *Iron ore.*	12	52N	119E
Berikul'skiy: RSFSR. *Gold* (Kiya Valley Field), arsenic.*	15	55N	88E
Berislav: Ukrainian SSR.	9	47N	33E
Berkovitsa: Bulgaria	11	43N	23E
BERLIN: Germany. *Area 344 square miles. Total population 1953: 3,481,000. Russian Sector (East Berlin). Area 156 sq. miles. Pop. 1948: 1,195,888 (i.e. 37% of the total) in the districts of Mitte, Prenzlauer, Berg, Friedrichshain, Treptow, Köpenick, Lichtenberg, Weissensee and Pankow. Capital of German Democratic Republic. Centre of Russian administrative organization in East Germany is at Karlshorst. Inland port. Eng., chemicals, pharmaceuticals.*	6	53N	13E
Bernburg: E. Germany. *Pop. 1946: 53,400. Potash*.*	6	52N	12E
Beshenkovichi: Byelorussian SSR.	8	55N	30E
Beshkent: Uzbek SSR.	17	39N	65E
Beslan: RSFSR.	12	43N	45E
Bet-Pak-Dala Steppe: USSR.	17	46N	72E
Beuthen *see* Bytom.			
Bezdan: Yugoslavia	10	46N	19E

	Page	Lat.	Long.
Bezhetsk: RSFSR. *Eng.**	8	58N	37E
Bezhitsa: RSFSR. *Pop. 1954: 95,000. Elec. power, eng., steel, chemicals.*	8	53N	34E
Bezmein: Turkmen SSR.	16	38N	58E
Biała Podlaska: Poland	7	52N	23E
Białogard [*Ger.* Belgard]: Poland	6	54N	16E
Białowieża: Poland.	7	53N	24E
Białystok: Poland. *Pop. 1950: 65,700. Eng., textiles, chemicals, sawmilling.*	7	53N	23E
Bibeşti: Romania	11	45N	24E
Bibi-Eybat: Azerbaydzhan SSR. *Petroleum.*	†	40N	30E
Bicaz: Romania	11	47N	26E
Bicske: Hungary	10	47N	19E
Bielawa [*Ger.* Langenbielau]: Poland	6	51N	16E
Bielsk: Poland	7	53N	23E
Bielsko Biała: Poland. *Textiles, sawmilling.*	6	50N	19E
Bihać: Yugoslavia	10	45N	16E
Bihor, *mtns.*: Romania	11	47N	23E
Bijeljin: Yugoslavia	10	45N	19E
Bijelo Polje: Yugoslavia..	10	43N	20E
Bikin: RSFSR.	19	47N	134E
Bíle Karpaty, *mtn.*: Czech.	6	49N	18E
Bilgoraj: Poland	7	50N	23E
Bilimbay: RSFSR. *Iron & steel.*	13	57N	60E
Bilisht: Albania.	11	41N	21E
Binagady: Azerbaydzhan SSR. *Petroleum.*	†	41N	50E
Binz: E. Germany	6	54N	14E
Birakan: RSFSR. *Antimony, sawmilling.*	19	49N	132E
Birobidzhan: RSFSR. *Textiles.*	19	49N	133E
Birsk: RSFSR.	13	55N	56E
Birzhay: Lithuanian SSR.	‡5	56N	24E
Biryakovo: RSFSR.	8	60N	42E
Biskupiec [*Ger.* Bischofsburg]: Poland. *Elec. power.*	7	54N	21E
Bistrica: Yugoslavia	10	46N	14E
Bistriţa: Romania	11	47N	25E
Bitola [Monastir]: Yugoslavia	11	41N	21E
Bitterfeld: E. Germany. *Pop. 1946: 32,800. Lignite, elec. power, textiles*, steel, aluminium smelting, chemicals.*	6	52N	12E
Biysk: RSFSR. *Pop. 1954: 100,000. Textiles, eng.*	15	53N	85E
Bjelovar: Yugoslavia	10	46N	17E
Black Sea	11	—	—
Blagodarnoye: RSFSR.	9	45N	43E
Blagodatnoye: Kazakh, SSR.	15	51N	73E
Blagoevgrad[Gorna Dzhumaya]: Bulgaria	11	42N	23E
Blagoveshchensk: RSFSR.	13	55N	56E
Blagoveshchensk: RSFSR. *Pop. 1954: 100,000. Eng., sawmilling.*	19	50N	128E
Blaj: Romania	11	46N	24E
Blansko: Czechoslovakia. *Iron.*	6	49N	16E
Błaszki: Poland	6	52N	18E
Blatna: Czechoslovakia	6	49N	14E
Blatnitsa: Bulgaria	11	44N	28E
Błażowa: Poland	7	50N	22E
Błonie: Poland	7	52N	21E
Blyava: RSFSR. *Copper*, gold*, silver*. (See Mednogorsk.)*	13	52N	58E
Bobolice: Poland	6	54N	16E
Bobrinets: Ukrainian SSR.	9	48N	32E
Bobrka: Ukrainian SSR.	7	50N	24E
Bobrov: RSFSR.	9	51N	40E
Bobruysk: Byelorus. SSR. *Pop. 1954: 100,000. Eng., sawmilling.*	7	53N	29E
Bochnia: Poland	7	50N	20E
Bocşa-Montană: Romania	11	45N	22E
Bodaybo: RSFSR. *Gold, eng.*	19	58N	114E
Bogachevka: RSFSR. *Petroleum.*	†	75N	160E
Bogdanovich: RSFSR. *Coal*.*	13	57N	62E
Bogodukhov: Ukrainian SSR.	9	50N	35E
Bogoroditsk: RSFSR. *Lignite* (Shchekino Group), iron ore*..*	8	54N	38E
Bogorodsk: RSFSR.	8	56N	44E
Bogotol: RSFSR.	15	56N	90E
Boguchany: RSFSR.	19	58N	97E
Boguchar: RSFSR.	9	50N	40E
Bogurayev: RSFSR.	9	48N	41E
Boguslav: Ukrainian SSR.	9	50N	31E
Bokombayevskoye: Kirgiz SSR.	17	42N	77E
Bokovo-Antratsit: Ukrainian SSR. *Anthracite.*	9	48N	39E
Bokovskaya: RSFSR.	9	49N	42E
Boksitogorsk: RSFSR. *Bauxite.*	8	59N	34E
Bolesławiec [*Ger.* Bunzlau]: Poland. *Pop. 1939: 22,455; 1946: 3,145. Copper, anhydrite mined at nearby Wizow.*	6	51N	16E

	Page	Lat.	Long.
Bolekhov: Ukrainian SSR.	7	49N	24E
Bolgrad: Ukrainian SSR.	11	46N	29E
Bolitinu Din Vale: Romania	11	44N	26E
Bolkhov: RSFSR.	8	53N	36E
Bolnisi: Georgian SSR.	12	41N	45E
Bologoye: RSFSR.	8	58N	34E
Bolokhovo: RSFSR. *Lignite* (Shchekino Group)*	8	54N	38E
Bolotnoye: RSFSR.	15	56N	85E
Bol'shaya Aleksandrovka: Ukrainian SSR.	9	47N	33E
Bol'shaya Belozerka: Ukr. SSR.	9	47N	35E
Bol'shaya Gribanovka: RSFSR.	9	52N	42E
Bol'shaya Martynovka: RSFSR.	9	47N	42E
Bol'shaya Pisarevka: Ukr. SSR.	9	50N	35E
Bol'shaya Vishera: RSFSR.	8	59N	32E
Bol'shaya Viska: Ukr. SSR.	9	48N	32E
Bol'shevik I.: RSFSR.	19	79N	102E
Bol'shinka: RSFSR.	9	49N	41E
Bol'shoy Tokmak: Ukr. SSR.	9	47N	36E
Bomnak: RSFSR. *Gold.*	19	55N	129E
Bondari: RSFSR.	8	53N	42E
Bor: RSFSR.	8	56N	44E
Bor: Yugo. *Copper, copper-concentrating plant, smelting and refining, gold, silver.*	†	44N	22E
Borca: Romania	11	47N	26E
Borislav [*Pol.* Borysław]: Ukr. SSR. *Pop. 1954: 50,000. Petroleum*, natural gas*, petroleum refining* (West Ukrainian Field).*	7	49N	23E
Borisoglebsk: RSFSR. *Pop. 1954: 60,000.*	9	51N	42E
Borisov: Byelorussian SSR. *Chemicals.*	7	54N	28E
Borisovka: Kazakh SSR.	17	43N	68E
Borisovo-Sudskoye: RSFSR.	8	60N	36E
Borispol': Ukrainian SSR.	9	50N	31E
Borki: Ukrainian SSR.	9	50N	36E
Borogontsy: RSFSR.	19	63N	131E
Borovan: Bulgaria	11	43N	24E
Borovichi: RSFSR. *Textiles*, eng., paper.*	9	58N	34E
Borovsk: RSFSR.	8	55N	36E
Borovsk: RSFSR.	13	60N	57E
Borovskoye: Kazakh SSR.	13	53N	64E
Borsa: Romania	11	48N	25E
Borshchev: Ukrainian SSR.	7	49N	26E
Borzhomi: Georgian SSR.	12	42N	43E
Borzna: Ukrainian SSR.	9	51N	32E
Borzya: RSFSR. *Tin (see Olovyannaya).*	12	51N	116E
Bosanska Gradiška: Yugoslavia	10	45N	17E
Bosanski Brod: Yugoslavia. *Petroleum refining*.*	†	43N	18E
Bosanski Novi: Yugoslavia	10	45N	16E
Boshnyakovo: RSFSR. *Coal* (Uglegorsk area).*	†	49N	142E
Boskovice: Czechoslovakia	6	49N	16E
Bosovka: Ukrainian SSR.	9	51N	32E
Botevgrad: Bulgaria	11	43N	24E
Botoşani: Romania	11	48N	27E
Boyarka-Budayevka: Ukr. SSR.	9	50N	30E
Bozshchakul': Kazakh SSR. *Copper.*	15	53N	75E
Brac, i.: Yugoslavia	10	43N	16E
Brád: Romania	11	46N	23E
Brăila: Romania. *Pop. 1948: 95,514 incl. suburbs. Inland port. Steel, chemicals.*	11	45N	28E
Brandenburg: E. Ger. *Pop. 1946: 70,632. Elec. power, eng., textiles, chemicals*	6	52N	12E
Brandýs: Czechoslovakia	6	50N	15E
Braniewo [*Ger.* Braunsberg]: Poland. *Pop. 1939: 21,142; 1946: 1,373.*	7	45N	20E
Bransk: Poland	7	53N	E23
Braşov: Romania *Petroleum refining, eng*, textiles.*	11	46N	26E
Braslav: Byelorus. SSR.	‡5	55N	27E
Bratislava: Czech. *Pop. 1947: 172,664. River port. Textiles, chemicals, eng., petroleum refining.*	6	48N	17E
Bratsk: RSFSR. *River port Iron ore, sawmilling.*	19	56N	101E
Bratslav: Ukrainian SSR.	9	49N	29E
Brčko: Yugoslavia	10	45N	19E
Břeclav: Czechoslovakia	6	49N	17E
Bredy: RSFSR. *Anthracite.*	13	52N	60E
Bregovo: Bulgaria	11	44N	23E
Breslau *see* Wrocław			
Brest [Brest-Litovsk, *Pol.* Brześć-nad-Bugiem]: Byelorussian SSR. *Pop. 1954: 80,000, Textiles.*	7	52N	24E
Brest Oblast: Byelorus. SSR.	EP	52N	25E
Bretcu: Romania	11	46N	26E
Breytovo: RSFSR.	8	58N	38E
Brežica: Yugoslavia	10	46N	16E

* Located by symbol, but not named, on special topic map. ‡ See Inset map. **EP** Front Endpaper map.
† On relevant special topic map only. For list of topics see p. VIII. Place names in [square] brackets are former, or alternative, names.
↓ See appendix on p.134

	Page	Lat.	Long.
Březnice: Czechoslovakia	6	49N	14E
Breznik: Bulgaria	11	43N	23E
Brezno: Czechoslovakia	6	49N	20E
Brichany [Briceni]: Mold. SSR.	7	48N	27E
Brno: Czech. *Pop. 1947: 273,127. Eng., textiles.*	6	49N	17E
Brod: Yugoslavia	10	45N	18E
Brod: Yugoslavia	11	41N	21E
Brodarevo: Yugoslavia	10	43N	20E
Brodnica: Poland	6	53N	20E
Brodokalmak: RSFSR.	13	56N	62E
Brody: Ukrainian SSR.	7	50N	25E
Bromberg *see* Bydgoszcz			
Broşteni: Romania	11	45N	23E
Brovary: Ukr. SSR. *Chemicals.*	9	50N	31E
Brusartsi: Bulgaria	11	44N	23E
Bryansk: RSFSR. *Pop. 1954: 90,000. Elec. power*, saw-milling, eng., textiles, chemicals, phosphates.*	8	53N	34E
Bryansk Oblast: RSFSR.	EP	53N	34E
Bryanskiy: Ukr. SRR. *Coal* (Donbass Field).*	9	48N	39E
Bryukhovetskaya: RSFSR.	9	46N	39E
Brza Palanka: Yugoslavia	11	44N	22E
Brzeg [*Ger.* Brieg]: Poland. *Pop. 1939: 31,000; 1946: 8,000. Chemicals.*	6	51N	17E
Brześć Kujawski: Poland	6	53N	19E
Brześć-nad-Bugiem *see* Brest			
Brzeziny: Poland	6	52N	20E
Bucha: Ukr. SSR. *Chemicals.*	9	50N	30E
Buchach: Ukrainian SSR.	7	49N	26E
BUCHAREST [Bucureşti]: Romania. *Pop. 1952: 1,042,000, incl. suburbs. River port. Textiles, chemicals and pharmaceuticals, sawmilling, petroleum refining, engineering*.*	11	44N	26E
Buciumeni: Romania	11	45N	26E
Budafok: Hungary	10	47N	19E
BUDAPEST: Hungary. *Pop. 1949: 1,058,288 (1,600,000 incl. suburbs). River port. Steel, textiles, chemicals, eng. petroleum refining.*	10	47N	19E
Budennovka: Kazakh SSR.	13	51N	53E
Budennovka: Ukrainian SSR.	9	47N	38E
Budennyy: Kirgiz SSR.	17	43N	73E
Budennovsk: RSFSR. *Textiles.*	9	45N	44E
Budesti: Bulgaria	11	45N	24E
Budesti: Bulgaria	11	44N	26E
Budogoshch: RSFSR.	8	59N	32E
Budweis *see* České Budějovice			
Bug, R.: Ukrainian SSR.	22	48N	30E
Bugojno: Yugoslavia	10	44N	18E
Bugul'ma: RSFSR. *Petroleum.*	13	55N	53E
Buguruslan: RSFSR. *Petroleum.*	13	54N	52E
Buhuşi: Romania	11	47N	27E
Buinsk: RSFSR.	14	55N	48E
Bukachacha: RSFSR. *Coal.*	12	53N	116E
Bukhara: Uzbek SSR. *Pop. 1954: 60,000. Textiles.*	16	40N	65E
Bukhara Oblast: Uzbek SSR.	EP	41N	64E
Bukuka: RSFSR. *Tungsten.*	12	51N	117E
Bulun: RSFSR.	19	71N	127E
Bumbeşti Jiu: Romania	11	45N	23E
Bunjani: Yugo *Petroleum*.*	†	46N	16E
Buotama, R.: RSFSR. *Iron ore.*	†	61N	127E
Burdalyk: Turkmen SSR.	16	39N	65E
Bureya: RSFSR.	19	50N	130E
Bureya, R.: RSFSR. *Coal*, gold*, molybdenum, tungsten (see Umal'tinskiy).*	23	51N	132E
Burg: E. Germany	6	52N	12E
Burgas: Bulgaria, *Port. Copper, eng., chemicals, textiles.*	11	42N	27E
Burli: Kazakh SSR.	13	51N	53E
Burlyu-Tobe: Kazakh SSR.	17	47N	79E
Burrel: Albania	11	42N	20E
Burshtyn: Ukrainian SSR.	7	49N	25E
Buryat Mongol ASSR.: RSFSR.	EP	54N	110E
Busk: Ukrainian SSR.	7	50N	25E
Buturlinovka: RSFSR.	9	51N	41E
Butysh: RSFSR.	13	56N	54E
Buurdu: Kirgiz SSR. *Lead.*	†	43N	75E
Buy: RSFSR. *Chemicals, saw-milling.*	8	58N	42E
Buyaga: RSFSR.	19	60N	127E
Buyan: Ukrainian SSR.	9	50N	30E
Buynaksk: RSFSR.	12	43N	47E
Buzachi Penin: USSR.	16	45N	53E
Buzău: Romania	11	45N	27E
Buzet: Yugoslavia	10	46N	14E
Buziaş: Romania	11	46N	22E
Buzovny: Azerbaydzhan SSR. *Petroleum.*	†	41N	49E
Buzuluk: RSFSR.	14	53N	52E
Byala: Bulgaria	11	43N	28E
Byandovan: Azer. SSR.	12	40N	49E

	Page	Lat.	Long.
Bydgoszcz [*Ger.* Bromberg]: Poland. *Pop. 1950: 159,800. River port. Eng., textiles, chemicals, lignite*.*	6	53N	18E
Byelorussian: SSR.	EP	54N	37E
Bykhov: Byelorussian SSR. *Chemicals.*	7	54N	30E
Bykovo: RSFSR.	9	50N	45E
Byrranga Mts.: RSFSR.	19	75N	100E
Bytom [*Ger.* Beuthen]: Poland. *Pop. 1950: 121,000. Coal, lead, zinc, eng.*	6	50N	19E
Bytosh': RSFSR.	8	54N	34E
Bytów [*Ger.* Bütow]: Poland.	6	54N	17E
Čačak: Yugoslavia	11	44N	20E
Čadca: Czechoslovakia	6	49N	19E
Cakovec: Yugoslavia	10	46N	16E
Calafat: Romania	11	44N	23E
Călăraşi: Romania	11	44N	27E
Calbe: E. Germany. *Lignite, chemicals, textiles*, iron.*	6	52N	12E
Călimăneşti: Romania	11	45N	24E
Călineşti: Romania	11	45N	24E
Câmpeni: Romania	11	46N	23E
Câmpina: Romania. *Elec. power, iron, petroleum refining, chemicals.*	11	45N	25E
Câmpulung: Romania	11	45N	25E
Câmpulung Moldava: Romania	11	48N	26E
Caracal: Romania	11	44N	24E
Caransebeş: Romania	11	45N	22E
Caraomer: Romania	11	44N	28E
Carei: Romania	11	48N	22E
Carpathian Mtns.: USSR	7	48N	24E
Čáslav: Czechoslovakia	6	50N	15E
Caspian Sea	16	—	—
Caucasus Mtns. *see* Greater, and Lesser Caucasus			
Cazin: Yugoslavia	10	45N	16E
Cegléd: Hungary	10	47N	20E
Celje: Yugoslavia. *Lignite, zinc smelter.*	10	46N	15E
Celldömölk: Hungary	10	47N	17E
Central Russian Uplands: RSFSR.	9	52N	36E
Central Siberian Plateau: RSFSR.	19	66N	107E
Cerknica: Yugoslavia	10	46N	14E
Cernavoda: Romania	11	44N	28E
Cerrik: Albania. *Petroleum* (see Stalin [Kuçove])*	†	41N	20E
Česká Třebová: Czechoslovakia. *Engineering, textiles*.*	6	50N	16E
Česká Lípa: Czechoslovakia	6	51N	14E
České Budějovice [*Ger.* Budweis]: Czechoslovakia. *Pop. 1947: 38,194. Anthracite.*	6	49N	14E
Český Krumlov: Czech. *Saw-milling.*	†	48N	14E
Český Těšín *see* Cieszyn			
Cetatea: Romania	11	44N	23E
Cetinje: Yugoslavia	10	42N	19E
Chaadeyevka: RSFSR.	8	53N	46E
Chadan: RSFSR. *Copper, asbestos, coal*.*	19	51N	91E
Chagan-Uzan: RSFSR. *Mercury*	†	50N	88E
Chagoda [Belyy Bychev]: RSFSR. *Elec. power*.*	8	59N	35E
Changyrtash: Kirgiz SSR. *Petroleum.*	17	41N	73E
Chany: RSFSR.	15	55N	77E
Chany, L.: RSFSR.	15	55N	78E
Chapayevsk [Ivashchenkovo]: RSFSR. *Pop. 1954: 80,000. Chemicals.*	14	53N	50E
Chaplinka: Ukrainian SSR.	9	46N	34E
Chaplino: Ukrainian SSR.	9	48N	36E
Chaplygin: RSFSR.	8	53N	40E
Chapoma: RSFSR.	18	66N	39E
Chardzhou [Leninsk-Turkmen-skiy]: Turkmen SSR. *Pop. 1954: 70,000. Textiles, chemicals.*	16	39N	63E
Chardzhou Oblast: Turkmen SSR.	EP	39N	62E
Charsk: Kazakh SSR.	17	50N	81E
Chartoriysk: Ukrainian SSR.	7	51N	26E
Charyshskoye: RSFSR.	15	51N	84E
Chasel'ka: RSFSR.	18	65N	81E
Chashniki: Byelorussian SSR.	8	55N	29E
Chasov Yar: Ukrainian SSR.	9	48N	38E
Chausy: Byelorussian SSR.	8	54N	31E
Chauvay: Kirgiz SSR. *Mercury, antimony.*	‡17	40N	72E
Cheb [*Ger.* Eger]: Czech. *Eng., textiles.*	6	50N	12E
Chebarkul': RSFSR.	13	54N	60E
Cheboksary: RSFSR. *Textiles, hydro-elec*.*	14	56N	47E

	Page	Lat.	Long.
Chekalin: RSFSR.	8	54N	36E
Chekhov [*Jap.* Noda]: RSFSR. *Petroleum, coal* (Sinegorsk Field).*	19	47N	142E
Chekmagush: RSFSR.	13	55N	55E
Chekunda: RSFSR. *Coal.*	19	51N	132E
Cheleken: Turkmen SSR. *Petroleum, chemicals.*	16	40N	53E
Chelkar: Kazakh SSR.	14	50N	52E
Chelkar: Kazakh SSR. *Chemicals.*	16	48N	60E
Chełm: Poland	7	51N	23E
Chełmno [*Ger.* Kulm]: Poland. *Engineering.*	6	53N	18E
Chełmża [*Ger.* Kulmsee]: Poland	6	53N	19E
Chelyabinsk: RSFSR. *Pop. 1954: 712,000. Elec. power, iron & steel, ferro-alloys, non ferrous metals refining, eng., chemicals, petroleum refining, textiles.*	13	55N	61E
Chelyabinsk Oblast: RSFSR.	EP	55N	60E
Chelyuskin, C.: RSFSR.	19	77N	95E
Chemdal'sk: RSFSR.	19	60N	103E
Chemnitz: E. Germany. *Pop. 1946: 290,200. Coal, lignite, textiles, eng.*, chemicals.*	6	51N	13E
Cherdyn': RSFSR.	13	60N	56E
Cheremkhovo: RSFSR. *Pop. 1954: 100,000. Coal, eng., chemicals.*	12	53N	104E
Cheremukhovo: RSFSR. *Bauxite*.*	†	60N	60E
Cherepanovo: RSFSR.	15	54N	83E
Cherepovets: RSFSR. *Pop. 1954: 50,000. Steel, eng., sawmills.*	8	59N	38E
Cherkassy: Ukrainian SSR. *Pop.1954: 53,000.*	9	49N	32E
Cherkassy Oblast: Ukr. SSR.	EP	49N	32E
Cherkess AO.: Stavropol' Kray, RSFSR.	EP	44N	42E
Cherkessk: RSFSR. *Chemicals, eng.*	9	44N	42E
Cherlak: RSFSR.	15	54N	75E
Chermoz: RSFSR. *Steel.*	13	59N	56E
Chern': RSFSR.	8	53N	37E
Chernaya: RSFSR.	18	70N	89E
Chernevo: RSFSR.	8	59N	28E
Chernigov: Ukr. SSR. *Pop. 1954: 60,000. Textiles, chemicals.*	9	51N	31E
Chernigov Oblast: Ukr. SSR.	EP	51N	34E
Chernikovsk: RSFSR. *Pop. 1954: 225,000. Eng., chemicals, petroleum refining, saw-milling.*	13	55N	56E
Chernobyl': Ukrainian SSR.	9	51N	30E
Chernogorsk: RSFSR. *Coal.*	15	54N	91E
Chernomorskoye: Ukr. SSR.	9	46N	33E
Chernovtsy [*Rom.* Cernăuţi]: Ukrainian SSR. *Pop. 1954: 200,000. Textiles, eng.*, chemicals, sawmills.*	7	48N	26E
Chernovtsy Oblast: Ukr. SSR.	EP	48N	27E
Chernushka: RSFSR.	13	57N	56E
Chernyakhov: Ukrainian SSR.	7	50N	29E
Chernyakhovsk [*Ger.* Insterburg]: Lith. SSR. *Chemicals, textiles*.*	7	55N	22E
Chernyanka: RSFSR.	9	51N	38E
Cherny Mys: RSFSR.	15	56N	80E
Chernyshevskaya: RSFSR.	9	49N	42E
Chernyshkovskiy: RSFSR.	9	48N	42E
Chernyy Ostrov: Ukr. SSR.	7	50N	27E
Chernyy Yar: RSFSR.	16	48N	46E
Cherskiy Range: RSFSR.	19	65N	145E
Chertkovo: RSFSR.	9	49N	40E
Cherven': Byelorussian SSR.	7	54N	28E
Chervyanka: RSFSR.	19	58N	100E
Chesnokovka: RSFSR. *Eng.*	15	53N	84E
Chiatura: Georgian SSR. *Manganese.*	12	42N	43E
Chiili: Kazakh SSR.	17	44N	67E
Chikachevo: RSFSR.	8	57N	30E
Chikin-Sala: RSFSR.	9	46N	44E
Chikishlyar: Turkmen SSR.	16	38N	54E
Chilia Veche: Romania	11	45N	29E
Chilik: Kazakh SSR.	17	44N	78E
Chimbay: Uzbek SSR.	16	43N	60E
Chimion: Uzbek SSR.	‡17	40N	72E
Chimkent: Kazakh SSR. *Pop. 1954: 120,000. Chemicals, eng., textiles, lead smelting.*	17	42N	70E
Chinaz: Uzbek SSR.	17	41N	69E
Chiragidzor: Azer. SSR.	12	40N	46E
Chirchik: Uzbek SSR. *Pop. 1939: 10,000; 1954: 70,000. Chemicals, hydro-elec*.*	17	41N	70E
Chirpan: Bulgaria	11	42N	25E

---

* Located by symbol, but not named, on special topic map.   ‡ See Inset map.   EP Front Endpaper map.
† On relevant special topic map only. For list of topics see p. VIII.   Place names in [square] brackets are former, or alternative, names.
⌐ See appendix on p.134

	Page	Lat.	Long.
Chirinda: RSFSR	19	67N	100E
Chishmy: RSFSR	13	55N	55E
Chişineu-Criş: Romania	11	47N	22E
Chistopol: RSFSR	14	55N	50E
ⵏChistyakovo: Ukr. SSR. *Pop. 1954: 100,000. Coal.*	9	48N	38E
Chita: RSFSR. *Pop. 1954: 300,000. Coal, eng., chemicals, sawmilling.*	12	52N	114E
Chita Oblast: RSFSR	EP	54N	115E
ⵏChkalov ⵏ: RSFSR. *Pop. 1954: 225,000. Eng., textiles.*	13	52N	55E
ⵏChkalov Oblast: RSFSR	EP	52N	55E
Chkalovsk: RSFSR	8	57N	43E
Choceň: Czechoslovakia	6	50N	16E
Chodzież: Poland	6	53N	17E
Choibalsan [Bayan Tumen]: Mongolian People's Republic	19	48N	115E
Chojna: Poland	6	53N	14E
Chojnice [*Ger.* Könitz]: Poland	6	54N	18E
Chojnów [*Ger.* Haynau]: Poland. *Pop. 1939: 11,114; 1946: 5,467.*	6	51N	16E
Chokhatauri: Georgian SSR	12	42N	42E
Chomutov: Czechoslovakia. *Coal, eng.*	6	50N	13E
Chop [*Czech.* Čop, *Hung.* Csap]: Ukrainian SSR	7	48N	22E
Chortkov: Ukrainian SSR	7	49N	26E
Chorukh-Dayron: Tadzhik SSR. *Tungsten.*	17	40N	70E
Chorzów [Królewska Huta, *Ger.* Königshütte]: Poland. *Pop. 1950: 142,000. Coal, iron & steel, chemicals*, eng.*	6	50N	19E
Choszczno [*Ger.* Arnswalde]: Poland. *Pop. 1939: 13,960; 1946: 2,052.*	6	53N	16E
Chrudim: Czechoslovakia	6	50N	16E
Chrzanów: Poland	6	50N	20E
Chu: Kazakh SSR	17	44N	74E
Chu, R.: USSR	22	45N	73E
Chud (Peipus), L.	‡5	59N	27E
Chudnov: Ukrainian SSR	7	50N	28E
Chudovo: RSFSR	8	59N	32E
Chugunash: RSFSR	15	53N	88E
Chugush Mt.: RSFSR	12	42N	40E
Chuguyev: Ukrainian SSR	9	50N	37E
Chukchi Sea: RSFSR	19	70N	175E
Chukchi NO.: RSFSR	EP	67N	170E
Chukhloma: RSFSR	8	59N	42E
Chulak-Tau: Kazakh SSR. *see* Kara-Tau.			
Chul'man: RSFSR	19	57N	125E
Chulym: RSFSR	15	55N	81E
Chuna, R.: RSFSR	23	58N	98E
Chupa: Karelo-Finnish SSR. *Mica.*	†	67N	33E
Chusovoy: RSFSR. *Pop. 1954: 100,000. Iron & steel.*	13	58N	58E
Chust: Uzbek SSR	‡17	41N	71E
Chuvash ASSR.: RSFSR	EP	55N	47E
Ciechanów: Poland	7	53N	21E
Ciechocinek: Poland	6	53N	19E
Cieszyn [*Ger.* Teschen, *Czech.* Český Těšín]: Poland. *Engineering. (Town divided between Poland and Czechoslovakia.)*	6	50N	19E
Cilibia: Romania	11	45N	27E
Cioara: Romania	11	44N	28E
Clejani: Romania	11	44N	26E
Cluj: Romania. *Pop. 1948: 117,915. Textiles, uranium, eng.**	11	47N	24E
Cobadin: Romania	11	44N	28E
Codlea: Romania	11	46N	25E
Cogealac: Romania	11	45N	29E
Comana: Romania	11	44N	26E
Constanţa: Romania. *Pop. 1948: 78,586. Chief Romanian Black Sea port.*	11	44N	29E
Corabia: Romania	11	44N	24E
Corovodë: Albania	11	41N	20E
Coşteşi: Romania	11	45N	25E
Cószeg: Hungary	10	47N	16E
Cottbus: E. Ger. *Pop. 1946: 60,100. Textiles.*	6	52N	14E
Craiova: Romania. *Pop. 1948: 84,574, incl. suburbs. Textiles*, chemicals, eng.**	11	44N	24E
Cres, i.: Yugoslavia	10	45N	14E
Crimea Oblast: Ukrainian SSR	EP	45N	34E
Crimmitschau: E. Ger. *Textiles* (in Central German textiles group)*	†	51N	12E
Črnomelj: Yugoslavia	10	45N	15E
Csepel: Hungary. *Petroleum refining, aluminium.*	†	47N	14E
Csepreg: Hungary	10	47N	17E
Csongrád: Hungary	11	47N	20E
Csorna: Hungary	10	48N	17E

	Page	Lat.	Long.
Csurgó: Hungary	10	46N	17E
Cugir: Romania	11	46N	23E
Cuprija: Yugoslavia	11	44N	21E
Curteade Arges: Romania	11	45N	25E
Cuzgun: Romania	11	44N	28E
Czarnków: Poland	6	53N	17E
Czempin: Poland	6	52N	17E
Czeremcha: Poland	7	53N	23E
Czersk: Poland	6	54N	18E
Częstochowa: Poland. *Pop. 1950: 118,900. Iron & steel, textiles.*	6	51N	19E
Człuchów [*Ger.* Schlochau]: Poland. *Pop. 1939: 6,029; 1946: 3,711.*	6	54N	17E
Czorsztyn: Poland	7	49N	20E
Dąbie [*Ger.* Altdamm]: Poland	6	53N	15E
Dąbrowa Górnicza: Poland. *Coal, zinc, iron ore.*	7	50N	21E
Dagestan ASSR.: RSFSR	EP	43N	47E
Dagestanskiye Ogni: RSFSR	12	42N	48E
Dagomys: RSFSR	12	44N	40E
Dakovica: Yugoslavia	11	42N	20E
Dakovo: Yugoslavia	10	45N	18E
Dalmatovo: RSFSR	13	56N	63E
Danilov: RSFSR	8	58N	40E
Danilov Grad: Yugoslavia	10	42N	19E
Danilovka: Kazakh SSR. *Gold* (Stepnyak Group).*	15	53N	70E
Dankov: RSFSR	8	53N	39E
Danube, R.	22	44N	27E
Danzig *see* Gdańsk.			
Darabani: Moldavian SSR	7	48N	26E
Darasun: RSFSR. *Gold, arsenic.*	12	52N	116E
Darganata: Turkmen SSR	16	40N	62E
Darłowo [*Ger.* Rügenwalde]: Poland	6	54N	16E
Darnitsa: Ukrainian SSR	9	50N	31E
Daruvar: Yugoslavia	10	46N	17E
Darvaza: Turkmen SSR. *Sulphur, chemicals.*	16	40N	58E
Dashava: Ukr. SSR. *Natural gas.*	7	49N	24E
Dashkesan: Azer. SSR. *Iron ore, cobalt.*	12	41N	46E
Daugavpils [*Rus.* Dvinsk]: Latvian SSR. *Pop. 1954: 60,000. Textiles, eng.*	‡5	56N	26E
David-Gorodok: Byelorussian SSR.	7	52N	27E
Davydovka: RSFSR	8	51N	40E
Debal'tsevo: Ukrainian SSR	9	48N	38E
Debar: Yugoslavia	11	42N	20E
Dębica: Poland	7	50N	21E
Dęblin: Poland	7	51N	22E
Dębno [*Ger.* Neudamm]: Poland	6	53N	15E
Debrecen: Hungary. *Pop. 1941: 125,936. Eng.*, chemicals.*	11	47N	22E
Děčín [*Ger.* Tetschen]: Czech. *Chemicals.*	6	51N	14E
Deda: Romania	11	47N	25E
Dedovichi: RSFSR	8	58N	30E
Dedovsk: RSFSR	8	56N	37E
Degtyarsk: RSFSR. *Copper, pyrites.*	13	57N	60E
Dej: Hungary. *Petroleum refining.*	11	47N	24E
Dekhkanabad: Uzbek SSR	17	38N	67E
Delčevo: Yugoslavia	11	42N	23E
Delitzsch: E. Germany	6	51N	13E
Delnice: Yugoslavia	10	45N	15E
Delvinë: Albania	11	40N	20E
Delyatin: Ukrainian SSR	7	48N	25E
Demidov: RSFSR	8	55N	32E
Demmin: E. Ger. *Engineering.*	6	54N	13E
Demyansk: RSFSR	8	58N	32E
Dem'yanskoye: RSFSR	14	60N	70E
Denau: Uzbek SSR	16	38N	68E
Derazhnya: Ukrainian SSR	7	49N	28E
Derbent: RSFSR. *Petroleum* (Izberbash Group), chemicals, textiles.*	12	42N	48E
Derbeshinskiy: RSFSR	13	56N	54E
Derecske: Hungary	11	47N	22E
Dergachi: RSFSR	14	51N	49E
Dergachi: Ukrainian SSR	9	50N	36E
Derventa: Yugoslavia	10	45N	18E
Derzhavinskoye: Kazakh SSR.	14	51N	66E
Dessau: E. Ger. *Pop. 1946: 88,300. Engineering.*	6	52N	12E
Deta: Romania	11	45N	21E
Detchino: RSFSR	8	55N	36E
Deva: Romania	11	46N	23E
Dévaványa: Hungary	11	47N	21E
Devin: Bulgaria	11	42N	24E
Deynau: Turkmen SSR	16	39N	63E
Dikson: RSFSR. *Port.*	18	73N	81E

	Page	Lat.	Long.
Dimitrovgrad [Kamenets]: Bulgaria. *Coal, elec. power, chemicals.*	11	42N	26E
Dimitrovgrad [Tsaribrod]: Yugoslavia. *Coal.*	11	43N	23E
Dimitrovo [Pernik]: Bulgaria. *Coal, iron & steel.*	11	43N	23E
Dinaric Alps: Yugoslavia	10	44N	17E
Disna: Byelorussian SSR.	8	56N	28E
Divnoye: RSFSR	9	46N	43E
Dmitriyev L'govskiy: RSFSR. *Chemicals.*	9	52N	35E
Dmitrov: RSFSR. *Engineering.*	8	56N	38E
Dmitrovka: Ukrainian SSR	9	47N	36E
Dmitrovka-Orlovskiy: RSFSR.	9	52N	35E
Dneprodzerzhinsk: Ukrainian SSR. *Pop. 1954: 75,000. Iron & steel, eng.*, chemicals.*	9	48N	35E
Dneproges: Ukrainian SSR. *Hydro-elec. power (largest installation in Europe until 1954). Suburb of Zaporozh'ye.*	9	48N	35E
Dnepropetrovsk: Ukrainian SSR. *Pop. 1954: 525,000. River port. Iron & steel, eng., chemicals, sawmilling, elec. power*	9	48N	35E
Dnepropetrovsk Oblast: Ukr. SSR.	EP	48N	35E
Dnieper, R.: Ukrainian SSR.	22	49N	33E
Dniester, R.: Ukrainian SSR.	22	49N	26E
Dno: RSFSR.	8	58N	30E
Dobele: Latvian SSR.	5	57N	23E
Döbeln: E. Ger. *Engineering.*	6	51N	13E
Doboj: Yugoslavia	10	45N	18E
Dobřany: Czech. *Coal*.*	6	50N	13E
Dobre Miasto: Poland	7	54N	20E
Dobri Voinikovo: Bulgaria	11	43N	27E
Dobrljin: Yugo. *Sawmilling.*	†	49N	17E
Dobromil': Ukrainian SSR	7	50N	23E
Dobrovol'sk [*Ger.* Pillkallen]: RSFSR.	7	54N	22E
Dobrush: Byelorussian SSR.	9	52N	31E
Dobryanka: RSFSR. *Steel.*	†	59N	56E
Dobryanka: Ukrainian SSR.	9	52N	31E
Dobrzyń: Poland	6	53N	19E
Dobšiná: Czech. *Asbestos, iron ore, copper.*	7	49N	20E
Dokshitsy: Byelorussian SSR.	8	55N	28E
Dokshukino: RSFSR.	12	43N	44E
Dolgaya: RSFSR.	9	52N	38E
Dolgorukovo: RSFSR.	9	52N	38E
Dolina: Ukr. SSR. *Petroleum refining*, chemicals*, natural gas*.*	7	49N	24E
Dolinsk [*Jap.* Ochiai]: RSFSR. *Paper.*	†	47N	143E
Dolinskaya: Ukrainian SSR.	9	48N	33E
Dolni-Chiflik: Bulgaria	11	43N	28E
Dolný Kubín: Czechoslovakia	6	49N	19E
Domanevka: Ukrainian SSR.	9	47N	31E
Domažlice: Czech. *Engineering.*	6	49N	13E
Dombarovskiy: RSFSR. *Coal.*	13	51N	60E
Dombóvár: Hungary	10	46N	18E
Dömitz: E. Ger. *Chemicals.*	6	53N	11E
Domodedovo: RSFSR.	8	56N	38E
Don, R.: USSR.	22	50N	41E
Donets, R.: USSR.	22	49N	38E
Donji Milanovač	11	44N	22E
Donji Vakuf: Yugoslavia	10	44N	18E
Donskoy: RSFSR. *Lignite* (Uzlovaya Group, Moscow Lignite Basin).*	†	54N	38E
Donskoye: Kazakh SSR.	14	50N	58E
Dorogobuzh: RSFSR. *Lignite*.*	8	55N	33E
Dorohoi: Romania.	11	48N	26E
Dorohusk: Ukrainian SSR.	7	51N	24E
Dosatuy: RSFSR.	12	51N	119E
Doschatoye: RSFSR.	8	56N	42E
Dossor: Kazakh SSR. *Petroleum.*	16	47N	53E
Drăgăşani: Romania	11	45N	24E
Drăgăneşti: Romania	11	44N	26E
Dravograd: Yugo. *Hydroelec*., petro eum refining.*	10	47N	15E
Drawsko [*Ger.* Dramburg]: Poland	6	54N	16E
Dresden: E. Ger. *Pop. 1946: 467,966. River port. Coal, chemicals, eng. (incl. optical & precision instr.). (Dresden china made at Meissen, 14 miles N.).*	6	51N	14E
ⵏDrissa: Byelorussian SSR.	8	56N	28E
Driniš: Yugoslavia	10	44N	16E
Drogichin: Byelorussian SSR.	7	52N	25E
Drogobych [*Pol.* Drohobycz]: Ukr. SSR. *Pop. 1954: 50,000. Petroleum*, petroleum refining*.*	7	49N	24E
ⵏDrogobych Oblast: Ukr. SSR.	EP	49N	23E
Druya: Byelorussian SSR.	8	56N	28E

* Located by symbol, but not named, on special topic map.
† On relevant special topic map only. For list of topics see p. VIII.
ⵏ See appendix on p. 134
‡ See Inset map.   EP Front Endpaper map.
Place names in [square] brackets are former, or alternative, names.

114

	Page	Lat.	Long.
Druzhkovka: Ukrainian SSR. Iron & steel* (Donbass)	9	48N	38E
Druzhnaya Gorka: RSFSR.	8	59N	30E
Dryanovo: Bulgaria	11	43N	26E
Dubna: RSFSR.	8	54N	37E
Dubno: Ukrainian SSR.	7	50N	26E
Dubossary: Moldavian SSR.	11	47N	29E
Dubovka: RSFSR..	9	49N	45E
Dubovka: Moscow, RSFSR. Lignite*. (Uzlovaya Group)..	†	54N	38E
Dubovskoye: RSFSR.	9	47N	43E
Dubrovitsa: Ukrainian SSR.	7	52N	27E
Dubrovka: RSFSR.	8	60N	31E
Dubrovo: RSFSR.	9	52N	38E
Dubrovnik [Ragusa]: Yugo. Port..	10	43N	18E
Dubrovno: Byelorussian SSR..	8	55N	30E
Duderstadt: E. Germany	6	52N	10E
Dudinka: RSFSR. Port for Noril'sk.	18	70N	86E
Dudorovskiy: RSFSR.	8	54N	35E
Due see Aleksandrovsk-Sakhalinskiy.			
Dugi Otok, i.: Yugoslavia	10	44N	15E
Dugna: RSFSR.	8	54N	37E
Dugo Selo: Yugo. Petroleum* (Sava River Field)	10	46N	16E
Dukla: Poland	7	50N	22E
Dŭlgopol: Bulgaria	11	43N	27E
Dulovo: Bulgaria	11	44N	27E
Dumbrăveni: Romania	11	46N	24E
Dunaföldvár: Hungary	10	47N	19E
Dunayevtsy: Ukrainian SSR.	7	49N	27E
Durbe: Latvian SSR.	‡5	57N	21E
Durge Nor, lake: M.P.R..	19	47N	93E
Durmitor, mtn.: Yugoslavia	10	43N	19E
Durdevac: Yugoslavia	10	46N	17E
Durrës [It. Durazzo]: Albania. Chief Albanian Port.	10	41N	19E
Duszniki Zdrój [Ger. Bad Reinerz]: Poland	6	50N	16E
Dve-mogili: Bulgaria	11	44N	26E
Dvůr Králové: Czech. Textiles..	6	50N	16E
D'yakovskaya: RSFSR.	8	60N	41E
Dyat'kovo: RSFSR.	8	54N	34E
Dyatlovo: Byelorussian SSR.	7	53N	25E
Dykh-Tau, mtn.: RSFSR.	12	43N	43E
Dymer: Ukrainian SSR.	9	51N	30E
Diósgyőr: Hungary	11	48N	21E
Dyubit: RSFSR.	8	58N	41E
Dzagidzor: Armenian SSR.	12	41N	45E
Dzamun Üude: Mongolian People's Republic	19	44N	111E
Dzan-Bulak: Mongolian People's Republic. Coal.	†	47N	116E
Dzarichi: Byelorussian SSR.	7	52N	29E
Dzaudzhikau see Ordzhonikidze.			
Dzerzhinsk [Kaidanovo]: Byelorussian SSR. Pop. 1954: 160,000.		54N	27E
Dzerzhinsk [Rastyapino]: RSFSR. Pop. 1954: 220,000. Chemicals..	8	56N	44E
Dzerzhinsk: Ukr. SSR. Coal* (Donbass).	9	48N	38E
Dzhalal-Abad: Kirgiz SSR. Engineering.	17	41N	73E
ᴸ Dzhalal-Abad Oblast: Kirgiz SSR.	EP	42N	73E
Dzhambul [Aulie-Ata]: Kazakh SSR. Pop. 1954: 120,000. Chemicals.	17	43N	72E
Dzhambul Oblast: Kazakh SSR.	EP	45N	72E
Dzhankoy: Ukrainian SSR.	9	46N	34E
Dzhardzhan: RSFSR.	19	69N	124E
Dzhebel: Turkmen SSR.	16	40N	55E
Dzhekonda: RSFSR. Gold* (Aldan Field).	19	58N	125E
Dzhetygara: Kazakh SSR. Gold, arsenic.	13	52N	61E
Dzhezdy: Kaz. SSR. Manganese.	17	48N	67E
Dzhezkazgan: Kazakh SSR. Copper, copper smelting and refining at Bol'shoy Dzhezkazgan.	17	48N	68E
Dzhirgalan: Kirgiz SSR. Lignite.	17	43N	79E
Dzhirgalantu see Jirgalantu			
Dzhirgatal': Tadzhik SSR.	17	39N	71E
Dzhizak: Uzbek SSR.	17	40N	68E
Dzhul'fa: Azerbaydzhan SSR.	12	39N	46E
Dzhusaly: Kazakh SSR. Engineering.	16	45N	65E
Działdowo [Ger. Soldau]: Poland	7	53N	20E
Dzierżoniów [Ger. Reichenbach]: Poland	6	51N	17E
Dzun Bulak: Mongolian People's Republic	19	47N	115E
East Kazakhstan Oblast: Kazakh SSR.	EP	49N	84E

	Page	Lat.	Long.
Eberswalde: E. Germany. Pop. 1946: 30,200. Chemicals, steel.	6	53N	14E
Echmiadzin: Armenian SSR.	12	40N	44E
Ege Khaya: RSFSR. Tin, tungsten.	19	67N	135E
Eger: Hungary	11	48N	20E
Eilenburg: E. Germany. Chemicals, eng.*	6	51N	12E
Eisenach: E. Germany. Pop. 1946: 51,834. Engineering*..	6	51N	10E
Eisleben: E. Germany. Copper smelting.	6	52N	12E
Ekhabi: RSFSR. Petroleum.	†	55N	127E
Ekibastuz : Kazakh SSR. Coal.	15	52N	75E
Ekimchan: RSFSR. Gold*.	19	53N	133E
Elbasan: Albania. Pop. 1945: 14,918. Chromium, iron, sawmilling.	11	41N	20E
Elbe, R.: Germany	22	53N	11E
Elbląg [Ger. Elbing]: Poland. Pop. 1939: 85,952; 1946: 20,924. Port. Eng., elec. power*.	7	54N	19E
El'brus, mtn.: Georgian SSR.	12	43N	42E
El'dikan: RSFSR. Gold*.	19	62N	135E
Elektrogorsk: RSFSR. Elec. power.	†	56N	38E
Elektrostal': RSFSR. Pop. 1954: 50,000. Steel, eng.	8	56N	38E
Elena: Bulgaria	11	43N	26E
El'gen-Ugol': RSFSR.	19	63N	152E
El'gyay: RSFSR.	19	62N	117E
Elk: Poland. Sawmilling.	7	54N	22E
Elkhovo: Bulgaria	11	42N	26E
El'ton: RSFSR. Salt.	16	49N	47E
El'va: Estonian SSR.	‡5	58N	26E
Emba: Kazakh SSR. Petroleum.	16	49N	58E
Emba, R.: Kazakh SSR. Petroleum.	22	48N	56E
Emine, C.: Bulgaria	11	43N	28E
Emmaste: Estonian SSR.	‡5	59N	22E
Endybal'sk: RSFSR. Lead-zinc, molybdenum.	19	65N	130E
Engel's [Pokrovsk]: RSFSR. Pop. 1954: 80,000. Former cap. Volga German ASSR. Chemicals, petroleum refinery*.	16	51N	46E
Enurmino: RSFSR.	19	67N	173W
Eplény: Hungary. Bauxite, manganese.	†	47N	18E
Erd: Hungary	10	47N	19E
Erentsab: Mongolian People's Republic.	19	50N	115E
Erfurt: E. Germany. Pop. 1946: 190,500. Eng., textiles.	6	51N	11E
Erseke: Albania.	11	40N	21E
Ertil': RSFSR.	9	52N	41E
Estonian SSR.	EP	58N	25E
Esztergom: Hungary	10	48N	19E
Etropole: Bulgaria	11	43N	24E
Evenki NO.: Krasnoyarsk Kray, RSFSR.	EP	65N	100E
Eyshishkes [Eišiškes]: Lith. SSR.	7	54N	25E
Fadd: Hungary	10	46N	19E
Faddey I.: RSFSR.	19	76N	144E
Făgăras: Romania	11	46N	25E
Făget: Romania	11	46N	22E
Fakiya: Bulgaria	11	42N	27E
Fălciu: Romania	11	46N	28E
Faleshty [Rom. Fălešti]: Moldavian SSR.	11	48N	28E
Fălticeni: Romania	11	47N	26E
Fastov: Ukrainian SSR.	9	50N	30E
Fatezh: RSFSR.	9	52N	36E
Fayansovyy: RSFSR.	8	54N	34E
Felsőgalla: Hungary. Lignite, aluminium works, chemicals..	10	48N	18E
Feodosiya: Ukr. SSR. Port.	9	45N	35E
Fergana [Skobelev]: Uzbek SSR. Hydro-elec., petroleum refining, textiles.	‡17	40N	72E
Fergana Oblast: Uzbek SSR.	EP	41N	73E
Feteşti: Romania	11	44N	28E
Fier: Albania	10	41N	20E
Fil'akovo: Czech. Coal, lignite.	6	48N	20E
Filiaşi: Romania	11	44N	24E
Finsterwalde: E. Ger. Lignite*, chemicals.	6	52N	14E
Fiume see Rijeka			
Floreshty: Moldavian SSR.	11	48N	28E
Foča: Yugoslavia	10	43N	19E
Focşani: Romania. Chemicals..	11	46N	27E
Fojnica: Yugoslavia	10	44N	18E
Folteşti: Romania	11	46N	28E
Fonyód: Hungary	10	47N	18E
Forró: Hungary	7	48N	21E
Forst: E. Germany. Chemicals, elec. power.	6	52N	14E

	Page	Lat.	Long.
Fort Shevchenko: Kazakh SSR.	12	45N	50E
Fosforitnyy: RSFSR. Phosphates.	8	55N	39E
Frankfurt: E. Germany. Pop. 1946: 51,577. River port. Engineering.	6	52N	14E
Franz Joseph Land: RSFSR.	18	80N	55E
Freiberg: E. Germany. Lead-zinc, tin mining & smelting.	6	51N	13E
Freital: E. Germany. Steel (at Döhlen).	6	51N	14E
Frolovo: RSFSR. Natural gas..	9	50N	44E
Frombork [Ger. Frauenburg]: Poland	7	54N	20E
Frumusica: Romania.	11	48N	27E
Frunze [Kadamdzhay]: Kirgiz SSR. Antimony..	‡17	40N	72E
FRUNZE [Pishpek]: Kirgiz SSR. Pop. 1954: 155,000. Eng. textiles, elec. power*.	17	43N	75E
Frunze Oblast: Kirgiz SSR.	EP	43N	75E
Frunzovka: Ukrainian SSR.	11	47N	30E
Frýdek: Czech. Coal, textiles*.	6	50N	18E
Fulnek: Poland	6	50N	18E
Furmanov: RSFSR.	8	57N	41E
Furmanovo: Kazakh SSR.	16	50N	50E
ᴸ Fürstenberg: E. Ger. Lignite (see Stalinstadt).	6	52N	15E
Fürstenwalde: E. Germany. Chemicals.	6	52N	14E
Fusca: Romania	11	45N	26E
Gąbin: Poland	6	52N	20E
Gabrovo: Bulgaria. Textiles.	11	43N	25E
Gacko: Yugoslavia	10	43N	18E
Gadyach: Ukrainian SSR.	9	50N	34E
Gadzhi-Gasan: Azer. SSR. Petroleum.	†	40N	50E
Găeşti: Romania	11	45N	25E
Gagra: Georgian SSR.	12	43N	40E
Galaţi: Romania. Pop. 1948: 80,411. Port (naval base), eng.	11	45N	28E
Galich: RSFSR.	8	58N	42E
Galište: Yugoslavia	11	41N	22E
Gánt: Hungary. Bauxite.	10	47N	18E
Gantsevichi: Byelorussian SSR.	7	53N	26E
Ganyushkino: RSFSR.	12	46N	49E
Gardelegen: E. Germany.	6	53N	11E
ᴸ Garm Oblast: Tadzhik SSR.	EP	39N	72E
Garwolin: Poland	7	52N	22E
Gasan-Kuli: Turkmen SSR.	16	38N	54E
Gastello [Jap. Nairo]: RSFSR. Coal* (Uglegorsk group).	†	49N	143E
Gătaia: Romania	11	45N	21E
Gatchina: RSFSR. Pop. 1954: 50,000. Sawmilling.	8	60N	30E
Gaurdak: Turkmen SSR. Chemicals, sulphur	17	38N	66E
Gavrilov: RSFSR.	8	56N	40E
Gavrilov-Yam: RSFSR.	8	57N	40E
Gayduk: RSFSR.	9	45N	38E
Gayny: RSFSR.	13	60N	54E
Gaysin: Ukrainian SSR.	9	49N	30E
Gayvoron: Ukrainian SSR.	9	48N	30E
Gbely: Czech. Petroleum*.	6	49N	17E
Gdańsk [Ger. Danzig]: Poland. Pop. 1950: 191,000. Port. Engineering.	6	54N	18E
Gdov: RSFSR.	8	59N	28E
Gdynia: Poland. Pop. 1950: 117,700. Port.	6	54N	19E
Gelendzhik: RSFSR.	9	44N	38E
Gel'myazov: Ukrainian SSR.	9	50N	32E
Genichesk: Ukrainian SSR.	9	46N	35E
General Toshevo: Bulgaria	11	44N	28E
Geokchay: Azerbaydzhan SSR.	12	41N	48E
Geokmaly: Azer. SSR. Petroleum	†	40N	50E
Geok-Tepe: Turkmen SSR.	16	38N	58E
George Land: RSFSR.	18	80N	45E
Georgian SSR.	EP	42N	43E
Georgiyevka: Kazakh SSR.	17	43N	75E
Georgiyevsk: RSFSR.	12	44N	44E
Gera: E. Ger. Pop. 1946: 101,100. Textiles, chemicals, eng.	6	51N	12E
Gergebil': RSFSR.	12	43N	47E
Gevgeli: Yugoslavia	11	41N	22E
Gheorgheni: Romania	11	47N	26E
Gherla: Romania	11	47N	24E
Ghimeş: Romania	11	47N	26E
Gidrotorf: RSFSR. Elec. plant*.	8	56N	43E
Gigant: RSFSR.	9	46N	41E
Gimoly: Karelo-Fin. SSR. Iron.	18	63N	32E
Guirgui: Romania. River port..	11	44N	26E
Gizel: RSFSR.	12	43N	44E
Gizhduvan: Uzbek SSR.	16	40N	65E
Gizhiga: RSFSR.	19	62N	160E
Gizycko: Poland	7	54N	22E

---

* Located by symbol, but not named, on special topic map.
† On relevant special topic map only. For list of topics see p. VIII.
ᴸ See appendix on p.134

‡ See Inset map. EP Front Endpaper map.
Place names in [square] brackets are former, or alternative, names.

Page Lat. Long.

Gjinokastër [Gk. Argyro-
kastron]: Albania . . 11 40N 20E
Glamoč: Yugoslavia . . 10 44N 17E
Glauchau: E. Ger. *Engineering.* 6 51N 12E
Glatz see Kłodzko.
Glazov: RSFSR. *Textiles.* . 13 58N 53E
Glina: Yugoslavia . . . 10 45N 16E
Glinka: RSFSR. . . . 8 55N 33E
Gliwice [Ger. Gleiwitz]: Poland.
Pop. 1950: 128,000. *Chemi-*
cals, iron & steel. . . 6 50N 19E
Głogów [Ger. Glogau]: Poland.
Pop. 1936: 33,495; 1946:
1,681. *Elec.power.* . . 6 52N 16E
Głubczyce [Ger. Leobschütz]:
Poland. Pop. 1939: 13,505;
1946: 5,020. . . . 6 50N 18E
Glubokiy: RSFSR. *Engineering.* 9 48N 40E
Glubokoye: Byelorussian SSR.. 8 55N 28E
Glubokoye: Kazakh SSR.
Copper. 15 50N 82E
Glukhov: Ukrainian SSR. . 9 52N 34E
Glussk: Byelorussian SSR. . 7 53N 29E
Gniezno [Ger. Gnesen]: Poland.
Engineering. . . . 6 53N 18E
Gnivan': Ukrainian SSR. . 7 61N 28E
Gnjilane: Yugoslavia . . 11 42N 22E
Godech: Bulgaria . . . 11 43N 23E
Golaya Pristan': Ukr. SSR. . 9 46N 33E
Gołdap: Poland . . . 7 54N 22E
Goleniów [Ger. Gollnow]:
Poland 6 54N 15E
Golosnovka: RSFSR. . . 9 52N 38E
Golpa-Zschornewitz: E. Ger.
Elec.power. . . . † 51N 12E
Golubovka: Ukr. SSR. *Coal**
(Donbass). . . . † 49N 39E
Golyshmanovo: RSFSR.. . 14 56N 68E
Gomel': Byelorus. SSR. *Pop.*
1954: 160,000. River port.
Eng., paper, sawmilling. . 9 52N 31E
Gomel' Oblast: Byelorus. SSR.. EP 53N 30E
Gora Chen, mtn.: RSFSR. . 19 65N 141E
Goradiz: Azerbaydzhan SSR. 12 39N 47E
Goragorskiy [Gorskiy]: RSFSR.
*Petroleum**. . . . 12 43N 45E
Gorbatov: RSFSR. . . 8 56N 43E
Gori: Georgian SSR. *Textiles..* 12 42N 44E
Goris: Armenian SSR. . 12 39N 46E
Gorki: Byelorussian SSR. . 8 54N 31E
Gor'kiy [Nizhniy Novgorod]:
RSFSR. Pop. 1954: 900,000.
River port. Eng., chemicals,
petroleum refining, steel,
textiles, elec.power. . 8 56N 44E
Gor'kiy Oblast: RSFSR. . EP 57N 45E
Gorlice: Poland . . . 7 50N 21E
Görlitz [Pol. Zgorzelec]: E. Ger.
Pop. 1946: 88,700. *Eng.,*
*textiles**, chemicals, lignite.* 6 51N 15E
Gorlovka: Ukr. SSR. *Pop.*
1954: 200,000. *Coal, chemi-*
cals, eng. 9 48N 38E
Gorna Oryakhovitsa: Bulgaria . 11 43N 26E
Gorno-Altay AO.: Altay Kray,
RSFSR. . . . . EP 51N 87E
Gorno-Altaysk [Oyrot-Tura]:
RSFSR. 15 52N 86E
Gorno-Badakhshan AO.:
Tadzhik SSR. . . . EP 38N 73E
Gornostayevka: Ukrainian SSR. 9 47N 34E
Gornyak: RSFSR. *Gold, lead-*
zinc. 15 51N 81E
Gornyatskiy: RSFSR. *Coal.* . † 68N 65E
Gornyy: RSFSR. *Oil shale.* . 14 52N 49E
Gornyy Balykley: RSFSR. . 9 49N 45E
Gorodenka: Ukrainian SSR. . 7 49N 25E
Gorodets: RSFSR. *Hydro-elec.* 8 57N 44E
Gorodishche [Pol. Horodyszcze]:
Byelorussian SSR. . . 7 53N 26E
Gorodishche: RSFSR. . . 8 53N 46E
Gorodishche: Ukrainian SSR. . 9 49N 31E
Gorodishchi: RSFSR. . . 8 56N 39E
Gorodnitsa: Ukrainian SSR. . 7 51N 27E
Gorodnya: Ukrainian SSR. . 9 52N 32E
Gorodok: Byelorussian SSR.
*Engineering**. . . . 8 55N 30E
Gorodok: RSFSR. *Tungsten.* . 12 50N 103E
Gorodok: Ukrainian SSR. . 7 49N 26E
Gorodok [Pol. Gródek Jagiel-
loński]: Ukrainian SSR. . 7 50N 24E
Gorokhovets: RSFSR. . . 8 56N 43E
Gorskoye: Ukrainian SSR.
Coal (Donbass). . . 9 49N 38E
Gorzów Wielkopolski [Ger.
Landsberg]: Poland. *Pop.*
1939: 48,053; 1946: 19,796. 6 53N 15E
Gospić: Yugoslavia . . 10 45N 15E
Gostivar: Yugoslavia . . 11 42N 21E
Gostomel': Ukrainian SSR. . 9 51N 30E
Gostyń: Poland . . . 6 52N 17E
Gostynin: Poland . . . 6 52N 19E

Page Lat. Long.

Gotha: E. Ger. *Pop. 1946:*
57,639. Eng. (incl. precision
instruments), chemicals. . 6 51N 11E
Gottwaldov [Zlín]: Czech.
Eng. (Bat'a Shoe factories). . 6 49N 18E
Gračac: Yugoslavia . . 10 44N 16E
Gračanica: Yugoslavia . . 10 45N 18E
Gradačac: Yugoslavia . . 10 45N 18E
Gradets: Bulgaria . . . 11 43N 27E
Gradizhsk: Ukrainian SSR. . 9 49N 33E
Graham Bell I.: RSFSR. . 18 82N 64E
Grajewo: Poland . . . 7 54N 22E
Grayvoron: RSFSR. . . 9 51N 36E
Great Fergana Canal: Uzbek
SSR.. . . . . ‡17 41N 72E
Greater Baku see Baku
Greater Caucasus: USSR. . 12 43N 45E
Greater Moscow see Moscow
Grebenkovskiy: Ukr. SSR.. . 9 50N 32E
Greifswald: E. Germany . 6 54N 13E
Gremyachinsk: RSFSR. *Coal**. 13 59N 58E
Grigoriopol': Moldavian SSR. . 11 47N 30E
Grigor'yevka: Ukrainian SSR.. 9 47N 31E
Grodno: Byelorus. SSR. *Pop.*
1954: 60,000. Eng., textiles,
chemicals. . . . 7 54N 24E
Grodno Oblast: Byelorus. SSR. EP 53N 25E
Grodzisk: Poland . . . 7 52N 20E
Grodzisk Wielkopolski: Poland 6 52N 16E
Grójec: Poland . . . 7 52N 21E
Gröningen: E. Ger. *Paper.* . † 52N 11E
Grossenhain: E. Germany . 6 51N 14E
Groznyy: RSFSR. *Pop. 1954:*
200,000. Petroleum, elec.
power*, petroleum refining,
chemicals, sawmilling, eng. . 12 43N 46E
Groznyy Oblast: RSFSR. . EP 44N 46E
Gruda: Yugoslavia . . 10 42N 18E
Grudziądz [Ger. Graudenz]:
Poland. *Sawmilling.* . 6 53N 19E
Gruža: Yugoslavia . . 11 44N 21E
Gruzino: RSFSR. . . 8 59N 32E
Gryazi: RSFSR. . . . 9 52N 40E
Gryazovets: RSFSR. . . 8 59N 40E
Grybów: Poland . . . 7 50N 21E
Gryfino: Poland . . . 6 53N 14E
Grykë: Albania . . . 10 41N 19E
Gubakha: RSFSR. *Pop. 1954:*
70,000. Coal, elec. power*,
chemicals. . . . 13 59N 58E
Gubin: E. Germany. *Textiles,*
*lignite**. 6 52N 15E
Gubkin: RSFSR. *Iron ore.* . 9 51N 38E
Gudauty: Georgian SSR. . 12 43N 41E
Gudermes: RSFSR. *Petroleum**
(Groznyy Field). . . 12 43N 46E
Gukovo: RSFSR. *Coal**
(Donbass). . . . 9 48N 40E
Gulbene: Latvian SSR. . 5 57N 27E
Gul'cha: Kirgiz SSR. *Copper..* 17 40N 74E
Gundorovka [Donetsk]:
RSFSR. *Coal.* . . . 9 48N 40E
Gura Humorului: Romania . 11 48N 26E

Gur'yev: Kazakh SSR. *Pop.*
1954: 80,000. Caspian port.
Petroleum refining, elec.
power, eng. . . . 16 47N 52E
Gur'yev Oblast: Kazakh SSR.. EP 45N 54E
Gur'yevsk: RSFSR. *Steel.* . † 54N 86E
Gurzuf: Ukrainian SSR. . 9 44N 34E
Gusev [Ger. Gumbinnen]:
RSFSR. 7 54N 22E
Gusinoozersk: RSFSR. *Lignite.* 12 51N 106E
Gus'-Khrustal'nyy: RSFSR. . 8 56N 41E
Guštanj: Yugoslavia. *Steel.* . 10 46N 15E
Güstrow: E. Germany . . 6 54N 12E
Gutay: RSFSR. *Gold, moly-*
denum. 19 50N 108E
Guzar: Uzbek SSR. . . 17 39N 66E
Gyda Penin.: RSFSR. . . 18 72N 80E
Gydan Range: RSFSR. . 19 63N 160E
Gyoma: Hungary . . . 11 47N 21E
Gyöngyös: Hungary. *Coal**. . 10 47N 19E
Győr: Hungary. *Pop. 1941:*
57,000. River port. Steel, eng.,
textiles, chemicals. . . 10 48N 18E
Gyueshevo: Bulgaria . . 11 42N 23E
Gyula: Hungary. *Engineering..* 11 47N 21E
Gyumush: Armenian SSR.
*Hydro-elec.** . . . 12 40N 45E
Gyuzdek: Azerbaydzhan SSR.
Petroleum. . . . † 40N 50E
Gzhatsk: RSFSR. . . 8 56N 35E

Hajdúböszörmeny: Hungary . 11 48N 21E
Hajdúnánás: Hungary . . 11 48N 22E
Hajnówka: Poland. *Sawmilling.* 7 53N 24E
Halberstadt: E. Ger. *Chemicals,*
*textiles**. 6 52N 11E
Haldensleben: E. Germany . 6 52N 12E
Hall I.: RSFSR. . . . 18 80N 60E

Page Lat. Long.

Halle: E. Germany. *Pop. 1946:*
278,400. Lignite, potash, eng.,
elec. power*, chemicals. . 6 52N 12E
Hangay Range: Mongolian
People's Republic . . 19 47N 100E
Han Pijesak: Yugoslavia . 10 44N 19E
Hara Usŭ, lake: Mongolian
People's Republic . . 19 48N 90E
Harkány: Hungary. *Sulphur.* . 10 46N 18E
Harlau: Romania . . . 11 47N 27E
Hârşova: Romania . . . 11 45N 28E
Hațeg: Romania . . . 11 46N 23E
Hatvan: Hungary . . . 10 48N 20E
Havelberg: E. Germany . . 6 53N 12E
Havlíčkův Brod: Czech. . 6 50N 16E
Heiligenstadt: E. Germany . 6 51N 10E
Hel: Poland 6 55N 19E
Hennigsdorf: E. Ger. *Eng., steel.* 6 53N 13E
Hercegnovi: Yugoslavia . 10 42N 18E
Herzberg: E. Germany . . 6 52N 13E
Heves: Hungary . . . 11 48N 20E
Hildburghausen: E. Germany . 6 50N 11E
Himarë: Albania . . . 10 40N 20E
Hirschberg see Jelenia Góra
Hlohovec: Czechoslovakia . 6 48N 18E
Hódmezővásárhely: Hungary.
Pop. 1941: 61,776. . . 11 46N 20E
Hodonín: Czech. *Petroleum.* . 6 49N 17E
Holič: Czechoslovakia . . 6 49N 17E
Horažd'ovice: Czechoslovakia . 6 49N 14E
Horšovský Týn: Czech. *Zinc.* . 6 49N 13E
Hradec Králové: Czech. *Pop.*
1947: 51,480. *Chemicals, eng.**
(photographic equip.), textiles.* 6 50N 16E
Hranice: Czechoslovakia . 6 49N 18E
Hrubieszów: Poland . . 7 51N 24E
Huedin: Romania . . . 11 47N 23E
Humenné: Czechoslovakia . 7 49N 22E
Humpolec: Czechoslovakia . 6 49N 15E
Hunedoara: Romania. *Iron &*
steel 11 46N 22E
Huşi: Romania . . . 11 47N 28E
Hvar, i.: Yugoslavia . . 10 43N 17E

Iaşi [Jassy]: Romania. *Pop.*
1948: 94,075. *Textiles,*
chemicals. . . . 11 47N 28E
Ibresi: RSFSR. . . . 14 55N 47E
Icha: RSFSR. . . . 19 56N 155E
Ichnya: Ukrainian SSR. . 9 51N 32E
Idrija: Yugoslavia. *Mercury.* . 10 46N 14E
Idrinskoye: RSFSR. . . 15 54N 92E
Idritsa: RSFSR. . . . 8 56N 29E
Idzhevan [Karavansarai]:
Armenian SSR. . . 12 41N 45E
Igarka: RSFSR. Port. *Saw-*
milling. 18 68N 87E
Ikhtiman: Bulgaria. *Asbestos..* 11 42N 24E
Ikryanoye: RSFSR. . . 12 46N 48E
Iksha: RSFSR. . . . 8 56N 38E
Ilanskiy: RSFSR. . . 19 56N 96E
Il'ava: Czechoslovakia . . 6 49N 18E
Iława [Ger. Deutsch Eylau]:
Poland 6 54N 20E
Ileanda: Romania . . . 11 47N 24E
Ilek: RSFSR. . . . 13 52N 53E
Ili [Iliysk]: Kazakh SSR.. . 17 44N 77E
Ili, R. China/USSR. . . 22 44N 79E
Ilia: Romania . . . 11 46N 23E
Il'ich: Kazakh SSR. . . 17 41N 68E
Il'inka: RSFSR. . . . 12 52N 107E
Illarionovo: Ukrainian SSR. . 9 48N 35E
Il'men, lake: RSFSR. . . 8 58N 31E
Ilovaysk: Ukrainian SSR. . 9 48N 38E
Ilovlinskaya: RSFSR. . . 9 49N 44E
Ilsenburg: E. Germany. *Copper*
smelting. . . . . † 51N 10E
Il'skiy: RSFSR. *Petroleum.* . 12 45N 39E
Iman: RSFSR. *Sawmilling.* . 19 46N 134E
Imeni Artema: Ukrainian SSR.
*Coal** (Donbass). . . † 56N 43E
Imeni Kalinina: Azer. SSR.
Petroleum. . . . † 40N 50E
Imeni Kominterna [Sosnin-
skaya]: RSFSR. . . 8 59N 32E
Imeni L. M. Kaganovicha:
Azer. SSR. *Petroleum.* . † 40N 50E
Imeni M. I. Kalinina: RSFSR.. 8 58N 45E
Imeni Stepana Razina: RSFSR. 8 55N 45E
Imeni Tsyurupy: RSFSR. . 8 55N 39E
Imeni V. M. Molotova: RSFSR.
*Eng.** Suburb of Gor'kiy. . † 56N 44E
Imishly: Azerbaydzhan SSR. . 12 40N 48E
Imotski: Yugoslavia . . 10 43N 17E
Imtonzha: RSFSR. *Tin, lead,*
zinc. 19 66N 130E
Inderborskiy: Kazakh SSR. . 16 49N 52E
Indiga: RSFSR. . . . 18 68N 49E
Indigirka, R.: RSFSR. . . 23 68N 147E
Ineu: Romania . . . 11 46N 22E
Innokent'yevskiy: RSFSR. . 19 48N 140E

* Located by symbol, but not named, on special topic map. ‡ See Inset map. EP Front Endpaper map.
† On relevant special topic map only. For list of topics see p. VIII. Place names in [square] brackets are former, or alternative, names.

116

	Page	Lat.	Long.
Inowrocław [Ger. Hohensalza]: Poland.	6	53N	18E
Inta: RSFSR.	18	66N	60E
Inya: RSFSR.	15	50N	87E
Inza: RSFSR. Chemicals, saw-milling*.	†	54N	46E
Inzer: RSFSR. Iron ore*.	13	54N	58E
Inzhavino: RSFSR.	9	52N	43E
Iolotan': Turkmen SSR.	16	37N	62E
Ionava: Lithuanian SSR..	‡5	55N	24E
Ioneşti: Romania	11	45N	24E
Ionishkis: Lith. SSR.	‡5	56N	23E
Ipatovo: RSFSR.	9	46N	43E
Irbit: RSFSR. Chemicals, eng.*	13	58N	63E
Irgiz: Kazakh SSR.	16	49N	61E
Irkutsk: RSFSR. Pop. 1954: 420,000. River port. Eng., elec. power, sawmilling, petroleum refining, chemicals.	12	52N	104E
Irkutsk Oblast: RSFSR.	EP	57N	105E
Irmino [Irminskiy Rudnik]: Ukrainian SSR.	9	49N	38E
Iron Gates, pass: Romania	11	46N	23E
Irpen': Ukrainian SSR.	9	50N	30E
Irtysh, R.: RSFSR.	22	57N	73E
Irtyshsk: Kazakh SSR.	15	53N	75E
Is [Sverdlovskiy Priisk]: RSFSR. Gold, platinum.	13	59N	60E
Isaccea: Romania.	11	45N	28E
Işalniţa: Romania	11	44N	24E
Isfara: Tadzhik SSR. Petroleum*, coal.	‡17	40N	71E
Isheyevka: RSFSR.	14	55N	48E
Ishim: RSFSR.	14	56N	69E
Ishim, R.: RSFSR.	22	54N	68E
Ishimbay: RSFSR. Pop. 1954: 50,000. Petroleum, petroleum refining.	13	54N	56E
Ishim Steppe: RSFSR.	15	55N	73E
Isil'-Kul': RSFSR.	15	55N	71E
Iskininskiy Kazakh SSR. Petroleum.	16	47N	53E
Iskitim: RSFSR.	15	55N	83E
Isperikh: Bulgaria	11	44N	27E
⁴Issyk-Kul' Oblast: Kirgiz SSR.	EP	42N	78E
Issyk-Kul', lake: USSR.	16	43N	77E
Istra [Voskresensk]: RSFSR.	8	56N	37E
Iturup, i.: RSFSR.	19	45N	147E
Ivailovgrad: Bulgaria	11	42N	26E
Ivangrad: Yugoslavia	10	43N	20E
Ivanichi: Ukrainian SSR.	7	51N	24E
Ivanić Grad: Yugoslavia	10	46N	16E
Ivanishchi: RSFSR	8	56N	40F
Ivanjica: Yugoslavia	11	44N	20E
Ivankov: Ukrainian SSR.	9	51N	30E
Ivankovo: RSFSR.	8	57N	37E
Ivanovo: Byelorussian SSR.	7	52N	25E
Ivanovo: RSFSR. Pop. 1954: 325,000. Textiles, eng.	8	57N	41E
Ivanovo Oblast: RSFSR.	EP	57N	42E
Ivanteyevka: RSFSR. Pop. 1954: 50,000.	8	56N	38E
Ivatsevichi: Byelorussian SSR.	7	53N	25E
Ivdel': RSFSR. Manganese, bauxite*, iron ore, coal.	13	61N	60E
Ivenets [Pol. Iwieniec]: Byelorussion SSR.	7	54N	27E
Ivot: RSFSR.	8	54N	34E
Iv'ye: Byelorussian SSR.	7	54N	26E
Izberbash: RSFSR. Petroleum.	12	43N	48E
Izdeshkovo: RSFSR.	8	55N	34E
Izhevsk: RSFSR. Pop. 1954: 225,000. Steel, eng.*	13	57N	53E
Izhma: RSFSR. Petroleum.	18	63N	54E
Izhma: RSFSR.	18	65N	54E
Izhma, R.: RSFSR.	22	64N	54E
Izmail: Ukrainian SSR. River port. Chemicals.	11	45N	29E
Izmalkovo: RSFSR.	9	53N	38E
Izobil'noye: RSFSR.	9	45N	42E
Izumrud: RSFSR. Beryllium.	13	57N	62E
Izvestiy Tsik Is.: RSFSR.	18	76N	82E
Izyaslav: Ukrainian SSR.	7	50N	27E
Izyum: Ukr. SSR. Engineering*.	9	49N	37E
Jablanac: Yugoslavia	10	45N	15E
Jablanica: Yugo. Hydro-elec.	10	44N	18E
Jablonec [Ger. Gablonz]: Czech. (Artificial jewellery.)	6	51N	15E
Jabłonow: Poland	6	54N	19E
Jablunkov: Czechoslovakia.	6	50N	19E
Jabukovac: Yugoslavia	11	44N	22E
Jáchymov [Ger. Joachimstal]: Czech. Pitchblende (uranium & radium prod.), lead, silver, nickel, cobalt.	6	50N	13E
Jajce: Yugoslavia. Chemicals, hydro-elec.*	10	44N	17E
Janja Lipa: Yugoslavia	11	44N	17E

	Page	Lat.	Long.
Jánoshalma: Hungary	10	46N	19E
Janów Lubelski: Poland	7	50N	22E
Janów Podlaski: Poland	7	52N	23E
Jarocin: Poland	6	52N	18E
Jaroměř: Czechoslovakia	6	50N	16E
Jarosław: Poland	7	50N	23E
Jasło: Poland. Petroleum	7	50N	22E
Jassy see Iaşi			
Jastrowie [Ger. Jastrow]: Poland	6	53N	17E
Jászárokszállás: Hungary	10	48N	20E
Jászberény: Hungary	10	47N	20E
Jawor [Ger. Jauer]: Poland	6	51N	16E
Jędrzejów: Poland	7	51N	20E
Jelenia Góra [Ger. Hirschberg]: Poland	6	51N	16E
Jelšava: Czech. Iron, magnesite.	7	49N	20E
Jena: E. Ger. Pop. 1946: 82,722. Eng. (incl. optical precision instruments), chemicals.	6	51N	12E
Jermenovci: Yugo. Petroleum.	11	45N	21E
Jesenice: Yugo. Iron & steel.	10	46N	14E
Jeseník: Czechoslovakia	6	50N	16E
Jibău: Romania	11	47N	23E
Jibhalanta (Uliassutai): Mongolian People's Republic	19	48N	97E
Jičín [Ger. Gitschin]: Czech.	6	49N	15E
Jihlava: Czech. Textiles, eng.*	6	49N	16E
Jindřichův Hradec: Czech.	6	49N	15E
Jirgalantu [Kobdo]: Mongolian People's Republic	19	48N	91E
Kaakhka: Turkmen SSR.	16	37N	60E
Kabardinian ASSR.: RSFSR.	EP	43N	43E
Kačanik: Yugoslavia	11	42N	21E
Kacha: Ukrainian SSR.	9	45N	34E
Kacha: RSFSR.	18	62N	56E
Kachanovo: RSFSR.	8	58N	28E
Kachuga: RSFSR. River port.	12	54N	106E
Kadiyevka [Sergo]: Ukr. SSR. Pop. 1954: 75,000. Coal, iron & steel* (Donbass), synthetic rubber.	9	49N	39E
Kadom: RSFSR.	8	55N	42E
Kaduy: RSFSR.	8	59N	38E
Kadyy: RSFSR.	8	58N	43E
Kadzharan: Armenian SSR. Copper, molybdenum.	12	39N	46E
Kadzhi-Say: Kirgiz SSR. Lignite.	17	42N	77E
Kafan: Armenian SSR. Copper, copper smelting.	12	39N	46E
Kagan [Novaya Bukhara]: Uzbek SSR.	16	40N	65E
⁴Kaganovich: RSFSR. Elec. power*.	†	55N	38E
⁴Kaganovich: RSFSR. Lignite* (Uzlovaya Group, Moscow Lignite Basin).	8	54N	38E
⁴Kaganovichabad: Tadzhik SSR.	17	37N	69E
⁴Kaganovichesk: Turkmen SSR.	16	39N	63E
⁴Kaganovichi: Ukrainian SSR..	9	51N	29E
Kagarlyk: Ukrainian SSR.	9	50N	31E
Kagul: Moldavian SSR.	11	46N	28E
Kainari: Moldavian SSR.	11	47N	29E
Kakhovka: Ukr. SSR. Eng., elec. power (site of 2nd Dnepr dam, scheduled for 1957).	9	47N	34E
Kalach: RSFSR.	9	50N	41E
Kalach: Turkmen SSR.	16	38N	65E
Kalachinsk: RSFSR.	15	55N	75E
Kalach-na-Donu: RSFSR.	9	49N	44E
Kalai-Khumb: Tadzhik SSR	17	39N	72E
Kalakan: RSFSR.	12	55N	116E
Kalanchak: Ukrainian SSR.	9	46N	33E
Kalarash [Rom. Călăraşi]: Moldavian SSR.	11	47N	28E
Kaliakra, C.: Bulgaria	11	45N	28E
Kalinin [Tver]: RSFSR. Pop. 1954: 250,000. Eng., textiles, chemicals, elec. power.	8	57N	36E
Kalinin Oblast: RSFSR.	EP	57N	35E
Kalininskoye: Ukrainian SSR.	9	47N	33E
Kalinkovichi: Byelorus. SSR. Chemicals.	7	52N	29E
Kalinovka: Ukrainian SSR.	7	49N	28E
Kalisz: Poland. Pop. 1950: 55,140.	6	53N	16E
Kallaste: Estonian SSR.	8	59N	27E
Kalmykovo: Kazakh SSR.	16	49N	52E
Kalocsa: Hungary	10	46N	19E
Kaluga: RSFSR. Pop. 1954: 100,000. River port. Hydro-elec., eng.*	8	55N	36E
Kaluga Oblast: RSFSR.	EP	54N	35E
Kalush: Ukrainian SSR. Potash, chemicals.	7	49N	24E

	Page	Lat.	Long.
Kalvariya: Lithuanian SSR.	7	54N	23E
Kal'ya: RSFSR. Gold, bauxite.	13	60N	60E
Kalyazin: RSFSR. Textiles.	8	57N	38E
Kama, R.: RSFSR. Large elec. power plant near Molotov.	22	57N	54E
Kama Hills: RSFSR.	13	58N	54E
Kambarka: RSFSR.	13	56N	54E
Kamchatka: RSFSR.	3	60N	160E
Kamen': RSFSR. Elec. power*, iron.	15	54N	81E
Kamenets: Byelorussian SSR.	7	52N	24E
Kamenets Podol'skiy: Ukrainian SSR. Pop. 1954: 50,000.	7	49N	27E
Kamenka: RSFSR.	8	57N	42E
Kamenka: RSFSR.	8	53N	44E
Kamenka: RSFSR.	9	52N	42E
Kamenka: RSFSR.	9	51N	40E
Kamenka: Ukrainian SSR.	9	49N	32E
Kamenka-Bugskaya: Ukr. SSR.	7	50N	24E
Kamenka-Dneprovskaya: Ukr. SSR.	9	48N	34E
Kamen'-Kashirskiy: Ukr. SSR.	7	52N	25E
Kamennogorsk [Fin. Antrea] RSFSR.	8	61N	29E
⁴Kamensk Oblast.: RSFSR.	EP	49N	41E
Kamenskiy [Grimm]: RSFSR.	9	51N	46E
Kamenskoye: RSFSR.	19	63N	165E
Kamensk-Shakhtinskiy [Kamensk]: RSFSR. Pop. 1954: 60,000. Chemicals.	9	48N	40E
Kamensk-Ural'skiy: RSFSR. Pop. 1954: 80,000. Iron, bauxite, aluminium, eng., steel.	13	56N	62E
Kamien Pomorski: Poland	6	54N	15E
Kamienna Góra [Ger. Landeshut]: Poland. Coal.	6	51N	16E
Kamnik: Yugoslavia.	10	46N	14E
Kamyshin: RSFSR. Textiles.	9	50N	45E
Kamyshlov: RSFSR.	13	57N	63E
Kamysh-Zarya: Ukr. SSR.	9	47N	37E
Kamyzyak: RSFSR.	12	46N	48E
Kanaker: Armenian SSR.	12	40N	45E
Kanash: RSFSR.	14	55N	47E
Kandagach: Kazakh SSR.	16	50N	57E
Kandakovo: RSFSR.	19	70N	152E
Kandalaksha: RSFSR. Port. Hydro-elec., aluminium refinery, sawmilling.	18	67N	33E
Kanev: Ukrainian SSR.	9	50N	31E
Kanevskaya: RSFSR.	9	46N	39E
Kangalasskiye Kopi: RSFSR. Lignite.	†	62N	130E
Kanibadam: Tadzhik SSR. Petroleum refinery*.	17	40N	70E
Kanin, C.: RSFSR.	18	68N	49E
Kansay: Tadzhik SSR. Lead-zinc.	17	40N	70E
Kansk: RSFSR. Pop. 1954: 60,000. Textiles, lignite, saw-milling.	19	56N	95E
Kant: Kirgiz SSR.	17	43N	75E
⁴Kantagi: Kazakh SSR. Lead, zinc. (Kentau Group.)	17	44N	68E
Kantemirovka: RSFSR.	9	50N	40E
Kaplice: Czechoslovakia	6	49N	14E
Kaposvár: Hungary.	10	46N	18E
Kapuvár: Hungary	10	48N	17E
Kara: RSFSR.	18	69N	65E
Karabakh Plateau: Azer. SSR.	12	40N	46E
Karabanovo: RSFSR.	8	56N	39E
Karabash: RSFSR. Copper mining & smelting, zinc.	13	55N	60E
Kara-Bogaz-Gol, lake: Turkmen SSR. Glauber's salt, chemicals.	16	41N	53E
Karbulak: Kazakh SSR.	17	45N	78E
Karachala: Azerbaydzhan SSR.	12	40N	49E
Karachev: RSFSR.	8	53N	35E
Karadag: Azer. SSR. Petroleum.	†	40N	50E
Karaga: RSFSR.	19	50N	73E
Karaganda: Kazakh SSR. Pop. 1926: 150; 1954: 420,000. Coal, eng.	17	50N	73E
Karaganda Oblast: Kazakh SSR.	EP	48N	70E
Karagayly: Kazakh SSR. Barium.	†	49N	76E
Karaginskiy I.: RSFSR.	19	59N	164E
Karaidel'skiy: RSFSR.	13	56N	57E
Kara-Kala: Turkmen SSR.	16	39N	56E
Kara-Kalpak ASSR.: Uzbek SSR.	EP	43N	57E
Karakas: Kazakh SSR.	17	48N	83E
Karakul': Turkmen SSR.	16	39N	64E
Kara-Kum: USSR.	16	39N	60E
Kara-Mazar: Tadzhik SSR. Lead-zinc.	17	41N	70E
Kara Sea: RSFSR.	18	72N	62E
Karasu: Kirgiz SSR.	‡17	41N	73E
Karasuk: RSFSR.	15	54N	78E
Kara-Tau: Kazakh SSR. Vanadium, phosphorite	17	43N	71E
Kara Tau mtns.: Kazakh SSR.	17	44N	68E

* Located by symbol, but not named, on special topic map.   ‡ See Inset map.   EP Front Endpaper map.
† On relevant special topic map only. For list of topics see p. VIII.   Place names in [square] brackets are former, or alternative, names.
⁴ See appendix on p.134

117

	Page	Lat.	Long.
Karaul: RSFSR.	18	70N	83E
Karayaz Steppe: Azer. SSR.	12	41N	46E
Karcag: Hungary	11	47N	21E
Kardeljevo: Yugoslavia	10	43N	18E
Karelo-Finnish SSR.	EP	65N	33E
Kargasok: RSFSR.	15	59N	81E
Kargat: RSFSR.	15	55N	80E
Kargopol': RSFSR.	18	62N	39E
Karintorf: RSFSR. *Elec.power.*	†	58N	50E
Karkaralinsk: Kazakh SSR. *Lead-silver.*	17	49N	76E
Karl Libknekht: RSFSR.	9	52N	35E
Karl-Marx-Stadt: E. Germany	6	51N	13E
Karlovac: Yugoslavia	10	45N	16E
Karlovka: Ukrainian SSR.	9	49N	35E
Karlovo: Bulgaria	11	43N	25E
Karlovy Vary [*Ger.* Karlsbad]: Czech. (*Noted health resort.*).	6	50N	13E
Karnilovka: Byelorussian SSR.	7	52N	29E
Karnobat: Bulgaria	11	43N	27E
Karpacz [*Ger.* Krummhübel]: Poland	6	51N	16E
Karpinsk [Bogoslovsk, Ugol'nyy]: RSFSR. *Lignite.*	13	60N	60E
Karpushikha: RSFSR. *Copper* zinc*.	†	58N	60E
Karsakpay: Kazakh SSR. *Elec.*, *copper smelting.*	17	48N	67E
Karsava: Latvian SSR.	‡5	57N	28E
Karshi [Bek-Budi]: Uzbek SSR.	17	39N	65E
Karskiye Vorota Strait: RSFSR.	18	70N	55E
Kartaly: RSFSR. *Anthracite, chromium (sometimes known as Varblyuzh'ya Gora deposit).*	13	53N	61E
Kartuzy [*Ger.* Karthaus]: Poland	6	54N	18E
Karviná [*Ger.* Karwin]: Czech. *Chemicals, coal*, iron* (Ostrava Group).*	†	50N	18E
Karyagino [Sardar]: Azer. SSR.	12	40N	47E
Karymskoye: RSFSR.	12	52N	114E
Kashary: RSFSR.	9	49N	41E
Kashin: RSFSR.	8	57N	38E
Kashira: RSFSR.	8	55N	38E
Kashka-Darya Oblast: Uzbek SSR.	EP	39N	66E
Kasimov: RSFSR. *Textiles, iron ore.*	8	55N	41E
Kasli: RSFSR.	13	56N	61E
Kaspi: Georgian SSR.	12	42N	44E
Kaspiysk [Dvigatel'stroy]: RSFSR.	12	43N	48E
Kaspiyskiy [Lagan']: RSFSR.	12	45N	47E
Kassansay: Uzbek SSR.	‡17	41N	72E
Kastornoye: RSFSR.	9	52N	38E
Katav-Ivanovsk: RSFSR. *Steel.*	13	55N	58E
Kataysk: RSFSR.	13	56N	63E
Katon Karagay: Kazakh SSR.	15	49N	86E
Katowice: Poland.	6	50N	19E
Katta-Kurgan: Uzbek SSR.	17	40N	66E
Katyk: Ukrainian SSR. *Coal* (Donbass).*	†	48N	38E
Kaunas [*Rus.* Kovno]: Lith. SSR. *Pop. 1954: 210,000. Chemicals, eng.*, textiles.*	‡5	55N	24E
Kavacha: RSFSR.	19	60N	170E
Kavadarci: Yugoslavia	11	41N	22E
Kavajë: Albania	10	41N	20E
Kavarna: Bulgaria	10	43N	28E
Kavkazskaya: RSFSR.	9	45N	41E
Kayakent: RSFSR. *Petroleum* (Izberbash Group).*	12	42N	48E
Kayshyadoris: Lith. SSR.	‡5	54N	24E
Kazachinskoye: RSFSR. *Thorium.*	15	58N	93E
Kazach'ye: RSFSR.	19	71N	136E
Kazakh SSR.	EP	49N	18E
Kazakh Uplands, USSR.	17	48N	75E
Kazalinsk: Kazakh SSR.	16	45N	62E
Kazan': RSFSR. *Pop. 1954: 630,000. River port. Eng. (incl. typewriters), chemicals, synthetic rubber, textiles, petroleum refining, paper, sawmilling.*	14	55N	49E
Kazandzhik: Turkmen SSR.	16	39N	56E
Kazanka: Ukrainian SSR.	9	48N	34E
Kazanlŭk: Bulg. (*Attar of roses.*)	11	42N	25E
Kazanovka: RSFSR. *Lignite* (Shchekino Group).*	8	54N	38E
Kazanshunkur: Kazakh SSR. *Gold*.*	17	50N	82E
Kazanskaya: RSFSR.	9	50N	41E
Kazarman: Kirgiz SSR.	17	41N	74E
Kazatin: Ukrainian SSR.	7	50N	29E
Kazbek, *mtn.*: Georgian SSR.	12	43N	44E
Kazhim: RSFSR. *Iron ore.*	13	60N	52E
Kazi-Magomed [Adzhikabul]: Azerbaydzhan SSR.	12	40N	49E
Kaztalovka: Kazakh SSR.	16	50N	49E
Kdyně: Czechoslovakia	6	49N	13E
Kecskemét: Hungary. *Pop. 1941: 87.000.*	10	47N	20E

	Page	Lat.	Long.
Kedabek: Azer. SSR. *Copper.*	12	41N	46E
Kegen': Kazakh SSR.	17	43N	80E
Kegums: Latvian SSR.	‡5	57N	24E
Kel'me [*Rus.* Kel'my]: Lith. SSR.	‡5	55N	23E
Kem': Karelo-Finnish SSR. *Port. Sawmilling.*	18	65N	35E
Kemerovo: RSFSR. *Pop. 1954: 366,000. Coal, elec., chemicals, textiles.*	15	55N	86E
Kemerovo Oblast: RSFSR.	EP	55N	86E
Kengir: Kazakh SSR. *Hydro-elec.*	17	38N	68E
Kenimekh: Uzbek SSR.	17	40N	65E
Ken-Tyube-Togay: Kazakh SSR. *Iron ore.*	†	44N	75E
Kępno [*Ger.* Kempen]: Poland.	6	51N	18E
Kerch': Ukr. SSR. *Pop. 1954: 150,000. Port. Petroleum, vanadium, steel, chemicals, elec. power.*	9	45N	36E
Kerchem'ya: RSFSR.	14	61N	54E
Kergez: Azer. SSR. *Petroleum.*	†	40N	50E
Kerki: Turkmen SSR.	17	38N	65E
Kerkichi: Turkmen SSR.	17	38N	65E
Kermine: Uzbek SSR.	17	40N	65E
Kesova Gora: RSFSR.	8	58N	38E
Kes'ma: RSFSR.	8	58N	37E
Keszthely: Hungary	10	47N	17E
Kętrzyn [*Ger.* Rastenburg]: Poland	7	54N	21E
Kežmarok: Czech. *Textiles.*	7	49N	20E
Keyla: Estonian SSR.	‡5	59N	24E
Khaapsalu [*Est.* Haapsalu]: Estonian SSR.	‡5	59N	23E
Khabarovsk: RSFSR. *Pop. 1954: 350,000. River port. Petroleum refining, eng. (incl. aircraft), sawmilling.*	19	48N	135E
Khabarovsk Kray: RSFSR.	EP	55N	140E
Khachmas: Azerbaydzhan SSR.	12	41N	49E
Khadabulak: RSFSR.	12	51N	116E
Khadkhal: Mongolian People's Republic. *Wool washing.*	12	51N	100E
Khadyzhensk: RSFSR.	9	44N	40E
Khakass AO.: Krasnoyarsk Kray, RSFSR.	EP	53N	90E
Khalilovo: RSFSR. *Iron ore, chromium, nickel, titanium, vanadium, cobalt.*	13	52N	58E
Khal'mer-Yu: RSFSR.	18	68N	65E
Khamar-Daban Range: RSFSR.	12	51N	105E
Khanty-Mansi NO.: RSFSR.	EP	62N	70E
Khanty-Mansiysk: Sawmilling..	14	61N	69E
Khapcheranga: RSFSR. *Tin, sawmilling.*	19	50N	113E
Kharik: RSFSR.	12	54N	102E
Khar'kov: Ukr. SSR. *Pop. 1954: 825,000. Eng., chemicals, paper.*	9	50N	36E
Khar'kov Oblast: Ukr. SSR.	EP	50N	36E
Kharlovka: RSFSR.	18	68N	38E
Kharlu: Kar.-Fin. SSR. *Paper..*	18	62N	31E
Kharmanli: Bulgaria	11	42N	26E
Kharovsk: RSFSR.	8	60N	40E
Khasavyurt: RSFSR.	12	43N	47E
Khashuri: Georgian SSR.	12	42N	44E
Khaskovo: Bulgaria. *Textiles.*	11	42N	26E
Khatanga: RSFSR.	19	72N	103E
Khatyrchi: Uzbek SSR.	17	40N	66E
Khaudag: Uzbek SSR. *Petroleum.*	†	38N	67E
Khaydarkan: Kirgiz SSR. *Mercury, antimony.*	‡17	40N	71E
Khazarasp: Turkmen SSR.	16	41N	61E
Kherson: Ukr. SSR. *Pop. 1954: 125,000. Port. Petroleum refining, eng., textiles.*	9	47N	33E
Kherson Oblast: Ukr. SSR.	EP	46N	33E
Khilok: RSFSR.	12	51N	111E
Khimki: RSFSR. *Pop. 1954: 100,000.*	8	56N	38E
Khislavichi: RSFSR.	8	54N	32E
Khiuma, *i.*: Est. SSR.	‡5	59N	23E
Khiva: Uzbek SSR. *Textiles*..*	16	41N	60E
Khizy: Azerbaydzhan SSR.	12	41N	49E
Khmel'nik: Ukrainian SSR.	7	50N	28E
Khmel'nitskiy [Proskurov]: Ukrainian SSR.	7	49N	27E
Khmel'nitskiy Oblast: Ukr. SSR.	EP	50N	28E
Khobso Gol, *lake*: Mongolian People's Republic.	12	51N	100E
Khodorov: Ukrainian SSR.	7	49N	24E
Khodyzhenskiy: RSFSR. *Petroleum* (Kuban' Field).*	†	45N	40E
Khodzheyli: Uzbek SSR. *Elec. power.*	16	42N	59E
Khokhol: RSFSR.	9	52N	39E
Kholbon: RSFSR.	12	52N	116E
Kholm: RSFSR.	8	57N	31E
Kholmsk [*Jap.* Maoka]: RSFSR. *Port.*	19	47N	142E

	Page	Lat.	Long.
Kholuy: RSFSR.	8	56N	42E
Khonu: RSFSR.	19	67N	143E
Khorinsk: RSFSR.	12	52N	110E
Khorog: Tadzhik SSR.	17	37N	72E
Khorol: Ukrainian SSR.	9	50N	33E
Khoseda-Khard: RSFSR.	18	67N	60E
Khotimsk: Byelorussian SSR.	8	54N	32E
Khotin: Ukrainian SSR.	7	49N	26E
Khot'kovo: RSFSR.	8	56N	38E
Khoyniki: Byelorussian SSR.	9	52N	30E
Khrapovitskaya: RSFSR.	8	56N	41E
Khrom-Tau: Kazakh SSR. *Chromium.*	14	50N	58E
Khudat: Azerbaydzhan SSR.	12	42N	49E
Khurdalan: Azerbaydzhan SSR. *Petroleum.*	†	40N	49E
Khust [*Czech.* Chust; *Hung.* Huszt]: Ukrainian SSR.	7	48N	23E
Khutor-Mikhaylovskiy: Ukr. SSR.	9	52N	34E
Khuzhir: RSFSR.	12	53N	107E
Khvalynsk: RSFSR.	14	53N	48E
Khvatovka: RSFSR.	14	53N	47E
Khvoynaya: RSFSR.	8	59N	34E
Kicevo: Yugoslavia	11	41N	21E
Kichiga: RSFSR.	19	60N	163E
Kielce: Poland. *Pop. 1950: 62,113. Eng., sawmilling.*	7	51N	21E
KIEV [Kiyev]: Ukr. SSR. *Pop. 1954: 900,000. River port. Eng. (incl. telephone and radio equipment), textiles, elec. power.*	9	50N	30E
Kiev Oblast: Ukrainian SSR.	EP	50N	33E
Kikinda: Yugoslavia	11	46N	20E
Kiliya [*Rom.* Chilia Nova]: Ukrainian SSR.	11	45N	29E
Kil'mez': RSFSR.	13	57N	51E
Kim [Santo]: Tadzhik SSR. *Petroleum.*	‡17	40N	70E
Kimovsk: RSFSR. *Lignite* (Uzlovaya Group).*	†	54N	39E
Kimry: RSFSR.	8	57N	37E
Kinel': RSFSR. *Petroleum.*	14	53N	50E
Kineshma: RSFSR. *Pop. 1954: 110,000. Textiles, chemicals, sawmilling*.*	8	57N	42E
Kingisepp: Estonian SSR.	‡5	58N	22E
Kingisepp: RSFSR. *Sawmilling.*	8	59N	29E
Kirensk: RSFSR.	19	58N	108E
Kireyevka: RSFSR. *Iron ore*, lignite* (Shchekino Group).*	8	54N	38E
Kirgiz Nor, *lake*: Mongolian People's Republic.	19	48N	95E
Kirgiz SSR.:	EP	42N	75E
Kirillov: RSFSR.	8	60N	38E
Kirillovka: Ukrainian SSR.	9	46N	35E
Kirishi: RSFSR.	8	59N	32E
Kirov: RSFSR.	8	54N	34E
Kirov [Vyatka]: RSFSR. *Pop. 1954: 225,000. Textiles, eng., sawmilling.*	14	59N	50E
Kirov Oblast: RSFSR.	EP	58N	50E
Kirovabad [Gandzha]: Azer. SSR. *Pop. 1954: 140,000. Petroleum, barytes, textiles, chemicals.*	12	41N	46E
Kirovakan [Karaklis]: Armenian SSR. *Pop. 1954: 55,000. Chemicals.*	12	41N	44E
Kirovgrad [Kalata]: RSFSR. *Pop. 1954: 50,000. Copper, zinc, pyrites, copper smelting, chemicals.*	13	57N	60E
Kirovo [Besh-Aryk]: Uzbek SSR.	‡17	40N	71E
Kirovograd: Ukr. SSR. *Pop. 1954: 125,000. Engineering.*	9	48N	32E
Kirovograd Oblast: Ukr. SSR.	EP	48N	32E
Kirovsk: Turkmen SSR.	16	38N	60E
Kirovsk [Khibinogorsk]: RSFSR. *Pop. 1954: 50,000. Apatite, nephelite (phosphates), chemicals.*	18	68N	34E
Kirovskiy: RSFSR.	12	46N	48E
Kirovskiy: RSFSR. *Gold*.*	†	45N	133E
Kirovskoye: Kirgiz SSR.	17	43N	72E
Kirs: RSFSR. *Steel.*	13	59N	52E
Kirsanov: RSFSR.	9	53N	43E
Kirya: RSFSR.	14	55N	47E
Kirzhach: RSFSR.	8	56N	39E
Kiselevsk: RSFSR. *Pop. 1954: 80,000. Coal.*	15	54N	87E
KISHINEV [*Rom.* Chişinău]: Moldavian SSR. *Pop. 1954: 170,000. Eng.*, sawmilling.*	11	47N	29E
Kiskőrös: Hungary.	10	47N	19E
Kiskunfélegyháza: Hungary	10	47N	20E
Kiskunhalas: Hungary	10	46N	19E
Kiskunmajsa: Hungary	10	44N	43E
Kislovodsk: RSFSR. *Pop. 1954: 70,000.*	12	44N	43E

* Located by symbol, but not named, on special topic map.
† On relevant special topic map only. For list of topics see p. VIII.
‡ See Inset map.
EP Front Endpaper map.
Place names in [square] brackets are former, or alternative, names.
↓ See appendix on p.134

118

	Page	Lat.	Long.
Kisújszállás: Hungary	11	47N	21E
Kisvárda: Hungary	7	48N	22E
Kitab: Uzbek SSR.	17	39N	67E
Kitoy: RSFSR.	12	53N	104E
Kivak: RSFSR.	19	64N	174W
Kiviyli [*Est.* Kivilõi]: Est. SSR. *Oil shale**.	8	59N	27E
Kiya: RSFSR.	18	68N	45E
Kiyevka: Kazakh SSR.	15	50N	72E
Kiyma: Kazakh SSR.	14	52N	68E
Kizel: RSFSR. *Pop. 1954: 63,000. Coal.*	13	59N	58E
Kizlyar: RSFSR.	12	44N	47E
Kizyl-Arvat: Turkmen SSR. *Engineering.*	16	39N	56E
Kizyl Atrek: Turkmen SSR.	16	38N	54E
Kladanj: Yugoslavia	10	44N	19E
Kladno: Czech. *Coal, iron & steel, eng. (semi-finished goods for further processing at Chomutov).*	6	50N	14E
Kladovo: Yugoslavia	11	45N	22E
Klatovy: Czech. *Textiles*.*	6	49N	13E
Klaypeda [Memel]: Lith. SSR. *Pop. 1954: 64,000. Port. Textiles, chemicals, paper.*	‡5	56N	21E
Klesov: Ukrainian SSR.	7	51N	27E
Kleszczele: Poland	7	53N	23E
Kletnya: RSFSR.	8	53N	33E
Kletsk: Byelorus. SSR. *Textiles*.*	7	53N	27E
Kletskaya: RSFSR.	9	49N	43E
Klimovichi: Byelorussian SSR. *Chemicals.*	8	54N	32E
Klimovo: RSFSR.	9	52N	32E
Klimovsk: RSFSR.	8	55N	38E
Klin: RSFSR. *Pop. 1954: 50,000. Eng., textiles*, elec. power*.*	8	56N	37E
Klintsy: RSFSR. *Textiles, elec. power*.*	8	53N	32E
Klisz: Poland	6	52N	18E
Ključ: Yugoslavia	10	44N	17E
Kłobuck: Poland	6	51N	19E
Kłodawa: Poland	6	52N	19E
Kłodzko [Glatz]: Poland, *Coal.*	6	50N	17E
Klos: Albania	11	41N	20E
Kloštar Ivanic: Yugo. *Petroleum* (Sava River Field).*	10	46N	16E
Kluchev: Byelorussian SSR.	8	54N	30E
Kluczbork [*Ger.* Kreuzburg]: Poland	6	51N	18E
Klukhori [Mikoyan-Shakhar]: Georgian SSR.	12	44N	42E
Klyastitsy: Byelorussian SSR.	8	56N	28E
Klyuchevsk: RSFSR. *Chromium.*	13	57N	61E
Klyuchi: RSFSR.	19	56N	161E
Knezha: Bulgaria	11	43N	24E
Knin: Yugoslavia	10	44N	16E
Knjaževac: Yugoslavia	11	44N	22E
Knyszyn [*Rus.* Knyshin]: Poland	7	53N	23E
Kobarid: Yugoslavia	10	46N	14E
Kobi: Azer. SSR. *Petroleum.*	†	40N	50E
Kobrin: Byelorussian SSR.	7	52N	24E
Kobuleti: Georgian SSR.	12	42N	42E
Kobyay: RSFSR.	19	63N	127E
Kočani: Yugoslavia	11	44N	20E
Koceljevo: Yogoslavia	10	44N	20E
Kočevje: Yugoslavia	10	46N	15E
Kochenevo: RSFSR.	15	55N	83E
Kochetovka: RSFSR.	8	53N	41E
Kochkar': RSFSR.	13	54N	61E
Kochumdek: RSFSR. *Lignite (uncertain exploitation).*	19	65N	93E
Kochura: RSFSR. *Iron ore.*	15	53N	88E
Kodyma: Ukrainian SSR.	9	48N	29E
Kokand: Uzbek SSR. *Pop. 1954: 82,000. Textiles, chemicals, eng.*	‡17	41N	71E
Kokan-Kishlak: Uzbek SSR.	‡17	41N	73E
Kokayty: Uzbek SSR. *Petroleum*.*	17	38N	68E
Kokchetav: Kazakh SSR. *Pop. 1954: 60,000.*	14	53N	70E
Kokchetav Oblast:Kazakh SSR.	EP	53N	70E
Kokhanovo: Byelorussian SSR.	8	54N	30E
Kokhma: RSFSR. *Textiles*.*	8	57N	41E
Kokhtla-Yarve: Estonian SSR. *Oil shale.*	‡5	59N	27E
Kokpekty: Kazakh SSR.	17	49N	82E
Koksovyy: RSFSR. *Coal* (Donbass).*	9	48N	41E
Koktash: Tadzhik SSR.	17	38N	69E
Kokuy: RSFSR.	12	52N	116E
Kok-Yangak: Kirgiz SSR. *Coal, elec. power.*	‡17	41N	73E
Kolanguy: RSFSR.	12	51N	116E
Kola Penin.: RSFSR.	18	68N	37E
▲Kolarovgrad: Bulgaria	11	43N	27E
Kol'chugino: RSFSR. *Engineering.*	8	56N	39E
Kolga Gulf: Est. SSR.	‡5	60N	23E
Kolguyev I: RSFSR.	18	69N	49E

	Page	Lat.	Long.
Kolín: Czech. *River port. Coal, iron, petroleum refining, eng*., chemicals.*	6	50N	15E
Kolka: Latvian SSR.	‡5	57N	22E
Kolki: Ukrainian SSR.	7	51N	26E
Kolno: Poland	7	53N	22E
Koło: Poland	6	52N	19E
Kołobrzeg [*Ger.* Kolberg]: Poland. *Pop. 1939: 36,616; 1946: 2,816. Port.*	6	54N	16E
Kolodnya: RSFSR.	8	55N	32E
Kologriv: RSFSR.	14	59N	44E
Kolomak: Ukrainian SSR.	9	50N	35E
▲Kolomna: RSFSR. *Pop. 1954: 120,000. Eng. (particularly locomotives).*	8	55N	39E
Kolomyya [*Pol.* Kolomyja]: Ukrainian SSR. *Petroleum refining*, eng.**	7	48N	25E
Kolonka: Ukrainian SSR.	9	45N	37E
Kolpashevo: RSFSR. *River port.*	15	58N	83E
Kolpino: RSFSR. *Pop. 1954: 50,000. Steel, eng.*	8	60N	30E
Kolpny: RSFSR.	9	52N	37E
Koltubanovskiy: RSFSR.	14	53N	52E
Koluton: Kazakh SSR. *Bauxite.*	14	52N	69E
Kolyberovo: RSFSR.	8	55N	39E
Kolyma, R.: RSFSR. *Gold.*	23	67N	153E
Kolyma Bay: RSFSR.	19	70N	160E
Kolyma Plain: RSFSR.	19	68N	155E
Kolyuchin, G. of: RSFSR.	19	67N	175W
Kolyvan': RSFSR. *Tungsten, copper, silver-lead.*	15	52N	82E
Kolyvan': RSFSR.	15	55N	83E
Komandor I.: RSFSR.	19	55N	166E
Komarichi: RSFSR.	9	52N	35E
Komárno: Czech. *River port. Textiles*.*	10	48N	18E
Komárom: Hungary	10	48N	18E
Komarovka: Ukrainian SSR. *Textiles.*	9	50N	36E
Komarovo: RSFSR. *Lignite.*	9	59N	34E
Komarovo: RSFSR. *Iron ore.*	13	54N	58E
Komi ASSR.: RSFSR.	EP	65N	55E
Kominternovskoye: Ukr. SSR.	9	47N	31E
Komi-Permyak NO.: Molotov Oblast, RSFSR.	EP	60N	54E
Kommunar [Bogomdarovanny]: RSFSR. *Gold*.*	15	54N	89E
Komrat [*Rom.* Comrat]: Moldavian SSR.	11	46N	29E
Komsomolets Bay: USSR.	16	45N	55E
Komsololets I.: RSFSR. *Uranium.*	19	81N	95E
Komsomol'sk: RSFSR. *Elec. power*.*	8	57N	40E
Komsomol'sk: RSFSR. *Pop. 1954: 150,000. River port. Petroleum refining, steel, eng. (aircraft), elec. power.*	19	51N	137E
Komyshnya: Ukrainian SSR.	9	50N	34E
Konakovo [Kuznetzovo]: RSFSR.	8	57N	37E
Kondinskoye: RSFSR.	14	63N	66E
Kondopoga: RSFSR. *Paper.*	†	63N	35E
Kondratyevskiy: Ukr. SSR. *Coal* (Donbass).*	†	48N	38E
Konevo: RSFSR.	8	60N	38E
Königsberg see Kaliningrad			
Königshütte see Chorzów			
Konin: Poland	6	52N	18E
Konispol: Albania	11	40N	20E
Konjic: Yugoslavia	10	44N	18E
Konosha: RSFSR. *Sawmilling.*	†	61N	40E
Konotop: Ukrainian SSR. *Pop. 1954: 50,000.*	9	51N	33E
Końskie: Poland	7	51N	20E
Konstantinovka: Ukr. SSR. *Pop. 1954: 125,000. Iron and steel, zinc smelting.*	9	48N	38E
Konstantinovskiy: RSFSR.	8	58N	40E
Konstantinovskiy: RSFSR.	9	48N	41E
Konstantinovskiy: RSFSR. *Petroleum refining.*	†	58N	40E
Konstantynów: Poland	6	52N	19E
Kopaonik, mtns.: Yugoslavia	11	43N	21E
Kopar: Yugoslavia	10	46N	14E
Kopani: Ukrainian SSR.	9	47N	32E
Kopatkevichi: Byelorus. SSR.	7	52N	29E
Kopet Dag, mtns.: USSR.	16	38N	58E
Kopeysk: RSFSR. *Pop. 1954: 75,000. Lignite.*	13	55N	62E
Koprivnica: Yugoslavia	10	46N	17E
Kopychintsy: Ukrainian SSR.	7	49N	26E
Korçë [Koritsa], Albania	11	41N	21E
Korcula, *i.*: Yugoslavia	10	43N	17E
Korenevo: RSFSR.	9	51N	35E
Korenovskaya: RSFSR.	9	45N	39E
Korepino: RSFSR.	13	61N	57E
Korets: Ukrainian SSR.	7	51N	27E
Korkino: RSFSR. *Pop. 1954: 50,000. Lignite.*	13	55N	61E

	Page	Lat.	Long.
Körmend: Hungary. *Sawmilling.*	10	47N	16E
Kornat, *i.*: Yugoslavia	10	44N	15E
Korneshty: Moldavian SSR.	11	47N	28E
Korneva [*Ger.* Zinten]: RSFSR.	7	54N	20E
Kornin: Ukrainian SSR.	9	50N	30E
Korocha: RSFSR.	9	51N	37E
Koronowo: Poland	6	53N	18E
Korosten': Ukrainian SSR. *Engineering.*	7	51N	28E
Korostyshev: Ukrainian SSR.	7	50N	29E
Korotoyak: RSFSR.	9	51N	39E
Korsakov [*Jap.* Otomari]: RSFSR. *Pop. 1954: 160,000. Port. Paper.*	19	47N	143E
Korsun'-Schevchenkovskiy: Ukrainian SSR.	9	49N	31E
Koryak NO.: RSFSR.	EP	62N	165E
Koryukovka: Ukrainian SSR.	9	52N	32E
Kosaya Gora: RSFSR. *Iron ore, lignite* (Shchekino Group), steel.*	8	54N	37E
Koschagyl': Kazakh SSR. *Petroleum.*	16	47N	54E
Kóscian: Poland	6	52N	17E
Kóscierzyna [*Ger.* Berent]: Poland.	6	54N	18E
Kosh-Agach: RSFSR.	15	50N	88E
Košice: Czech. *Pop. 1947: 58,080. Magnesite, chemicals, textiles, sawmilling, antimony nearby, eng.*	7	49N	21E
Kosino [Kosa] RSFSR.	14	58N	52E
Kosmach: Ukrainian SSR.	7	48N	25E
Kosov: Ukrainian SSR.	7	48N	25E
Kosovska Mitrovica: Yugo.	11	43N	21E
Kospash: RSFSR. *Coal.*	13	59N	58E
Kostelec: Czechoslovakia	6	50N	16E
Kostopol': Ukrainian SSR.	7	51N	26E
Kostroma Oblast: RSFSR.	EP	58N	45E
Kostroma: RSFSR. *Pop. 1954: 160,000. Textiles.*	8	58N	41E
Kostrzyn [Küstrin]: Poland. *Pop. 1939: 23,711; 1946: 634. Paper*.*	6	52N	14E
Kostyukovichi: Byelorus. SSR..	8	53N	32E
Kostyukovka: Byelorus. SSR.	9	52N	31E
Kos'ya: RSFSR. *Gold*, platinum.*	13	58N	60E
Koszalin [Köslin]: Poland. *Pop. 1939: 33,479; 1949: 17,115. Eng., textiles, sawmilling.*	6	54N	16E
Kőszeg: Hungary	10	47N	16E
Kotel: Bulgaria	11	43N	26E
Kotel'nich: RSFSR.	14	58N	48E
Kotel'nikovskiy: RSFSR.	9	48N	43E
Kotel'nyy I: RSFSR.	19	76N	138E
Köthen: E. Germany. *Lignite, textiles*.*	6	52N	12E
Kotka: Finland	8	60N	27E
Kotlas: RSFSR. *River port. Sawmilling.*	14	61N	47E
Kotor [*It.* Cattaro]: Yugo. *Port.*	10	42N	19E
Kotor Varoš: Yugoslavia	10	45N	17E
Kotovsk: RSFSR. *Chemicals.*	9	53N	42E
Kotovsk [Birzula]: Ukr. SSR.	11	48N	30E
Kounradskiy: Kazakh SSR. *Copper, molybdenum.*	17	47N	75E
Kovda: RSFSR. *Sawmilling.*	18	67N	33E
Kovel' [*Pol.* Kowel]: Ukr. SSR.	7	51N	25E
Kovno see Kaunas			
Kovrov: RSFSR. *Pop. 1954: 80,000. River port. Eng.*, textiles.*	8	56N	41E
Kovylkino: RSFSR.	8	54N	44E
Kovzhinskiy Zavod: RSFSR.	8	60N	37E
Koysug: RSFSR.	9	47N	40E
Koytash: Uzbek SSR. *Tungsten, molybdenum.*	17	40N	67E
Kozel'sk: RSFSR.	8	54N	36E
Kozelets: Ukrainian SSR.	9	51N	31E
Kozel'shchina: Ukr. SSR.	9	49N	34E
Kozhva: RSFSR. *Petroleum* (Pechora Group), sawmilling.*	18	65N	57E
Kozienice: Poland	7	52N	22E
Koźle [Cosel]: Poland	6	50N	18E
Kozlov, C.: RSFSR.	19	55N	162E
Kozlovo: RSFSR.	8	58N	36E
Kozlovshchina: Byelorus. SSR.	7	53N	25E
Kozluk: Yugoslavia	10	44N	19E
Kozmodem'yansk: RSFSR.	14	56N	46E
Kragujevac: Yugoslavia	11	44N	21E
Kraków: Poland. *Pop. 1950: 347,048. Chemicals, paper, eng., metallurgical centre at Nowa Huta..*	6	50N	20E
Kralupy: Czech. *Petroleum refining.*	6	50N	14E
Kramatorsk: Ukr. SSR. *Pop. 1954: 125,000. Iron & steel, eng.*	9	49N	38E
Kranj: Yugoslavia	10	46N	14E
Krapina: Yugoslavia	10	46N	16E

* Located by symbol, but not named, on special topic map.
† On relevant special topic map only. For list of topics see p. VIII.
▲ See appendix on p.134

‡ See Inset map.

EP Front Endpaper map.
Place names in [square] brackets are former, or alternative, names.

Place	Page	Lat.	Long.
Krapivino: RSFSR. *Coal* (*Kuzbass*).	15	55N	87E
Krasavino: RSFSR. *Textiles.*	14	61N	46E
Krasilov: Ukrainian SSR.	7	50N	27E
Krasino: RSFSR.	18	71N	55E
Kraslava: Latvian SSR.	‡5	56N	27E
Kraskino: RSFSR. *Lignite.*	19	42N	131E
Kraslice: Czechoslovakia.	6	50N	12E
Krasnaya Gorbatka: RSFSR.	8	56N	42E
Krasnaya Gorka: RSFSR.	13	55N	57E
Krasnaya-Shapochka: RSFSR. *Bauxite.*	†	60N	60E
Krasnik: Poland	7	51N	22E
Krasnoarmeysk [Voznesenskaya Manufaktura]: RSFSR.	8	56N	38E
Krasnoarmeysk [Golyy Karamysh, Baltser]: RSFSR.	9	51N	46E
Krasnoarmeysk: RSFSR.	9	49N	44E
Krasnoarmeyskaya: RSFSR.	9	45N	38E
Krasnoarmeyskiy: RSFSR.	9	47N	42E
KrasnoarmeyskiyRudinsk: Ukrainian SSR.	9	48N	37E
Krasnoarmeyskoye: RSFSR.	12	43N	46E
Krasnoarmeyskoye: Ukr. SSR. *Coal* (*Donbass*).	9	48N	37E
Krasnodar [Yekaterinodar]: RSFSR. *Pop. 1954: 250,000. Petroleum refining, eng., elec. power, textiles.*	9	45N	39E
Krasnodar Kray: RSFSR.	EP	46N	40E
Krasnodon [Sorokino]: Ukr. SSR. *Coal.*	9	48N	40E
Krasnofarfornyy: RSFSR.	8	59N	32E
Krasnogorovka: Ukr. SSR.	9	48N	38E
Krasnograd [Konstantinograd]: Ukrainian SSR.	9	49N	35E
Krasnogvardeyskiy: RSFSR.	13	57N	62E
Krasnogvardeyskoye: Ukr. SSR.	9	45N	34E
Krasnokamsk: RSFSR. *Pop. 1954: 50,000. Petroleum, petroleum refining, chemicals, paper.*	13	58N	56E
Krasnokutsk: Ukr. SSR.	9	50N	35E
Krasnolesnyy: RSFSR.	9	52N	40E
Krasnomayskiy: RSFSR.	8	58N	34E
Krasnoostrovskiy [Byerskiy]: RSFSR.	8	60N	28E
Krasno-Perekopsk: Ukr. SSR.	9	46N	34E
Krasnopol'ye: Ukrainian SSR.	9	48N	35E
Krasnopol'ye: Ukr. SSR.	9	51N	35E
Krasnoslobodsk: RSFSR.	8	54N	44E
Krasnotur'insk [Turinskiy]: RSFSR. *Pop. 1926: 5,602; 1947: 45,000. Aluminium refinery, copper, zinc.*	13	60N	60E
Krasnoufimsk: RSFSR.	13	57N	58E
Krasnoural'sk: RSFSR. *Copper, zinc, asbestos, pyrites, copper smelting, chemicals.*	13	58N	60E
Krasnousol'skiy: RSFSR. *Chemicals.*	13	54N	56E
Krasnovishersk: RSFSR. *Paper.*	13	60N	57E
Krasnovodsk: Turkmen SSR. *Port. Petroleum refining,eng.*	16	40N	53E
‖ Krasnovodsk Oblast: Turkmen SSR.	EP	40N	55E
Krasnoyarsk: RSFSR. *Pop. 1954: 325,000. River port. Petroleum refining, eng., textiles, sawmilling, paper, synthetic rubber.*	15	56N	93E
Krasnoyarsk Kray: RSFSR.	EP	60N	95E
Krasnoye: RSFSR.	9	53N	39E
Krasnoye: RSFSR.	8	54N	31E
Krasnoye Ekho: RSFSR.	8	56N	41E
Krasnoye Selo: RSFSR.	8	60N	30E
Krasnozavodsk [Zagorskiy]: RSFSR.	8	57N	38E
Krasnozerskoye: RSFSR.	15	54N	79E
Krasnystaw: Poland	7	51N	23E
Krasnyye Baki: RSFSR.	8	57N	45E
Krasnyy Bor: RSFSR. *Chemicals.*	8	60N	35E
Krasnyy Kholm: RSFSR.	8	58N	37E
Krasnyy Kholm: RSFSR.	13	52N	54E
Krasnyy Klyuch: RSFSR. *Paper.*	13	55N	57E
Krasnyy Kut: RSFSR.	14	51N	47E
Krasnyy Liman [Liman]: Ukrainian SSR.	9	49N	38E
Krasnyy Luch: Ukrainian SSR. *Pop. 1954: 60,000. Coal.*	9	48N	40E
Krasnyy-Sulin: RSFSR. *Pop. 1954: 75,000. Elec. power*, steel.*	9	48N	40E
Krasnyy Tekstil'shchik: RSFSR. *Textiles*.*	9	51N	46E
Krasnyy Tkach: RSFSR.	8	56N	39E
Krasnyy Yar: RSFSR.	12	46N	48E
Krasnyy Yar: RSFSR.	9	51N	45E
Kremenchug: Ukr. SSR. *Pop. 1954: 110,000. Eng., textiles.*	9	49N	34E
Kremenets [*Pol.* Krzemieniec]: Ukr. SSR. *Lignite, eng.**	7	50N	26E
Kremennaya [Novo-Glukhov]: Ukr. SSR. *Coal* (*Donbass*).	9	49N	38E
Krestovyy: Georgian SSR.	12	43N	45E
Kresttsy: RSFSR.	8	58N	32E
Kresty: RSFSR.	19	72N	102E
Kretinga: Lithuanian SSR.	‡5	56N	21E
Krichev: Byelorussian SSR.	8	54N	32E
Krinichki: Ukrainian SSR.	9	48N	34E
Kristinovka: Ukrainian SSR.	9	49N	30E
Krivorozh'ye: RSFSR. *Coal* (*Donbass*).	9	48N	40E
Krivoy Rog: Ukrainian SSR. *Pop. 1954: 250,000. Iron, iron ore, eng.*	9	47N	34E
Križevci: Hungary	10	46N	16E
Krk, *i.*: Yugoslavia	10	45N	15E
Krnov: Czech. *Textiles*.*	6	50N	18E
Krolevets: Ukrainian SSR.	9	52N	33E
Kroměříž: Czech. *Engineering.*	6	49N	17E
Kronotskiy Bay: RSFSR.	19	54N	161E
Kronshtadt: RSFSR. *Pop. 1954: 50,000. Port (naval base). Suburb of Leningrad.*	8	60N	30E
Kropachevo: RSFSR.	13	55N	58E
Kropotkin: RSFSR.	19	59N	115E
Kropotkin: RSFSR. *Pop. 1954: 50,000.*	9	45N	41E
Krosniewice: Poland	6	52N	19E
Krosno: Poland. *Petroleum refining*, sawmilling.*	7	50N	22E
Krosno [*Ger.* Crossen]: Poland.	7	52N	15E
Krotoszyn: Poland	6	52N	17E
‖ Kruglyakov: RSFSR.	9	48N	44E
Krujë: Albania. *Bauxite.*	10	42N	20E
Krumovgrad: Bulgaria. *Chromium.*	11	41N	26E
Krumovo: Bulgaria. *Iron ore.*	†	42N	25E
Krupanj: Yugoslavia. *Antimony, lead, antimony smelting.*	10	44N	19E
Krupina: Czech. *Magnesite.*	6	48N	19E
Krupki: Byelorussian SSR.	7	54N	29E
Krupnik: Yugoslavia	11	42N	23E
Kruševac: Yugoslavia	11	44N	21E
Kruševo: Yugoslavia	11	41N	21E
Krustpils: Latvian SSR.	‡5	56N	26E
Krutaya: RSFSR.	18	63N	55E
Krutikha: RSFSR.	15	54N	81E
Krymskaya: RSFSR	9	45N	38E
Krynica: Poland	7	49N	21E
Krzepice: Poland	6	51N	19E
Krzyz: Poland	6	53N	16E
Ksen'yevka: RSFSR.	12	54N	117E
Kuba: Azerbaydzhan SSR.	12	41N	48E
Kuban', R.: RSFSR.	22	45N	39E
Kubena, *lake*: RSFSR.	8	60N	39E
Kubrat: Bulgaria	11	44N	26E
Kučevo: Yugoslavia	11	44N	22E
Kucově *see* Stalin			
Kudymkar: RSFSR.	13	58N	55E
Kugitang: Turkmen SSR.	17	38N	67E
Kukës: Albania. *Iron, chromium.*	11	42N	20E
Kukisvumchorr: RSFSR. *Apatite, nepheline(phosphates).*	†	68N	34E
Kukmor: RSFSR. *Textiles.*	14	56N	51E
Kula: Bulgaria	10	44N	23E
Kula: Yugoslavia	10	46N	20E
Kulaly I: Caspian Sea	12	45N	50E
Kuldiga: Latvian SSR.	‡5	57N	22E
Kulebaki: RSFSR. *Pop. 1954: 50,000. Steel.*	8	55N	42E
Kulotino: RSFSR. *Textiles.*	8	58N	34E
Kul'sary: Kazakh SSR. *Petroleum* (*Emba Field*).*	16	47N	54E
Kultuk: RSFSR.	12	52N	104E
Kuludzhunskiy: Kazakh SSR. *Gold*.*	†	49N	83E
Kulunda: RSFSR.	15	53N	79E
Kulunda Steppe: RSFSR.	15	53N	78E
Kulyab: Tadzhik SSR.	17	38N	70E
Kulyab Oblast: Tadzhik SSR.	EP	38N	70E
Kumak: RSFSR.	13	51N	60E
Kumanovo: Yugoslavia	11	42N	22E
Kum-Bel': Kirgiz SSR. *Tungsten, molybdenum.*	17	42N	76E
Kum-Dag: Turkmen SSR. *Petroleum.*	†	38N	55E
Kumertau: RSFSR. *Coal.*	†	52N	56E
Kunashir, *i.*: RSFSR.	19	44N	146E
Kunda: Estonian SSR.	‡5	60N	27E
Kungrad: Uzbek SSR.	16	43N	59E
Kungur: RSFSR.	13	57N	57E
Kunszentmiklós: Hungary.	10	47N	19E
Kuntsevo: RSFSR. *Pop. 1954: 140,000. Textiles, eng.*	8	56N	38E
Kun'ya: RSFSR.	8	56N	31E
Kupino: RSFSR.	15	54N	78E
Kupishkis: Lith. SSR.	‡5	55N	25E
Kupyansk: Ukrainian SSR.	9	50N	38E
Kupyansk-Uzlovoy: Ukr. SSR.	9	50N	38E
Kura Lowlands: Azer. SSR.	12	40N	48E
Kura, R.: Azer. SSR.	22	40N	48E
Kuragino: RSFSR.	15	54N	93E
Kŭrdzhali: Bulgaria	11	42N	25E
Kureyka: RSFSR.	18	67N	88E
Kurgal'dzhino: Kazakh SSR.	15	50N	70E
Kurgan: RSFSR. *Pop. 1954: 75,000. Engineering.*	14	55N	65E
Kurgan Oblast: RSFSR.	EP	55N	65E
Kurgankaya: RSFSR.	9	45N	41E
Kurganovka [Zaboishchik]: RSFSR.	15	55N	86E
Kurgan-Tyube: Tadzhik SSR.	17	38N	69E
Kuril Is.: RSFSR.	19	45N	150E
Kuril'sk: RSFSR.	19	45N	148E
Kurlovskiy: RSFSR.	8	56N	41E
Kurmenty: Kirgiz SSR.	17	43N	78E
Kurovskoye: RSFSR.	8	56N	39E
Kursk: RSFSR. *Pop. 1954: 140,000. Eng.*, synthetic rubber, textiles, iron ore nearby (Kursk Magnetic Anomaly).*	9	52N	36E
Kursk Oblast: RSFSR.	EP	52N	36E
Kurskiy Zaliv: RSFSR.	7	55N	21E
Kuršumlija: Yugoslavia	11	43N	21E
Kurtamysh: RSFSR.	13	55N	64E
Kusa: RSFSR. *Iron ore, titanium*, eng.*	13	55N	59E
Kusary: Azerbaydzhan SSR.	12	41N	48E
Kushchevskaya: RSFSR.	9	46N	40E
Kushmurun: Kazakh SSR. *Lignite, bauxite.*	14	52N	65E
Kushnarenkovo: RSFSR.	13	55N	55E
Kushva: RSFSR. *Iron ore.*	13	58N	59E
Kustanay: Kazakh SSR. *Pop. 1954: 60,000.*	13	53N	64E
Kustanay Oblast: Kazakh SSR.	EP	52N	63E
Kutais: RSFSR.	9	45N	39E
Kutaisi: Georgian SSR. *Pop. 1954: 100,000. Chemicals, textiles, barium, eng., saw-milling.*	12	42N	43E
Kutna Hora: Czechoslovakia	6	50N	16E
Kutno: Poland	6	52N	19E
Kutulik: RSFSR.	12	53N	103E
Kuty: Ukrainian SSR. *Saw-milling.*	7	48N	25E
Kuuli-Mayak: Turkmen SSR.	16	40N	53E
Kuvandyk: RSFSR. *Bauxite.*	13	52N	58E
Kuvasay: Uzbek SSR.	‡17	40N	72E
Kuvshinovo: RSFSR.	8	57N	34E
Kuybyshev [Samara]: RSFSR. *Pop. 1954: 750,000. River port. Eng., sulphur, paper, petroleum refining.*	14	53N	50E
Kuybyshev: RSFSR.	15	55N	78E
Kuybyshev Oblast: RSFSR.	EP	53N	50E
Kuybyshevka-Vostochnaya: RSFSR.	19	51N	128E
Kuybyshevo: Kazakh SSR.	16	44N	52E
Kuybyshevo: RSFSR.	9	48N	39E
Kuybyshevo [Rishtan, Imeni Kuybysheva]: Uzbek SSR.	‡17	40N	71E
Kuybyshevskiy Zaton: RSFSR.	14	55N	50E
Kuyvastu: Estonian SSR.	‡5	58N	23E
Kuzino: RSFSR.	13	57N	59E
Kuznetsk: RSFSR.	14	53N	47E
Kwidzyn [*Ger.* Marienwerder]: Poland. *Elec. power*.*	6	54N	19E
Kyakhta: RSFSR.	12	50N	106E
Kyardla: Estonian SSR. *Textiles.*	‡5	59N	22E
Kyra: RSFSR.	19	50N	112E
Kyritz: E. Germany	6	53N	12E
Kyshtym: RSFSR. *Pop. 1954: 50,000. Copper smelting, chemicals*.*	13	56N	60E
Kysucké Nové Mesto: Czech.	6	49N	19E
Kytlym: RSFSR. *Gold, platinum.*	13	59N	59E
Kyusyur: RSFSR.	19	71N	127E
Kyustendil: Bulgaria	11	42N	23E
Kyzas: RSFSR.	15	52N	90E
Kyzyl: RSFSR.	15	52N	95E
Kyzyl-Kiya: Kirgiz SSR. *Lignite.*	‡17	40N	72E
Kyzyl-Kum: USSR.	16	43N	65E
Kzyl-Orda: Kazakh SSR. *Pop. 1954: 60,000. Engineering.*	17	45N	65E
Kzyl-Orda Oblast: Kazakh SSR.	EP	45N	65E
Kzyl-Tu: Kazkh SSR.	15	54N	72E
Labin: Yugoslavia	10	45N	14E
Labinot: Albania	10	41N	20E
Labinsk: RSFSR.	9	45N	41E
Labytnangi: RSFSR.	18	67N	66E
Lachinovo: RSFSR.	9	52N	38E
Ladan: Ukrainian SSR.	9	50N	33E
Ladoga, Lake: USSR.	8	61N	32E
Lakinskiy: RSFSR.	8	56N	40E

* Located by symbol, but not named, on special topic map.   ‡ See Inset map.   EP Front Endpaper map.
† On relevant special topic map only. For list of topics see p. VIII.   Place names in [square] brackets are former, or alternative, names.
‖ See appendix on p.134

120

	Page	Lat.	Long.
Lal'sk: RSFSR.	14	61N	48E
Lańcut: Poland	7	50N	22E
Landeshut see Kamienna Góra			
Landsberg see Gorzów Wielkopolski			
Langensalza: East Germany. Paper, eng.*	6	51N	11E
Lanškroun: Czechoslovakia.	6	50N	17E
Laptev Sea: RSFSR.	19	75N	125E
Laptev Strait: RSFSR.	19	73N	142E
Laptevo: RSFSR.	8	55N	38E
Lar'yak: RSFSR.	15	62N	80E
Lashma: RSFSR.	8	55N	41E
Lastovo, i.: Yugoslavia	10	43N	17E
Latnaya: RSFSR.	9	52N	39E
Latvian SSR:	EP	57N	25E
Lavdona: Latvian SSR	‡5	57N	26E
Lazarevac: Yugoslavia	11	44N	20E
Lazdiyay: Lithuanian SSR.	7	54N	24E
Łeba [Ger. Leba]: Poland	6	55N	18E
Lebedin: Ukrainian SSR.	9	51N	34E
Lebedyan': RSFSR.	8	53N	39E
Lębork [Ger. Lauenburg]: Poland.	6	53N	18E
Lebyazh'ye: RSFSR.	14	55N	66E
Łęczna: Poland	7	51N	23E
Łęczyca: Poland	6	52N	19E
Legnica [Ger. Liegnitz]: Poland. Pop. 1950: 55,940. Textiles.	6	51N	16E
Legovskiy: RSFSR.	9	49N	44E
Leipzig: E. Ger. Pop. 1946: 607,700. Eng., textiles, chemicals.	6	51N	12E
Lel'chitsy: Byelorussian SSR.	7	52N	28E
Lena, R.: RSFSR.	23	65N	125E
Lendava: Yugo. Petroleum.	10	47N	16E
Lenger: Kazakh SSR. Lignite.	17	42N	70E
Leninabad [Khodzhent] Tadzhik SSR. Pop. 1954: 60,000. Textiles.	17	40N	70E
Leninabad Oblast: Tadzhik SSR.	EP	40N	70E
Leninakan [Gumry, Aleksandropol']: Armenian SSR. Pop. 1954: 115,000. Textiles, eng.	12	41N	44E
Leningrad [St. Petersburg, Petrograd]: RSFSR. Pop. 1954: 3,150,000. Main port of USSR. Eng., petroleum refining, chemicals, textiles, synthetic rubber, steel, paper. Extensive suburbs. Important for its research and specialization in eng. and chemicals.	8	60N	30E
Leningrad Oblast: RSFSR.	EP	60N	30E
Leningradskaya: RSFSR.	9	46N	39E
Lenino: RSFSR.	8	56N	38E
Leninogorsk [Ridder]: Kazakh SSR. Pop. 1954: 80,000. Tin, lead, zinc, silver, elec. power*, lead smelting.	15	50N	84E
Lenin Peak: USSR.	17	39N	73E
Leninsk: RSFSR.	9	49N	45E
Leninsk [Assake] Uzbek SSR.	‡17	41N	72E
Leninskiy: RSFSR.	8	57N	46E
Leninsk-Kuznetskiy: RSFSR. Pop. 1954: 125,000. Coal, elec. power.	15	55N	86E
Leninskoye: Kazakh SSR.	17	42N	69E
Leninskoye: RSFSR. Antimony (concentrating plant).	19	48N	133E
Lenkoran': Azer. SSR.	12	39N	49E
Lentvaris: Lith. SSR.	‡5	54N	25E
Leonidovo [Jap. Kami-Shikuka]: RSFSR. Coal* (Uglegorsk Group).	†	49N	143E
Leovo [Rom. Leova]: Mold. SSR.	11	46N	28E
Lepel': Byelorussian SSR.	8	55N	29E
Lepsy: Kazakh SSR.	17	46N	79E
Lesh: Albania	10	42N	20E
Lesken: RSFSR. Sawmilling.	†	43N	44E
Lesko: Poland	7	49N	22E
Leskovac: Yugoslavia	11	43N	22E
Leskovik: Albania	11	40N	20E
Lesnoy: RSFSR. Sawmilling.	18	67N	35E
Lesogorsk [Jap. Nayoshi]: RSFSR. Coal* (Uglegorsk Group)	19	50N	142E
Lesogorskiy [Fin. Jääski]: RSFSR. Sawmilling*.	†	61N	29E
Lesopil'noye: RSFSR.	19	46N	135E
Lesozavodsk: RSFSR. Sawmilling.	19	45N	133E
Lesser Caucasus: USSR.	12	41N	45E
Leszno [Ger. Lissa]: Poland. Lignite, eng.	6	51N	16E
Letichev: Ukrainian SSR.	7	49N	28E
Letnerechenskiy: Karelo-Finnish SSR.	18	64N	35E
Letnyaya Stavka: RSFSR.	9	45N	44E
Leushi: RSFSR.	14	60N	66E
Levice: Czechoslovakia	6	48N	19E
Levikha: RSFSR. Copper, zinc.	13	58N	60E
Levoča: Czechoslovakia	7	49N	21E
Lev Tolstoy: RSFSR.	8	53N	39E
Leżajsk: Poland	7	50N	22E
Leznaya: Byelorussian SSR.	7	53N	26E
L'gov: RSFSR.	9	52N	35E
L'govskiy: RSFSR.	9	52N	35E
Libau see Liepaya			
Liberec [Ger. Reichenberg]: Czech. Pop. 1947: 52,798 (incl. suburbs). Textiles, chemicals.	6	51N	15E
Lichtenstein in Sachsen: E. Ger. Coal*, textiles*.	†51N		12E
Lida: Byelorussian SSR.	7	54N	25E
Lidzbark Warmiński [Ger. Heilsberg]: Poland	7	54N	21E
Liegnitz see Legnica			
Liepaya [Ger. Libau]: Latvian SSR. Pop. 1954: 90,000. Port. Steel, eng., chemicals.	‡5	56N	21E
Lifudzin: RSFSR. Tin.	19	44N	135E
Likhoslavl': RSFSR.	8	57N	36E
Likhovskoy [Likhaya]: RSFSR. Coal* (Donbass), eng.	9	48N	40E
Likhula: Estonian SSR.	‡5	58N	24E
Likino-Dulevo: RSFSR.	8	56N	39E
Limanowa: Poland	7	50N	20E
Limbazhi: Latvian SSR.	‡5	57N	24E
Liozno: Byelorussian SSR.	8	55N	30E
Lipetsk: RSFSR. Pop. 1954: 140,000. Iron ore, iron, eng., ferro-alloys.	9	53N	40E
Lipetsk Oblast: RSFSR.	EP	53N	39E
Lipiany [Ger. Lippehne]: Poland.	6	53N	15E
Lipiya: RSFSR.	8	56N	42E
Lipkany [Rom. Lipcani]: Mold. SSR.	7	48N	27E
Lipno: Poland	6	53N	19E
Lipova: Romania	11	46N	22E
Lipovets: Ukrainian SSR.	9	49N	29E
Lisichansk: Ukrainian SSR. Pop. 1954: 40,000. Iron & steel, coal* (Donbass).	9	49N	38E
Liski [Svoboda]: RSFSR.	9	51N	40E
Lisna, lake: Byelorus. SSR.	‡5	56N	28E
Lispeszentadorján: Hungary. Petroleum.*	10	47N	17E
Listonadovka: RSFSR.	9	51N	42E
Listvyanka: RSFSR.	12	52N	105E
Listvyanskiy: RSFSR. Coal.	†	54N	84E
Lithuanian SSR.:	EP	55N	25E
Litija: Yugo. Zinc smelting.	10	46N	15E
Litin: Ukrainian SSR.	7	49N	28E
Litoměřice: Czechoslovakia	6	50N	14E
Litomyšl: Czechoslovakia	6	50N	16E
Litovko: RSFSR.	19	49N	135E
Livani: Latvian SSR.	‡5	56N	26E
Livno: Yugoslavia	10	44N	17E
Livny: RSFSR. Iron*.	9	52N	38E
Ljubija: Yugoslavia. Iron ore.	10	45N	16E
Ljubinje: Yugoslavia	10	43N	18E
Ljubljana [Ger. Laibach]: Yugo. Pop. 1953: 138,211. Engineering.	10	46N	14E
Ljubuški: Yugoslavia	10	43N	18E
Llixhë: Albania. Sulphur.	†	41N	20E
Lobez [Ger. Labes]: Poland	6	54N	16E
Lobva: RSFSR.	13	59N	61E
Lodeynoye Pole: RSFSR.	8	61N	34E
Łódź: Poland. Pop. 1950: 622,500. Eng., textiles, paper.	6	52N	20E
Logoysk: Byelorussian SSR.	7	54N	28E
Loin: Bulgaria	11	44N	23E
Lok-Batan: Azerbaydzhan SSR. Petroleum.*	†	40N	50E
Lokhvitsa: Ukrainian SSR.	9	50N	33E
Loknya: RSFSR.	8	57N	30E
Lokot': RSFSR.	9	52N	34E
Lom: Bulgaria. Lignite.	†	43N	22E
Lomonosov: RSFSR.	8	60N	30E
Lomonosovo: RSFSR.	8	56N	32E
Łomża: Poland	7	53N	22E
Lopandino: RSFSR.	9	52N	35E
Louny: Czechoslovakia	6	50N	14E
Losinj, i.: Yugoslavia	10	45N	14E
Lovászi: Hungary. Petroleum (natural gas).	†	47N	17E
Lovech: Bulgaria	11	43N	25E
Lovozero: RSFSR. (Loparite).	18	68N	34E
Lower Amur Oblast: RSFSR.	EP	55N	135E
Lower Tunguska, R.: RSFSR.	23	64N	94E
Łowicz: Poland	6	52N	20E
Loyev: Byelorussian SSR.	9	52N	31E
Lozovaya: Ukrainian SSR.	9	49N	36E
Lubaczów: Poland	7	50N	23E
Lubán [Ger. Lauban]: Poland. Lignite, textiles*.	6	51N	15E
Lubana, lake: Latvian SSR.	‡5	57N	27E
Lubartów: Poland	7	51N	23E
Lubawa: Poland	6	54N	20E
Lübben: E. Germany	6	52N	14E
Lubin [Ger. Lüben]: Poland	6	51N	16E
Lublin: Poland. Pop. 1950: 111,000. Engineering.	7	51N	22E
Lubliniec: Poland	6	51N	19E
Lubny: Ukrainian SSR. Eng.*, chemicals.	9	50N	33E
Lubsko [Ger. Sommerfeld]: Poland	6	52N	15E
Lučenec: Czechoslovakia. Magnesite, textiles, sawmilling.	6	48N	20E
Luckenwalde: E. Ger. Elec. power, textiles*, eng.*, chemicals.	6	52N	13E
Ludwigslust: E. Germany	6	53N	12E
Ludza: Latvian SSR.	8	56N	28E
Luga: RSFSR. Pop. 1954: 50,000.	8	59N	30E
Lugoj: Romania	11	46N	22E
Lugovoy: Kazakh SSR.	17	43N	73E
Lugskaya Gulf: RSFSR.	‡5	60N	28E
Lukachek: RSFSR. Gold.	19	53N	132E
Lukovit: Bulgaria, Hydro-elec..	11	43N	24E
Lukhovitsy: RSFSR.	8	55N	39E
Lukovnikovo: RSFSR.	8	57N	34E
Łuków: Poland	7	52N	22E
Lukoyanov: RSFSR.	8	55N	44E
Luninets [Pol. Łuniniec]: Byelorussian SSR.	7	52N	27E
Lunino: RSFSR.	8	54N	45E
Lun'yevka: RSFSR. Coal* (Kizel' Group)	13	59N	58E
Lupeni: Romania. Textiles*.	11	45N	23E
Lupków: Poland	7	49N	22E
Lushnjë: Albania.	10	41N	20E
Lutsk [Pol. Łuck]: Ukr. SSR.	7	51N	25E
L'vov [Pol. Lwów, Ger. Lemberg]: Ukr. SSR. Pop. 1954: 425,000. Eng., textiles, chemicals, petroleum refining, elec. power, sawmilling.	7	50N	24E
L'vov Oblast: Ukrainian SSR.	EP	50N	24E
Lwówek: Poland	6	51N	16E
Lyady: RSFSR.	‡5	59N	28E
Lyakhov Is.: RSFSR.	19	74N	142E
Lyangar: Uzbek SSR. Tungsten.	†	40N	66E
Lyskovo: RSFSR.	8	56N	45E
Lys'va: RSFSR. Pop. 1954: 80,000. Steel.	13	58N	58E
Lyuban': Byelorussian SSR.	7	53N	28E
Lyuban': RSFSR.	8	59N	32E
Lyubar: Ukrainian SSR.	7	50N	28E
Lyubashevka: Ukrainian SSR.	9	48N	30E
Lyubcha: Byelorussian SSR.	7	54N	26E
Lyubertsy: RSFSR. Pop. 1954: 125,000.	8	56N	38E
Lyubeshov: Ukrainian SSR.	7	52N	26E
Lyubino: RSFSR. Pop. 1954: 100,000.	8	56N	38E
Lyubinskiy: RSFSR.	15	55N	73E
Lyuboml': Ukrainian SSR.	7	51N	24E
Lyubotin: Ukrainian SSR.	9	50N	36E
Lyudinovo: RSFSR.	8	54N	34E
Madan: Bulgaria. Lead-zinc*.	†	41N	25E
Madona: Latvian SSR.	‡5	57N	26E
Magadan: RSFSR. Pop. 1954: 50,000. Port. Engineering.	19	60N	151E
Magadan Oblast: RSFSR.	EP	65N	155E
Magdagachi: RSFSR.	19	54N	125E
Magdeburg: E. Ger. Pop. 1946: 236,326. River port. Potash, lignite, eng., chemicals.	6	52N	12E
Magnitka: RSFSR. Iron ore*, titanium, vanadium, zinc.	13	55N	60E
Magnitnaya Mt.: RSFSR.	13	53N	59E
Magnitogorsk: RSFSR. Pop. 1954: 300,000. Iron ore, elec. power*, iron & steel, eng., chemicals.	13	53N	59E
Mago: RSFSR.	19	54N	140E
Majdanpek: Yugoslavia. Gold*.	11	44N	22E
Makanchi: Kazakh SSR.	17	47N	82E
Markarakskiy: RSFSR. Gold* (Kiya Valley Field)	15	56N	88E
Makarov [Jap. Shirutoru]: RSFSR. Coal*.	19	48N	142E
Makarska: Yugoslavia	10	43N	17E
Makar'yev: RSFSR.	8	58N	44E
Makat: Kazakh SSR. Petroleum* (Emba Field).	16	48N	53E
Makeyevka: Ukr. SSR. Pop. 1954: 275,000. Iron & steel, eng., coal.	9	48N	38E
Makhachkala: RSFSR. Pop. 1954: 110,000. Port. Petroleum refining*, chemicals, textiles, eng.	12	43N	48E
Makharadze]Ozurgety]: Georgian SSR. Textiles.	12	42N	42E
Makinsk: Kazakh SSR. Engineering.	15	53N	70E
Makó: Hungary	11	46N	20E

---

* Located by symbol, but not named, on special topic map.
† On relevant special topic map only. For list of topics see p. VIII.
⊥ See appendix on p.134

‡ See Inset map.

Place names in [square] brackets are former, or alternative, names.

EP Front Endpaper map.

	Page	Lat.	Long.
Maków [Mazowiecki]: Poland .	7	52N	21E
Maków [Podhalański]: Poland.	7	49N	19E
Maksatikha: RSFSR.	8	58N	36E
Maksimkin Yar: RSFSR.	15	59N	87E
Makushino: RSFSR.	14	55N	67E
Malacky: Czech. *Petroleum.*	6	48N	17E
Malaya: RSFSR.	19	68N	152E
Malaya Belozerka: Ukr. SSR.	9	47N	35E
Malaya Kheta: RSFSR. *Petroleum.*	18	69N	84E
Malaya Serdoba: RSFSR.	8	52N	45E
Malaya Sopcha: RSFSR. *Nickel** (*Monchegorsk deposit*).	†	68N	33E
Malaya Viska: Ukr. SSR.	9	49N	32E
Malaya Vishera: RSFSR.	8	59N	32E
Malbork (Marienburg): Poland. *Elec. power, sawmilling.*	6	54N	19E
Malé Karpaty, *mtns.*: Czech.	6	48N	17E
Malgobek: RSFSR. *Petroleum.*	12	44N	44E
Malin: Ukr. SSR. *Sawmilling.*	9	51N	29E
Małkinia: Poland	7	53N	22E
Malko Turnovo: Bulgaria	11	42N	28E
Malmyzh: RSFSR.	14	56N	50E
Maloarkhangel'sk: RSFSR.	9	52N	36E
Malomal'sk: RSFSR. *Gold*, platinum.*	13	59N	60E
Malorita: Byelorussian SSR.	7	52N	24E
Malo-Uchalinskiy: RSFSR. *Manganese.*	13	54N	60E
Maloyaroslavets: RSFSR.	8	55N	36E
Malye Derbety: RSFSR.	9	48N	45E
Mama: RSFSR. *Mica.*	19	58N	113E
Mamison: Georgian SSR.	12	43N	44E
Mamlyutka: Kazakh SSR.	14	55N	68E
Mamonovo [*Ger.* Heiligenbeil]: RSFSR.	7	54N	20E
Mamontovo: RSFSR.	15	53N	81E
Mandal Gobi: Mongolian People's Republic.	19	46N	106E
Mănecui Ungureni: Romania .	11	45N	26E
Manevichi: Ukrainian SSR.	7	51N	26E
Mangalia: Romania	11	43N	28E
Mangit: Uzbek SSR.	16	42N	60E
Mangyshlak Penin.: USSR. *Manganese, phosphates (see Tauchik).*	16	44N	52E
Manoylin: RSFSR.	9	49N	43E
Mansfeld: E. Germany. *Copper.*	†	51N	11E
Manturovo: RSFSR. *Sawmilling, oil shale.*	8	58N	45E
Manych Depression: RSFSR.	9	47N	42E
Mărăşeşti: Romania	11	46N	27E
Marcali: Hungary	10	47N	17E
Mardakyany: Azer. SSR. *Petroleum*.*	†	40N	50E
Marek: Bulgaria	11	42N	23E
Marevo: RSFSR.	8	57N	32E
Marganets: Kazakh SSR. *see Dzhezdy*	17	48N	67E
Marganets [Gorodishche, Komintern]: Ukr. SSR. *Manganese** (*Nikopol' Group*)	9	48N	35E
Margelan: Uzbek SSR. *Textiles.*	‡17	40N	72E
Mărghita: Romania	11	47N	22E
Mari ASSR.: RSFSR.	EP	57N	48E
Marianské Lazně [*Ger.* Marienbad]: Czech. *Antimony.* (*Noted health resort.*)	6	50N	13E
Maribor [*Ger.* Marburg]: Yugo. *Pop. 1953: 77,124. Engineering.*	10	47N	16E
Mariinsk: RSFSR.	15	56N	88E
Mar'ina Gorka: Byelorus. SSR.	7	54N	28E
Mar'inka: Ukrainian SSR.	9	48N	38E
Mar'ino: RSFSR.	9	51N	37E
Maritime Kray: RSFSR.	EP	45N	135E
Maritsa [Simeonovgrad]: Bulg.	11	42N	26E
Mariyampole: Lithuanian SSR. *Textiles*.*	‡5	54N	23E
Mariyets: RSFSR.	14	56N	50E
Markakol', *lake*: Kazakh SSR.	15	49N	86E
Markha: RSFSR.	19	63N	119E
Markovo: RSFSR.	19	65N	170E
Marsyaty: RSFSR. *Manganese.*	13	60N	60E
Martyyakha: RSFSR.	18	70N	68E
Mary [Merv]: Turkmen SSR. *Pop. 1954: 80,000. Textiles.*	16	38N	62E
Mary Oblast: Turkmen SSR.	EP	37N	62E
Mar'yevka: Kazakh SSR.	14	54N	67E
Mashtagi: Azer. SSR. *Petroleum.*	†	41N	50E
Maslyanino: RSFSR.	15	55N	84E
Maslovo: RSFSR.	13	60N	60E
Matay: Kazakh SSR.	17	46N	79E
Matcha [Madrushkent]: Tadzhik SSR.	17	39N	70E
Mátészalka: Hungary	7	48N	22E
Matochkin Shar Strait: RSFSR.	18	73N	56E
Matveyev-Kurgan: RSFSR.	9	48N	39E
Maykain: Kazakh SSR. *Gold, copper.*	15	52N	76E

	Page	Lat.	Long.
Maykop: RSFSR. *Pop. 1954: 70,000. Petroleum** (*Kuban' Field*), *eng.*	9	45N	40E
Maykor: RSFSR.	13	59N	56E
Mayli-Say: Kirgiz SSR. *Uranium, vanadium, bauxite.*	‡17	41N	72E
Mayno-Pyl'gino: RSFSR.	19	63N	177E
Mayskiy: RSFSR. *Gold.*	19	52N	129E
Mayskiy *see* Krasnokamsk			
Mazheykyay: Lith. SSR.	‡5	56N	22E
Mazirbe: Latvian SSR.	‡5	57N	22E
Mazul'skiy: RSFSR. *Manganese.*	14	56N	90E
Mazuria: Poland	7	54N	21E
Medak: Yugoslavia	10	44N	16E
Medgidia: Romania	11	44N	28E
Mediaş: Romania	11	46N	24E
Mednogorsk: RSFSR. *Copper (with gold, silver), copper smelting, chemicals.*	13	52N	58E
Medvedevka: RSFSR.	13	55N	59E
Medvedovskaya: RSFSR.	9	45N	39E
Medvezhi Is.: RSFSR.	19	71N	161E
Medvezh'yegorsk [*Fin.* Karhumäki]: Karelo-Finnish SSR. *Copper, sawmilling*.*	18	63N	34E
Medyn': RSFSR.	8	55N	36E
Medzhibozh: Ukrainian SSR.	7	49N	27E
Meerane: E. Ger. *Pop. 1946: 27,700. Textiles*.*	6	51N	12E
Meganom, C.: Ukr. SSR.	9	45N	35E
Megri: Armenian SSR.	12	39N	46E
Meiningen: E. Germany .	6	51N	10E
Meissen: E. Germany. *Pop. 1946: 48,900. Chemicals (famous china works, sometimes known as Dresden china).*	6	51N	14E
Melekess: RSFSR.	14	54N	50E
Meleuz: RSFSR.	13	53N	56E
Melitopol': Ukr. SSR. *Pop. 1954: 110,000. Engineering.*	9	47N	35E
Melnik: Bulgaria	11	42N	23E
Mělník: Czechoslovakia	6	50N	14E
Mel'nikovo: RSFSR.	15	57N	84E
Melovoye: Ukrainian SSR.	9	49N	40E
Memel *see* Klaypeda			
Mena: Ukrainian SSR.	9	52N	32E
Menzelinsk: RSFSR.	13	56N	53E
Mercurea: Romania	11	46N	24E
Mercurea Ciuc: Romania	11	46N	26E
Merefa: Ukrainian SSR.	9	50N	36E
Merke: Kazakh SSR.	17	43N	73E
Merković: Yugoslavia	10	43N	18E
Merseburg: E. Ger. *Lignite, chemicals.*	6	51N	12E
Mersrag: Latvian SSR.	‡5	57N	23E
Meshchanitsy: Byelorus. SSR.	8	54N	28E
Meshchera: RSFSR.	8	55N	40E
Meshchovsk: RSFSR.	8	54N	35E
Meshkovskaya: RSFSR.	9	50N	41E
Metlika: Yugoslavia	10	46N	15E
Mezdra: Bulgaria	11	43N	24E
Mezen': RSFSR. *Port. Sawmilling.*	18	66N	45E
Mezhdurech'ye [Shali]: RSFSR.	12	43N	46E
Mezhirech'ye: Ukrainian SSR.	11	48N	26E
Mezhirichi [*Pol.* Międzyrzec]: Ukrainian SSR.	12	51N	27E
Mežica: Yugoslavia. *Lead.*	†	47N	15E
Mezőberény: Hungary	11	47N	21E
Mezőhegyes: Hungary	11	46N	21E
Mezőkövesd: Hungary	11	48N	21E
Mezőtúr: Hungary	11	47N	21E
Mgachi: RSFSR.	19	52N	142E
Miass: RSFSR. *Pop. 1954: 50,000. Gold, eng.*	13	55N	60E
Miastko [*Ger.* Rummelsburg]: Poland	6	54N	17E
Michurin [Tsarevo]: Bulgaria	11	42N	28E
Michurinsk: RSFSR. *Pop. 1954: 90,000.*	8	53N	40E
Miechów: Poland	7	50N	20E
Międzylesie [*Ger.* Mittelwalde]: Poland	6	50N	17E
Międzyrzec: Poland	7	52N	23E
Międzyrzecz [*Ger.* Meseritz]: Poland	7	53N	16E
Mielec: Poland	7	50N	21E
Migulinskaya: RSFSR.	9	50N	41E
Mikalovce: Czechoslovakia	7	49N	22E
Mikhailovgrad: Bulgaria	11	43N	23E
Mikhaylov: RSFSR.	8	54N	39E
Mikhaylovka: RSFSR.	9	50N	43E
Mikhaylovka: Ukrainian SSR. *Coal** (*Donbass*).	†	48N	39E
Mikhaylovka: Ukrainian SSR.	9	47N	35E
Mikhaylovskiy: RSFSR.	15	52N	80E
Mikhaylovskoye: RSFSR.	9	45N	42E
Mikoyan: RSFSR. *Tin.*	†	49N	131E
Mikoyanovka: RSFSR.	9	50N	36E
Mikulov: Czechoslovakia	6	49N	17E
Mil'kovo: RSFSR.	19	55N	158E
Millerovo: RSFSR.	9	49N	40E

	Page	Lat.	Long.
Milyutinskaya: RSFSR.	9	49N	42E
Mindszent: Hungary	11	46N	20E
Mineral'nyye Vody: RSFSR.	12	44N	43E
Mingechaur: Azerbaydzhan SSR. *Hydro-elec.*	12	41N	47E
MINSK: Byelorus. SSR. *Pop. 1954: 320,000. Eng., textiles, elec. power.*	7	54N	28E
Minsk Mazowiecki: Poland	7	52N	22E
Minsk Oblast: Byelorus. SSR.	EP	58N	27E
Minusinsk: RSFSR. *Coal*.*	15	54N	92E
Min'yar: RSFSR. *Engineering*.*	13	55N	58E
Mirgalimsay: Kazakh SSR. *Lead-zinc.* (*Kentau Group.*)	17	43N	68E
Mirgorod: Ukrainian SSR.	9	50N	34E
Mironovka: Ukrainian SSR.	9	50N	31E
Miroşi: Romania	11	44N	25E
Mirzaani: Georgian SSR. *Petroleum, petroleum refining.*	12	41N	46E
Mishkino: RSFSR.	13	55N	56E
Miskolc: Hungary. *Pop. 1941: 77,362. Steel, paper, eng.*	11	48N	21E
Mistek: Czechoslovakia .	6	50N	18E
Miłicz: Poland	6	52N	17E
Mittweida: E. Ger. *Textiles*, eng.**	6	51N	13E
Mizil: Romania	11	45N	26E
Mizoch: Ukrainian SSR.	7	50N	26E
Mizur: RSFSR.	12	43N	44E
Mladá Boleslav: Czechoslovakia. *Chemicals.*	6	50N	15E
Mladenovac: Yugoslavia	11	44N	21E
Mława: P. and	7	53N	20E
Mljet, *i.*: Yugoslavia	10	43N	17E
Mogilev: Byelorus. SSR. *Pop. 1954: 90,000. Eng., textiles.*	8	54N	30E
Mogilev Oblast: Byelorussian SSR.	EP	54N	30E
Mogilev-Podol'skiy: Ukr. SSR.	11	48N	28E
Mogocha: RSFSR. *Gold.*	12	54N	120E
Mogochin: RSFSR. *Sawmilling.*	15	58N	83E
Mogzon: RSFSR.	12	52N	112E
Mohács: Hungary. *Textiles, eng.*	10	46N	19E
Moineşti: Romania. *Petroleum, petroleum refining.*	†	46N	27E
Mointy: Kazakh SSR. *Lead-zinc.*	17	47N	74E
Mokraya Ol'khovka: RSFSR.	9	50N	45E
Moldavian SSR.	EP	46N	29E
Moldova-Nouă: Romania	11	45N	22E
Molochansk: Ukrainian SSR.	9	47N	36E
Molodechno [*Pol.* Molodeczno]: Byelorussian SSR.	7	54N	27E
Molodechno Oblast: Byelorus. SSR.	EP	55N	27E
Molodoy Tud: RSFSR.	8	56N	34E
Molotov: RSFSR. *Pop. 1954: 450,000. River port. Eng., chemicals, copper refining, petroleum refinery, textiles, hydro-elec.* (*Kama Station*).	13	58N	56E
Molotovabad: Kirgiz SSR.	‡17	40N	72E
Molotovo: Georgian SSR.	12	42N	44E
Molotovo: RSFSR.	9	52N	40E
Molotov Oblast: RSFSR.	EP	60N	58E
Molotovsk [Sudostroy]: RSFSR. *Pop. 1954: 70,000.*	18	65N	40E
Molotovsk: RSFSR.	14	58N	50E
Molotovskoye: RSFSR.	9	46N	42E
Moma Mts.: RSFSR.	19	65N	147E
Momchilgrad: Bulgaria	11	43N	25E
Monastyrshche: Ukr. SSR.	9	49N	30E
Monchegorsk: RSFSR. *Nickel, copper.*	18	68N	33E
Monor: Hungary	10	47N	19E
Mor: Hungary	10	47N	18E
Moravská Ostrava [*Ger.* Mährisch Ostrau]: Czech. *Pop. 1947: 790,285. Coal, iron & steel, eng., petroleum refining, chemicals.*	6	50N	18E
Moravská Třebová: Czech.	6	49N	16E
Moravské Budějovice: Czech.	6	49N	15E
Mordovinian ASSR.: RSFSR.	EP	54N	45E
Morozovsk: RSFSR. *Engineering*.*	9	48N	42E
Morshansk: RSFSR. *Chemicals.*	8	53N	42E
Morshin [*Pol.* Morszyn]: Ukrainian SSR.	7	49N	24E
Mor'ye: RSFSR.	8	60N	31E
Morzhenga: RSFSR.	8	60N	40E
Mosal'sk: RSFSR.	8	54N	35E
MOSCOW [Moskva]: RSFSR. *Pop. 1954: 5,250,000. Capital of the USSR. Port (linked by river and canal to the Baltic, White Sea, Black Sea, and the Caspian). Textiles, steel, eng., petroleum refining, chemicals.*			

(Cont. on next page.)

---

* Located by symbol, but not named, on special topic map.
† On relevant special topic map only. For list of topics see p. VIII.
⌐ See appendix on p.134

‡ See Inset map.

EP Front Endpaper map.
Place names in [square] brackets are former, or alternative, names.

MOSCOW—NOVOCHERKASSK

	Page	Lat.	Long.
MOSCOW (cont.) Suburbs incl. within the metropolitan area: Babushkin, Kuntsevo, Lyubertsy, Lyublino, Mytishchi, Perovo, Tushino. Moscow produces approx. 16% of the industrial output of the USSR.	8	56N	38E
Moscow Oblast: RSFSR.	EP	57N	38E
Moskal'vo: RSFSR. Petroleum refining.	19	54N	142E
Moskovskoye: RSFSR.	9	51N	40E
Mosonmagyaróvár: Hungary	10	48N	17E
Mospino: Ukrainian SSR.	9	48N	38E
Most [Ger. Brüx]: Czech. Lignite, chemicals.	6	50N	14E
Mostar: Yugoslavia. Bauxite, lignite, aluminium plant.	10	43N	18E
Mostiska: Ukrainian SSR.	7	50N	23E
Mosty: Byelorussian SSR.	7	53N	25E
Motol': Byelorussian SSR.	7	52N	26E
Mozdok: RSFSR.	12	44N	45E
Mozhaysk: RSFSR.	8	56N	36E
Mozhga: RSFSR. Chemicals, sawmilling.	13	57N	52E
Mozyr': Byelorussian SSR. Sawmilling.	7	52N	29E
Mramor Brdo: Yugoslavia. Petroleum*.	†	45N	17E
Mstera: RSFSR.	8	56N	42E
Mstibovo: Byelorussian SSR.	7	53N	24E
Mstislavl': Byelorussian SSR.	8	54N	32E
Mtsensk: RSFSR.	8	53N	36E
Mtskheta: Georgian SSR.	12	41N	45E
Mühlhausen: E. Ger. Pop. 1946: 48,013. Barytes, paper, eng.*	6	51N	11E
Mukachevo [Czech. Mukačevo, Hung. Munkacs]: Ukr. SSR. Pyrites.	7	48N	23E
ı Mukhtuya: RSFSR.	19	61N	115E
Mukhu, i. & strait: Est. SSR.	‡5	58N	23E
Mumra: RSFSR.	12	46N	48E
Mundybash: RSFSR.	15	53N	87E
Murashi: RSFSR.	14	59N	49E
Muravlyanka: RSFSR.	8	54N	41E
Murgab [Pamirskiy Post]: Tadzhik SSR.	17	38N	74E
Murmansk: RSFSR. Pop. 1954: 160,000. Port. Eng., elec. power.	18	69N	34E
Murmansk Oblast: RSFSR.	EP	68N	35E
Murmino: RSFSR. Textiles.	8	55N	40E
Murom: RSFSR. Pop. 1954: 50,000. Textiles. eng.	8	56N	42E
Murygino: RSFSR.	14	59N	49E
Muskau [Bad Muskau]: E. Ger. Lignite*.	6	52N	15E
Muslyumovo: RSFSR.	13	55N	53E
Mustve: Est. SSR.	‡5	59N	27E
Mutoray: RSFSR.	19	61N	100E
Muya: RSFSR.	19	57N	115E
Muynak: Uzbek SSR. Port.	16	44N	59E
Muyun Kum: USSR.	17	44N	70E
ı Muztor: Kirgiz SSR.	17	42N	73E
Myar'yamaa: Est. SSR.	‡5	59N	24E
Myatlevo: RSFSR.	8	55N	36E
Mydzk: Ukr. SSR. Copper.	†	51N	26E
Myshega: RSFSR.	8	54N	37E
Myślibórz [Ger. Soldin]: Poland.	6	53N	15E
Mysłowice: Poland. Coal*.	†	50N	19E
Mytishchi: RSFSR. Pop. 1954: 110,000.	8	56N	38E
Myysakyula [Moisakula]: Estonian SSR.	5	58N	25E
Naberezhnyye Chelny: RSFSR.	13	56N	52E
Náchod: Czech. Textiles*.	6	50N	16E
Nădlac: Romania	11	46N	21E
Nadushita: Moldavian SSR.	7	48N	28E
Nadvornaya: Ukrainian SSR.	7	49N	25E
Naftalan: Azer. SSR. Petroleum.	12	41N	46E
Nagorno-Karabakh AO.: Azer. SSR.	EP	40N	48E
Nagornskiy: RSFSR. Coal* (Gubakha).	13	59N	58E
Nagornyy: RSFSR. Gold, mica.	19	56N	125E
Nagor'ye: RSFSR.	8	57N	38E
Nagykanizsa: Hungary	10	46N	17E
Nagykáta: Hungary	10	47N	20E
Nagykörös: Hungary	10	47N	20E
Nakhichevan': Azer. SSR.	12	39N	45E
Nakhichevan' ASSR.: Azer. SSR.	EP	39N	45E
Nakhodka: RSFSR. Pop. 1954: 50,000. Port. Engineering.	19	43N	133E
Nakło: Poland	6	53N	18E
Nalaykha: Mongolian People's Republic. Coal.	†	48N	107E
Nal'chik: RSFSR. Pop. 1954: 50,000.	12	43N	44E
Naliboki: Byelorussian SSR.	7	54N	26E
Namangan: Uzbek SSR. Pop. 1954: 90,000. Textiles*.	‡17	41N	72E
ı Namangan Oblast: Uzbek SSR.	EP	42N	73E
Námestovo: Czechoslovakia	6	49N	19E
Namysłow [Ger. Namslau]: Poland	6	51N	18E
Napas: RSFSR.	15	60N	82E
Naroch, L.: Byelorussian SSR.	‡5	55N	26E
Narodichi: Ukrainian SSR.	7	51N	29E
Naro-Fominsk: RSFSR. Textiles.	8	55N	37E
Narovlya: Byelorussian SSR.	9	52N	30E
Narva: Est. SSR. Textiles, eng.*, hydro-elec.	8	59N	28E
Nar'yan-Mar: RSFSR. Port. Sawmilling.	18	68N	53E
Narym: RSFSR.	15	59N	82E
Naryn: Kirgiz SSR.	17	41N	76E
Naryn: RSFSR.	19	50N	97E
Naryshkino: RSFSR.	8	53N	36E
Năsăud: Romania	11	47N	24E
Našice: Yugoslavia	10	45N	18E
Nasielsk: Poland	7	53N	21E
Nauen: E. Ger. Chemicals.	†	53N	13E
Naumburg an der Saale: E. Ger. Chemicals.	†	51N	12E
Navarin, C.: RSFSR.	19	62N	179E
Navlya: RSFSR.	8	53N	34E
Navoloki: RSFSR.	8	58N	42E
Nazarovka: RSFSR.	8	54N	42E
Nazarovo: RSFSR. Lignite.	15	56N	90E
Nebit-Dag [Neftedag]: Turkmen SSR. Petroleum, petroleum refining.	16	40N	55E
Nebolchi: RSFSR.	8	59N	34E
Nefteabad: Tadzhik SSR. Petroleum*.	‡17	40N	71E
Neftechala [Imeni 26 Bakinskikh Komissarov]: Azer. SSR. Petroleum, chemicals.	12	39N	49E
Neftegorsk: RSFSR. Petroleum.	9	44N	40E
Neftyannya: RSFSR.	9	44N	40E
Negoiul, mtn.: Romania	11	46N	24E
Negoreloye: Byelorussian SSR.	7	54N	27E
Negotin: Yugoslavia	11	44N	22E
Negotino: Yugoslavia	11	41N	22E
Nehoiu: Romania	11	45N	26E
Nekhvoroshcha: Ukrainian SSR.	9	49N	35E
Nelidovo: RSFSR. Lignite.	8	56N	33E
Nel'kan: RSFSR.	19	58N	136E
Nel'ma: RSFSR.	19	48N	139E
Neman [Ger. Ragnit]: RSFSR.	7	55N	22E
Nemirov: Ukrainian SSR.	9	49N	29E
Nenets NO.: Arkhangel'sk Oblast, RSFSR.	EP	68N	55E
Nepa: RSFSR.	19	59N	108E
Nerchinsk: RSFSR.	12	52N	116E
Nerchinskiy Zavod: RSFSR.	12	52N	119E
Nerekhta: RSFSR.	8	57N	40E
Nero, lake: RSFSR.	8	57N	40E
Nes': RSFSR.	18	67N	45E
Nesebŭr: Bulgaria	11	43N	28E
Nesterov: Lithuanian SSR.	7	55N	23E
Nesterov [Ger. Stallupönen, Ebenrode]: RSFSR.	7	50N	24E
Nesvizh: Byelorussian SSR.	7	53N	27E
Neubrandenburg: E. Germany. Engineering.	6	54N	13E
Neuruppin: E. Ger. Chemicals.	6	53N	13E
Neustadt: E. Germany	6	54N	11E
Neustrelitz: E. Ger. Engineering*.	6	53N	13E
Nevel': RSFSR.	8	56N	30E
Nevel'sk [Jap. Honto]: RSFSR. Engineering.	19	47N	142E
Never [Larinsky]: RSFSR. Gold.	19	54N	124E
Nevinnomyssk: RSFSR. Textiles.	9	45N	42E
Nevrokop: Bulgaria	11	42N	24E
Nev'yansk: RSFSR. Gold*, platinum.	13	57N	60E
Ney-To, lake: RSFSR.	18	70N	72E
Neya: RSFSR.	8	58N	44E
Nezhin: Ukrainian SSR. Pop. 1954: 50.000.	9	51N	32E
Nicaj: Albania	10	42N	20E
Nidzica: Poland.	7	53N	20E
Nikel' [Fin. Kolosjoki]: RSFSR. Nickel.	18	69N	30E
Nikitinka: Kazakh SSR. Gold*.	17	50N	82E
Nikitinka: RSFSR.	8	56N	33E
Nikitovka: Ukrainian SSR. Mercury, antimony.	9	48N	38E
Nikolayev: Ukr. SSR. Pop. 1954: 225,000. Port. Eng. (2nd largest shipbuilding centre in USSR.), petroleum refining.	9	47N	32E
Nikolayev Oblast: Ukr. SSR.	EP	47N	32E
Nikolayevka: Ukrainian SSR.	9	47N	33E
Nikolayevsk: RSFSR. Pop. 1954: 50,000. Port (naval base). Iron ore, eng., petroleum refining.	19	53N	140E
Nikolayevskiy: RSFSR.	9	50N	45E
Nikol'sk: RSFSR.	14	60N	45E
Nikol'skaya: RSFSR.	14	54N	46E
Nikol'skiy Khutor: RSFSR.	8	53N	46E
Nikol'skoye: RSFSR.	8	59N	38E
Nikol'skoye: RSFSR.	16	48N	46E
Nikopol': Bulgaria	11	44N	25E
Nikopol': Ukr. SSR. Pop. 1954: 100,000. Manganese (mined nearby at Marganets), eng., steel.	9	48N	34E
Nikšić: Yugoslavia	10	43N	19E
Niš: Yugoslavia. Pop. 1953: 60,677. Textiles.	11	43N	22E
Nitra [Ger. Neutra]: Czech.	6	48N	18E
Nitra Pravna: Czechoslovakia	6	49N	19E
Nivskiy: RSFSR. Hydro-elec.*	†	67N	32E
Nizhne Angarsk [Kozlovo]: RSFSR. Lake & river port.	12	56N	109E
Nizhne Duvanka: Ukr. SSR.	9	50N	38E
Nizhnegorskiy [Seytler]: Ukr. SSR.	9	45N	35E
Nizhne-Ilimsk: RSFSR. Iron ore.	19	57N	103E
Nizhne Kamchatsk: RSFSR.	19	56N	162E
Nizhne Kolymsk: RSFSR.	19	68N	160E
Nizhne Troitskiy: RSFSR.	13	54N	54E
Nizhneudinsk: RSFSR.	19	55N	99E
Nizhniy Baskunchak: RSFSR. Salt.*	†	47N	45E
Nizhniye Kresty: RSFSR.	19	69N	157E
Nizhniye Sergi: RSFSR. Iron & steel.	13	57N	59E
Nizhniy Lomov: RSFSR.	8	54N	44E
Nizhniy Pyandzh: Tadzhik SSR.	17	37N	68E
Nizhniy Tagil: RSFSR. Pop. 1926: 38,849; 1954: 400,000. Iron ore, iron & steel, eng., chemicals, sawmilling.	13	58N	60E
Nizhnyaya Salda: RSFSR. Asbestos* (Asbest Group of mines), iron & steel.	13	58N	62E
Nizhnyaya Tura: RSFSR. Iron & steel.	13	59N	60E
Nogay Steppe: RSFSR.	12	44N	46E
Nogaysk: Ukrainian SSR.	9	47N	36E
Noginsk: RSFSR. Pop. 1954: 125,000. Textiles.	8	56N	38E
Noginsk: RSFSR.	19	65N	90E
Nordhausen: E. Germany. Pop. 1946: 40,400. Potash, eng.*	6	51N	11E
Nordvik: RSFSR. Port. Petroleum, salt.	19	74N	111E
Nori: RSFSR.	18	66N	72E
Noril'sk: RSFSR. Pop. 1954: 125,000. Nickel, platinum, copper, coal, cobalt, gold, non-ferrous metal smelting.	18	69N	88E
North Dvina, R. RSFSR.	22	63N	43E
North Kazakhstan Oblast: Kazakh SSR.	EP	54N	69E
North Osetian ASSR.	EP	43N	44E
North Sos'va, R.: RSFSR.	22	64N	63E
Noshul': RSFSR.	14	60N	50E
Novabad [Shulmak]: Tadzhik SSR.	17	39N	70E
Nova Varoš: Yugoslavia	10	43N	20E
Novaya Chigla: RSFSR.	9	51N	40E
Novaya Kasanka: Kazakh SSR.	16	49N	50E
Novaya Ladoga: RSFSR.	8	60N	32E
Novaya Lyalya: RSFSR. Paper.	13	59N	61E
Novaya Odessa: Ukr. SSR.	9	47N	32E
Novaya Sibir' I: RSFSR.	19	75N	150E
Novaya Slobodka: RSFSR.	8	57N	43E
Novaya Ushitsa: Ukr. SSR.	7	49N	27E
Novaya Vodolaga: Ukr. SSR.	9	50N	36E
Novaya Vyzhva: Ukr. SSR.	7	51N	24E
Novaya Zemlya: RSFSR. Lead & zinc, copper (exploitation uncertain).	18	75N	60E
Nova Zagora: Bulgaria	11	42N	26E
Nové Zámky: Czech.	6	48N	18E
Novgorod: RSFSR. Pop. 1954: 50,000. Eng., sawmilling*.	8	59N	31E
Novgorod Oblast: RSFSR.	EP	58N	33E
Novgorod-Severskiy: Ukr. SSR.	9	52N	33E
Novi: Yugoslavia	10	45N	15E
Novi Pazar: Bulgaria	11	43N	27E
Novi Pazar: Yugoslavia	11	43N	21E
Novi Sad: Yugoslavia. Pop. 1953: 83,223. River port. Coal.	10	45N	20E
Novo-Aleksandrovskaya: RSFSR.	9	46N	41E
Novo-Annenskiy: RSFSR.	9	50N	43E
Novo-Arkhangel'sk: Ukr. SSR.	9	49N	31E
Novoasbest [Krasnouralskiy Rudnik]: RSFSR. Asbestos* (Asbest Group of Mines).	13	58N	60E
Novo-Aydar: Ukrainian SSR.	9	49N	39E
Novobogatinskoye: Kazakh SSR. Petroleum.	16	47N	51E
Novocherkassk: RSFSR. Pop. 1954: 90,000. Eng., chemicals.	9	47N	40E

* Located by symbol, but not named, on special topic map. ‡ See Inset map. EP Front Endpaper map.
† On relevant special topic map only. For list of topics see p. VIII. Place names in [square] brackets are former, or alternative, names.
ı See appendix on p.134

* Located by symbol, but not named, on special topic map.   ‡ See Inset map.   **EP** Front Endpaper map.
† On relevant special topic map only.   For list of topics see p. VIII.   Place names in [square] brackets are former, or alternative, names.
↓ See appendix on p.134

124

	Page	Lat.	Long.
Oster: RSFSR.	8	54N	33E
Oster: Ukrainian SSR.	9	51N	32E
Ostróda: Poland	6	54N	20E
Ostrog: Ukrainian SSR.	7	50N	26E
Ostrogozhsk: RSFSR.	9	51N	39E
Ostropol': Ukrainian SSR.	7	50N	28E
Ostrołęka: Poland	7	53N	22E
Ostrov: Romania	11	45N	28E
Ostrov: RSFSR.	8	57N	28E
Ostrovnoye: RSFSR.	19	68N	164E
Ostrowiec: Poland. *Lignite, iron ore, iron.*	7	51N	21E
Ostrów Mazowiecka: Poland	7	53N	22E
Ostrów Wielkopolski: Poland	6	52N	18E
Ostrzeszów: Poland	6	51N	18E
Osveya: Byelorussian SSR.	8	56N	28E
Oświęcim [*Ger.* Auschwitz]: Poland. *Coal, elec. power. (Notorious concentration camp during Second World War.)*	6	50N	19E
Otepya: Estonian SSR.	5	58N	26E
Otočac: Yugoslavia	10	45N	15E
Otwock: Poland	7	52N	21E
Ovidiopol': Ukrainian SSR.	11	46N	30E
Ovruch: Ukrainian SSR.	7	51N	29E
Oyash: RSFSR.	15	55N	84E
Oymyakon: RSFSR.	19	63N	143E
Oytal: Kazakh SSR.	17	43N	73E
Ózd: Hungary. *Lignite, iron & steel.*	11	48N	20E
Ozernovskiy: RSFSR.	19	52N	157E
Ozery: RSFSR. *Engineering.*	8	55N	38E
Ozinki: RSFSR.	14	51N	50E
Ozorków: Poland	6	52N	19E
Pabianice: Poland *Eng., textiles*.	6	52N	19E
Pachelma: RSFSR.	8	53N	43E
Padina: Yugoslavia	11	45N	21E
Pag, *i.*: Yugoslavia	10	45N	15E
Pakhtusovo: RSFSR.	18	75N	59E
Pakrac: Yugoslavia	10	45N	17E
Paks: Hungary	10	47N	19E
Palanka: Yugoslavia	11	44N	21E
Paldiski [*Ger.* Baltischport]: Estonian SSR. *Port (naval base).*	‡5	59N	24E
Palekh: RSFSR.	8	57N	42E
Pallasovka: RSFSR.	16	50N	47E
Palvantash: Uzbek SSR. *Petroleum.*	‡17	41N	73E
Pamirs, *mtns.*: Tadzhik SSR.	17	37N	73E
Pamyati 13 Bortsov [Znamenskiy]: RSFSR.	15	56N	92E
Pamyat' Parizhskoy Kommuny: RSFSR.	8	56N	44E
Panagyurishte: Bulg. *Copper.*	11	42N	24E
Pančevo: Yugoslavia	11	45N	21E
Panciu: Romania	11	46N	27E
Panevezhis: Lith. SSR. *Pop. 1954: 60,000. Textiles.*	‡5	55N	24E
Panfilov: Kazakh SSR. *Hydro-elec.*	17	44N	80E
Panino: RSFSR.	9	52N	40E
Pankrat'yev I.: RSFSR.	18	76N	60E
Panyutino: Ukrainian SSR.	9	49N	36E
Pap: Uzbek SSR.	‡17	41N	71E
Pápa: Hungary	10	47N	17E
Parabel': RSFSR.	15	59N	82E
Paraćin: Yugoslavia	11	44N	22E
Parafiyeyevka: Ukrainian SSR.	9	51N	33E
Parakhino Poddub'ye: RSFSR.	8	58N	32E
Paramushir, *i.*: RSFSR.	19	50N	156E
Parchim: E. Germany	6	54N	12E
Parczew: Poland	7	52N	23E
Pardubice: Czech. *Petroleum refining, eng.* (radio and telephone equipment).	6	50N	16E
Parfino: RSFSR.	8	58N	32E
Pargolovo: RSFSR.	8	60N	30E
Parichi: Byelorussian SSR.	7	53N	29E
Parizhskaya Kommuna: Ukr. SSR. *Coal* (Donbass).	†	48N	39E
Paromnaya: RSFSR.	9	49N	45E
Paşcani: Romania	11	47N	27E
Pasewalk: E. Germany	6	54N	14E
Pashiya: RSFSR.	13	58N	58E
Pashkovskaya: RSFSR.	9	45N	39E
Pashkiy Perevoz: RSFSR.	8	60N	33E
Pasym [*Ger.* Passenheim]: Poland	7	54N	21E
Patos: Albania. *Petroleum.*	10	41N	20E
Pavelets: RSFSR.	8	54N	39E
Pavilosta: Latvian SSR.	‡5	57N	21E
Pavlikeni: Bulgaria	11	43N	25E
Pavlodar: Kazakh SSR. *Pop. 1954: 60,000. River port. Chemicals, sulphates.*	15	52N	77E
Pavlodar Oblast: Kazkh SSR.	EP	52N	75E
Pavlograd: Ukrainian SSR.	9	48N	36E

	Page	Lat.	Long.
Pavlogradka: RSFSR.	15	54N	73E
Pavlovo: RSFSR. *Eng.* (notably surgical instruments).	8	56N	43E
Pavlovo: RSFSR.	19	63N	115E
Pavlovsk: RSFSR.	15	54N	83E
Pavlovsk: RSFSR.	8	60N	30E
Pavlovsk: RSFSR. *Chemicals.*	9	50N	40E
Pavlovskaya: RSFSR.	9	46N	40E
Pavlovskiy: RSFSR.	13	58N	54E
Pavlovskiy-Posad: RSFSR. *Pop. 1954: 50,000. Textiles.*	8	56N	39E
Payde: Estonian SSR.	‡5	59N	25E
Payturma: RSFSR.	19	71N	96E
Pazardzhik: Bulgaria	11	42N	24E
Pazin: Yugoslavia	10	45N	14E
Peć: Yugoslavia	11	43N	20E
Pechenegi: Ukrainian SSR.	9	50N	37E
Pechenezhin: Ukrainian SSR.	7	48N	25E
Pechenga [*Fin.* Petsamo]: RSFSR. *Nickel, cobalt.*	18	70N	30E
Pechora: RSFSR. *Coal, petroleum.*	18	65N	58E
Pechora, R.: RSFSR.	22	66N	52E
Pechory: RSFSR.	8	58N	28E
Pécs: Hungary. *Pop. 1941: 73,000. Coal, chemicals.*	10	46N	18E
Peenemünde: E. Ger. (*Research centre for guided missiles established during 2nd World War*).	6	54N	14E
Pegin: Albania	10	41N	20E
Pehčevo: Yugoslavia	11	42N	23E
Peipus [Chud]. *L.*: RSFSR.	‡5	59N	27E
Pelagruža, *i.*: Yugoslavia	10	42N	16E
Peleduy: RSFSR. *Sawmilling.*	19	60N	113E
Pelhřimov: Czechoslovakia	6	49N	15E
Pelješac, *i.*: Yugoslavia	10	43N	18E
Pen'kovo: RSFSR.	8	54N	37E
Peno: RSFSR. *Sawmilling.*	8	57N	33E
Penza: RSFSR. *Pop. 1954: 250,000. Eng., paper.*	8	53N	45E
Penza Oblast: RSFSR.	EP	53N	45E
Penzhino: RSFSR.	19	63N	168E
Perechin: Ukr. SSR. *Chemicals.*	7	49N	22E
Pereginsko: Ukrainian SSR.	7	49N	24E
Perekop: Ukrainian SSR.	9	46N	34E
Pereshchino: Ukrainian SSR.	9	49N	35E
Pereslavl'-Zalesskiy: RSFSR.	8	57N	39E
Pereyaslav-Khmel'nitsy: Ukr. SSR.	9	50N	31E
Perleberg: E. Germany	6	53N	12E
Perovo: RSFSR.	8	56N	38E
Perušić: Yugoslavia	10	45N	15E
Pervoavgustovskiy: RSFSR.	9	52N	35E
Pervomaysk [Tashino]: RSFSR. *Iron.*	8	55N	44E
Pervomaysk: Ukrainian SSR.	11	48N	31E
Pervomaysk: Ukr. SSR. *Coal* (Donbass).	9	49N	39E
Pervomayskiy: RSFSR. *Bauxite.*	13	55N	59E
Pervomayskiy: Ukr. SSR. *Coal* (Donbass).	†	48N	38E
Pervoural'sk: RSFSR. *Pop. 1954: 60,000. Vanadium, vanadium concentrating plant.*	13	57N	60E
Peschanoye: RSFSR.	12	46N	45E
Peshkopije: Albania	11	42N	20E
Peshtera: Bulgaria	11	42N	24E
Peskovka: RSFSR.	13	59N	52E
Peskovka: Ukrainian SSR.	9	51N	30E
Pesochnoye: RSFSR.	8	58N	39E
Pestovo: RSFSR.	8	59N	36E
Pétfürdő: Hungary. *Petroleum refining.*	†	47N	18E
Petrich: Bulgaria	11	41N	23E
Petrikorka: Ukrainian SSR.	9	49N	34E
Petrikov: Byelorussian SSR.	7	52N	28E
Petrila: Romania. *Coal*.	11	46N	23E
Petrodvorets: RSFSR.	8	60N	30E
Petrokrepost': RSFSR. *Pop. 1954: 50,000. Chemicals.*	8	60N	31E
Petropavlovsk: Kazakh SSR. *Pop. 1954: 120,000. Engineering.*	14	55N	69E
Petropavlovskiy: RSFSR.	16	48N	46E
Petropavlovsk - Kamchatskiy: RSFSR. *Port (naval base).*	19	53N	159E
Petroşani: Romania. *Coal.*	11	45N	23E
Petrosiţa: Romania	11	45N	25E
Petrovac: Yugoslavia	11	44N	21E
Petrovsk: RSFSR. *Pop. 1954: 50,000. Steel.*	12	51N	109E
Petrovsk: RSFSR.	9	52N	45E
Petrovskaya: RSFSR.	9	45N	38E
Petrovskiy: RSFSR.	8	54N	37E
Petrovskoye: RSFSR.	9	45N	43E
Petrovskoye: RSFSR.	8	57N	39E
Petrov Val: RSFSR.	9	50N	45E
PETROZAVODSK [*Fin.* Petroskoi]: Karelo-Finnish SSR. *Pop. 1954: 135,000. Eng., mica, paper*.	18	62N	35E

	Page	Lat.	Long.
Petukhovo: RSFSR. *Engineering*.	14	55N	68E
Pevek: RSFSR.	19	70N	170E
Piatra Neamt: Romania. *Saw-milling, chemicals* (pharmaceuticals, soap).	11	47N	26E
Pichayevo: RSFSR.	8	53N	42E
Piešť any: Czechoslovakia	6	49N	18E
Pikalevo: RSFSR.	8	60N	34E
Piła [*Ger.* Schneidemühl]: Poland. *Pop. 1939: 45,791; 1946: 10,671. Textiles.*	6	53N	17E
Pil'khyn: RSFSR.	19	70N	175E
Pillau *see* Baltiysk			
Pilsen *see* Plzeň			
Pińczów: Poland	7	50N	21E
Pinega: RSFSR.	18	65N	44E
Pinsk: Byelorus. SSR. *Paper.*	7	52N	26E
Piotrków: Poland	6	51N	20E
Piran: Yugoslavia	10	46N	14E
Pirdop: Bulgaria	11	43N	24E
Pirna: E. Germany	6	51N	14E
Pirogovka: Ukrainian SSR.	9	52N	33E
Pirot: Yugoslavia	11	43N	23E
Pirsagat: Azerbaydzhan SSR.	12	40N	49E
Piryatin: Ukrainian SSR. *Engineering*.	9	50N	33E
Písek: Czechoslovakia. *Textiles.*	6	49N	14E
Pisz: Poland	7	54N	22E
Pitelino: RSFSR.	8	55N	42E
Piteşti: Romania. *Petroleum.*	11	45N	25E
Pit-Gorodok: RSFSR. *Gold* (Yenisey Field).	†	59N	91E
Pitkyaranta: Karelo-Finnish SSR. *Copper, paper.*	18	62N	32E
Piua Petri: Romania	11	45N	28E
Plăineşti: Romania	11	46N	27E
Planá: Czechoslovakia. *Mica.*	6	50N	13E
Planerskoye: Ukrainian SSR.	9	45N	35E
Plast: RSFSR. *Gold, arsenic.*	13	54N	61E
Plauen: E. Ger. *Pop. 1946: 88,300. Eng., textiles.*	6	50N	12E
Plav: Yugoslavia	10	43N	20E
Plavnica: Yugoslavia	10	42N	19E
Plavsk: RSFSR. *Engineering.*	8	54N	37E
Pleniţa: Romania	11	44N	23E
Plesetsk: RSFSR. *Sawmilling.*	18	63N	40E
Pleshchanitsy: Byelorus. SSR.	7	54N	28E
Pleszew [*Ger.* Pleschen]: Poland.	6	52N	18E
Pleven: Bulgaria. *Eng., textiles.*	11	43N	25E
Pliskov: Ukrainian SSR.	9	49N	29E
Pljevlja: Yugoslavia	10	43N	19E
Płock: Poland	6	53N	20E
Ploeşti: Romania. *Pop. 1948: 95,632. Petroleum, petroleum refining, eng.*	11	45N	26E
Płońsk: Poland	7	53N	20E
Ploskosh': RSFSR.	8	57N	31E
Płoty: Poland	6	54N	15E
Plovdiv: Bulgaria. *Pop. 1950: 127,000. Textiles*, chemicals.	11	42N	25E
Plunge: Lith. SSR.	‡5	56N	21E
Plyavinyas: Latvian SSR.	‡5	56N	26E
Plyusa: RSFSR.	8	58N	29E
Plzeň [*Ger.* Pilsen]: Czech. *Pop. 1947: 117,814. Coal, iron ore, steel, eng.* (Škoda works), chemicals*.	6	50N	14E
Pobedinskiy: RSFSR.	8	54N	40E
Pochep: RSFSR.	8	53N	34E
Pochinki: RSFSR.	8	54N	45E
Pochinok: RSFSR.	8	54N	32E
Podborov'ye: RSFSR.	8	60N	35E
Poddor'ye: RSFSR.	8	58N	31E
Poděbrady: Czechoslovakia	6	50N	15E
Podgaytsy: Ukrainian SSR.	7	49N	25E
Podgornoye: RSFSR.	9	50N	40E
Podgorodnoye: Ukr. SSR.	9	48N	35E
Podkagernaya: RSFSR.	19	60N	162E
Podol'sk: RSFSR. *Pop. 1954: 125,000. Eng.*, tin smelting.	8	56N	38E
Podporozh'ye: RSFSR.	8	61N	34E
Podravska Slatina: Yugoslavia.	10	46N	18E
Podu Iloaiei: Romania	11	47N	28E
Podvolochisk: Ukrainian SSR.	7	50N	26E
Pogar: RSFSR.	9	52N	33E
Pogradec: Albania	11	41N	21E
Pogrebishchenskiy: Ukr. SSR.	9	49N	29E
Pokhvistnevo: RSFSR. *Petroleum.*	14	54N	52E
Pokrovsk *see* Engel's			
Pokrovskiy Rudnik: RSFSR. *Cobalt.*	†	60N	60E
Pokrovskoye: Ukrainian SSR.	9	48N	36E
Pokrovsk-Ural'skiy: RSFSR. *Bauxite.*	13	60N	60E
Polessk [*Ger.* Labiau]: RSFSR.	7	55N	21E
Polevskoy: RSFSR. *Artificial cryolite, copper, magnesite, pyrites, cobalt, copper refining (see Severskiy).*	13	56N	60E
Polgár: Hungary	11	48N	21E

* Located by symbol, but not named, on special topic map.   ‡ See Inset map.   EP Front Endpaper map.
† On relevant special topic map only. For list of topics see p. VIII.   Place names in [square] brackets are former, or alternative, names.

	Page	Lat.	Long.
Pologi: Ukrainian SSR.	9	47N	36E
Polonnoye: Ukrainian SSR.	7	50N	28E
Polotsk: Byelorussian SSR.	8	55N	29E
Polovinka: RSFSR. *Coal* (*Kizel' Group*).	13	59N	58E
Poltava: Ukr. SSR. *Pop. 1954: 150,000. Eng., textiles.*	9	50N	34E
Poltava Oblast: Ukr. SSR.	EP	50N	35E
Poltava: RSFSR.	15	54N	72E
Polunochnoye: RSFSR. *Manganese.*	13	61N	60E
Pomorie: Bulgaria	11	43N	28E
Pomoshnaya: Ukrainian SSR.	9	48N	31E
Pomyt: RSFSR.	18	63N	69E
Ponizov'ye: RSFSR.	8	56N	31E
Ponoy: RSFSR.	18	67N	41E
Popasnaya: Ukrainian SSR.	9	49N	38E
Popel'niya: Ukrainian SSR.	9	50N	30E
Popigay: RSFSR.	19	72N	10E
Popovo: Bulgaria	11	43N	26E
Poprad: Czechoslovakia	7	49N	20E
Poreč: Yugoslavia	10	45N	14E
Porkhov: RSFSR.	8	58N	30E
Poronaysk [*Jap.* Shikuka]: RSFSR. *Paper.*	19	49N	143E
Port Il'icha: Azer. SSR.	12	39N	49E
Posevnaya: RSFSR.	15	54N	83E
Poshekhon'ye-Volodarsk: RSFSR.	8	58N	39E
Pössneck: E. Germany	6	51N	12E
Postojna: Yugoslavia	10	46N	14E
Postavy [*Pol.* Postawy]: Byelorussian SSR.	‡5	55N	27E
Poti: Georgian SSR. *Port.*	12	42N	42E
Potsdam: E. Germany. *Pop. 1946: 115,100. (Former garrison town. Allied agreements on Germany signed here in 1945).*	6	52N	13E
Povorino: RSFSR.	9	51N	42E
Poyarkovo: RSFSR.	19	50N	128E
Požarevac: Yugoslavia	11	45N	21E
Pozhva: RSFSR.	13	59N	56E
Požega: Yugoslavia	10	45N	17E
Poznań [*Ger.* Posen]: Poland. *Pop. 1950: 327,192. River port. Eng., iron founding, chemicals.*	6	52N	17E
Pragersko: Yugoslavia	10	46N	16E
PRAGUE [*Czech.* Praha]: Czech. *Pop. 1947: 922,284. River port. Eng., chemicals, petroleum refining.*	6	50N	14E
Prahovo: Yugoslavia	11	44N	23E
Praid: Romania	11	47N	25E
Praskoveya: RSFSR.	9	45N	44E
Pravdinsk: RSFSR. *Paper.*	8	56N	44E
Predivinsk: RSFSR.	19	57N	93E
Pregradnaya: RSFSR.	12	44N	41E
Přelouč: Czech. *Manganese, hydro-elec.*	6	50N	16E
Prenzlau: E. Ger. *Elec.power.*	6	53N	14E
Přerov: Czech. *Textiles*.	6	49N	17E
Preševo: Yugoslavia	11	42N	22E
Preslav: Bulgaria	11	43N	27E
Prešov: Czechoslovakia	7	49N	21E
Préspa Lake: Yugoslavia	11	41N	21E
Přeštice: Czech. *Iron ore.*	6	50N	13E
Priazovskoye: Ukrainian SSR.	9	47N	36E
Pribinjč: Yugoslavia	10	44N	18E
Priboj: Yugoslavia	10	44N	20E
Příbram: Czech. *Silver, lead, zinc, barium, antimony.*	6	50N	14E
Prichal'naya: RSFSR.	9	49N	45E
Priekule: Latvian SSR.	‡5	56N	22E
Prievidza: Czechoslovakia	6	49N	19E
Prijedor: Yugoslavia. *Coal nearby.*	10	45N	17E
Prijepolje: Yugoslavia	10	43N	20E
Prilep: Yugoslavia	11	41N	22E
Priluki: Ukrainian SSR.	9	51N	32E
Primorsk [*Ger.* Fischhausen]: RSFSR.	7	55N	20E
Primorsk [Kiovisto]: RSFSR.	8	60N	29E
Primorsko-Akhtarsk: RSFSR.	9	46N	38E
Pripet Marshes: Byelorus. SSR.	7	52N	28E
Pristina: Yugoslavia	11	43N	21E
Pritzwalk: E. Germany	6	53N	12E
Privolzhsk: RSFSR.	8	57N	41E
Privolzhskiy: RSFSR.	12	46N	48E
Privolzhskiy: RSFSR.	8	51N	46E
Prizren: Yugoslavia	11	42N	21E
Priyutnoye: RSFSR.	9	46N	44E
Prnjavor: Yugoslavia	10	45N	18E
Prokhladnyy: RSFSR.	12	44N	44E
Prokop'yevsk: RSFSR. *Pop. 1954: 250,000. Coal.*	15	54N	87E
Prokuplje: Yugoslavia	11	43N	22E
Proletariy: RSFSR.	8	58N	32E
Proletarsk: Ukr. SSR. *Coal* (*Donbass*).	†	49N	38E
Proletarskaya: RSFSR.	9	47N	42E

	Page	Lat.	Long.
Proletarskiy: RSFSR.	9	51N	36E
Pronsk: RSFSR.	8	54N	40E
Prostějov: Czechoslovakia	6	49N	17E
Prosyanaya: Ukrainian SSR.	9	48N	36E
Proszowice: Poland. *Sulphur.*	7	50N	20E
Provadiya: Bulgaria	11	43N	27E
Provideniya: RSFSR. *Port.*	19	65N	174W
Prudnik [*Ger.* Neustadt]: Poland	6	50N	18E
Pruszków: Poland. *Engineering*.	7	52N	21E
Prut, R.: Romania/USSR.	22	46N	28E
Pruzhany [*Pol.* Pružana]: Byelorussian SSR.	7	52N	24E
Przasnysz: Poland	7	53N	21E
Przedbórz: Poland	6	51N	20E
Przemyśl: Poland. *Chemicals.*	7	50N	23E
Przeworsk: Poland	7	50N	22E
Przheval'sk [Karakol]: Kirgiz SSR. *Port at Pristan' Przheval'sk nearby on lake Issyk-Kul'. Engineering.*	17	42N	78E
Pskent: Uzbek SSR.	17	41N	69E
Pskov: RSFSR. *Pop. 1954: 60,000. River port. Textiles, sawmilling.*	8	58N	28E
Pskov Oblast: RSFSR.	EP	58N	30E
Pszczyna: Poland	6	50N	19E
Ptich: Byelorussian SSR.	7	52N	29E
Ptuj: Yugoslavia	10	46N	16E
Puchezh: RSFSR.	8	57N	43E
Puck [*Ger.* Putzig]: Poland	6	55N	18E
Pudozh [Pudozhgora]: Karelo-Finnish SSR. *Titaniferous iron, titanium.*	18	62N	37E
Puești: Romania	11	46N	28E
Pugachev [Nikolayevsk] RSFSR	14	52N	49E
Pugachevo [*Jap.* Maguntanhama]: RSFSR.	19	48N	142E
Pukë: Albania. *Iron ore*, copper.*	10	42N	20E
Pula [*It.* Pola]: Yugoslavia. *Port (naval base).*	10	45N	14E
Puławy: Poland	7	51N	22E
Pułtusk: Poland	7	53N	21E
Pur, R.: RSFSR.	23	65N	79E
Purvomay: Bulgaria	11	42N	25E
Pushkin [Tsarskoye Selo]: RSFSR. *Pop. 1954: 50,000. Chemicals. Suburb of Leningrad.*	8	60N	30E
Pushkino: Azerbaydzhan SSR..	12	40N	48E
Pushkino: RSFSR. *Pop. 1954: 50,000. Eng.*, textiles*.*	8	56N	38E
Pushkino [Urbakh]: RSFSR.	16	51N	47E
Pushkinskiye Gory: RSFSR.	8	57N	29E
Püspökladány: Hungary	11	47N	21E
Pustoshka: RSFSR.	8	56N	29E
Puta: Azer. SSR. *Petroleum.*	†	39N	49E
Putnok: Hungary	7	48N	20E
Putoran, *mtns.:* RSFSR.	19	69N	95E
Puyko: RSFSR.	18	67N	70E
Pyarnu [Parnu]: Est. SSR. *Port. Textiles,eng.*.*	‡5	58N	24E
Pyatigorsk: RSFSR. *Pop.1954: 100,000. Eng. (radio equipment).*	12	44N	43E
Pyatikhatki: Ukrainian SSR.	†	48N	34E
Pyrkakai: RSFSR. *Tin.*	†	69N	175E
Pyrzyce: Poland	6	53N	15E
Pyshma: RSFSR.	13	57N	63E
Pytalovo: RSFSR.	8	57N	28E
Quedlinburg: E. Ger. *Pop. 1946: 35,100. Chemicals. (Known for its Harz cheese).*	6	52N	11E
Rab, *i.:* Yugoslavia	10	45N	15E
Rabka: Poland	7	50N	20E
Racibórz [*Ger.* Ratibor]: Poland. *Pop. 1939: 50,004; 1946: 19,065. Engineering.*	6	50N	18E
Ráckeve: Hungary	10	47N	19E
Rădăuți: Romania. *Engineering.*	11	48N	26E
Radnevo: Bulgaria	11	42N	26E
Radom: Poland. *Pop. 1950: 83,167. Engineering.*	7	51N	21E
Radomir: Bulgaria	11	42N	23E
Radomsko: Poland	6	51N	19E
Radomyshl': Ukr. SSR. *Textiles.*	9	51N	29E
Radoshkovichi [*Pol.* Radoszkowice]: Byelorussian. SSR..	7	54N	27E
Radoviš: Yugoslavia	11	42N	22E
Radul': Ukrainian SSR.	9	52N	31E
Radvilishkis: Lith. SSR.	‡5	56N	23E
Radymno: Poland	7	50N	23E
Radzymin: Poland	7	52N	21E
Radzyn: Poland	7	52N	23E
Rakhov: Ukrainian SSR.	7	48N	24E

	Page	Lat.	Long.
Rakitnoye: RSFSR.	9	51N	36E
Rakvere: Estonian SSR.	‡5	59N	26E
Ramenskoye: RSFSR. *Pop. 1954: 75.000.*	8	56N	38E
Rameshki: RSFSR.	8	58N	36E
Râmnicu Sărat: Romania. *Petroleum.*	11	45N	27E
Râmnicu Vâlcea: Romania	11	45N	24E
Ramon': RSFSR.	9	52N	39E
Rankovićevo [Kraljevo]: Yugo.	11	44N	21E
Raseynyay: Lith. SSR.	‡5	55N	23E
Rasskazovo: RSFSR. *Textiles.*	9	53N	42E
Raška: Yugoslavia	11	43N	21E
Rathenow: E. Germany. *Eng. (optical and precision instruments), chemicals.*	6	53N	12E
Ratno: Ukrainian SSR.	7	52N	24E
Ratta: RSFSR.	18	63N	84E
Rava-Russkaya [*Pol.* Rawa Ruska]: Ukrainian SSR. *Lignite.*	7	50N	24E
Rawa Mazowiecka: Poland	7	52N	20E
Rawicz: Poland	6	52N	17E
Raychikhinsk: RSFSR. *Pop. 1954: 50,000. Lignite.*	19	50N	129E
Rayevskiy: RSFSR.	13	54N	55E
Razdel'naya: Ukrainian SSR.	11	47N	30E
Razdolinsk: RSFSR. *Antimony, gold, magnesium.*	19	58N	94E
Razdol'noye: Ukrainian SSR.	9	46N	34E
Razelm, *lake:* Romania	11	45N	29E
Razgrad: Bulgaria	11	44N	26E
Razlog: Bulgaria	11	42N	22E
Recaş: Romania	11	46N	22E
Rechitsa: Byelorussian SSR. *Chemicals.*	9	52N	30E
Redkino: RSFSR.	8	56N	36E
Regar: Tadzhik SSR.	17	39N	68E
Reghin: Romania	11	47N	25E
Reichenbach: E. Ger. *Pop. 1946: 34,700. Eng.*, textiles*.*	6	51N	12E
Reichenberg *see* Liberec			
Rekinniki: RSFSR.	19	61N	164E
Remontnoye: RSFSR.	9	47N	44E
Reni: Romania	11	45N	28E
Repetek: Turkmen SSR.	16	39N	63E
Rep'yevka: RSFSR.	9	51N	39E
Resen: Yugoslavia	11	41N	21E
Reshetilovka: Ukrainian SSR.	9	49N	34E
Reşiţa: Romania. *Iron & steel, eng., coal.*	11	45N	22E
Rétság: Hungary	10	48N	19E
Revda: RSFSR. *Copper.*	13	57N	60E
Rezekne: Latvian SSR.	‡5	56N	27E
Rezh: RSFSR. *Asbestos* (*Asbest Group*), bauxite, nickel.*	13	57N	61E
Rezina [*Rom.* Rezina-Târg]: Moldavian SSR.	9	48N	29E
Rezna, *lake:* Latvia	5	56N	27E
Rgotina: Yugoslavia	11	44N	22E
Rhodope, *mtns.:* Bulgaria	11	42N	24E
Riesa: E. Germany. *Steel.*	6	51N	13E
RIGA: Latvian SSR. *Pop. 1954: 530,000. Port. Chemicals, paper, textiles, eng., elec. power.*	5	57N	24E
Rijeka [*It.* Fiume]: Yugo. *Pop. 1953: 75,112 (incl. Sušak). Port. Hydro-elec. eng., petroleum refining.*	10	45N	14E
Rila, *mtns.:* Bulgaria	11	42N	23E
Rimavska Sobota: Czech.	7	48N	20E
Rioni [Rionges]: Georgian SSR. *Hydro-elec.*	12	42N	43E
Risan: Yugoslavia	10	42N	19E
Rodensk: Byelorussian SSR.	8	54N	28E
Rodna-Veche: Romania	11	47N	25E
Rodniki: RSFSR.	8	57N	42E
Rogachev: Byelorussian SSR.	8	53N	30E
Rogan': Ukrainian SSR.	9	50N	37E
Rogatica: Yugoslavia	10	44N	19E
Rogatin: Ukrainian SSR.	7	49N	25E
Rokishkis [*Lith.* Rokiškis]: Lithuanian SSR.	‡5	56N	25E
Rokyčany: Czechoslovakia	6	50N	14E
Roman: Romania	11	47N	27E
Roman Kosh, *mtn.:* Ukr. SSR..	9	44N	34E
Romanovka: Moldavian SSR.	11	46N	29E
Romanovka: RSFSR.	9	52N	43E
Romanovka: RSFSR.	12	54N	112E
Romany: Azer. SSR. *Petroleum.*	†	40N	50E
Romny: Ukr. SSR. *Petroleum.*	9	51N	34E
Romodan: Ukrainian SSR.	9	50N	33E
Roshal': RSFSR.	8	56N	40E
Roșiorii de Vede: Romania	11	44N	25E
Rositsa: Bulgaria	11	44N	28E
Roslavl': RSFSR.	8	54N	33E
Rossosh': RSFSR.	9	50N	40E
Rostock: E. Ger. *Pop. 1946: 117,000. Port. Engineering.*	6	54N	12E
Rostov: RSFSR.	8	57N	40E

---

* Located by symbol, but not named, on special topic map.   ‡ See Inset map.   EP Front Endpaper map.
† On relevant special topic map only. For list of topics see p. VIII.   Place names in [square] brackets are former, or alternative, names.
⅄ See appendix on p.134

	Page	Lat.	Long.
Rostov-na-Donu: RSFSR. *Pop. 1954: 600,000. Port. Eng., elec. power, paper.*	9	47N	40E
Rostov Oblast: RSFSR.	EP	47N	41E
Roven'ki: RSFSR.	9	50N	39E
Roven'ki: Ukr. SSR. *Coal* (Donbass).*	9	48N	39E
Rovinj: Yugoslavia	10	45N	14E
Rovno [*Pol.* Równe]: Ukr. SSR. *Pop. 1954: 50,000. Textiles, paper.*	7	51N	26E
Rovno Oblast: Ukr. SSR.	EP	51N	32E
Rovnoye [Zel'man]: RSFSR.	16	51N	46E
Roya: Latvian SSR.	‡5	57N	23E
Rozaj: Yugoslavia	10	43N	20E
Rozhva: RSFSR.	13	59N	56E
Rozhishche: Ukrainian SSR.	7	51N	25E
Rožňava: Czech. *Antimony.*	7	49N	21E
Rozwadów: Poland	7	50N	22E
Rrubig *see* Lesh			
Rtishchevo: RSFSR. *Engineering*.*	9	52N	44E
Rubezhnoye: Ukr. SSR. *Pop. 1954: 50,000. Coal, chemicals.*	9	49N	38E
Rubtsovsk: RSFSR. *Pop.1954: 50,000. Engineering.*	15	52N	81E
Rudabánya: Hungary. *Iron ore.*	†	48N	21E
Rudensk: Byelorussian SSR.	7	54N	28E
Rudki: Ukrainian SSR.	7	50N	23E
Rudnichnyy: RSFSR. *Iron ore* (Bakal Group).*	†	55N	59E
Rudnichnyy: RSFSR. *Phosphates, chemicals.*	13	60N	52E
Rudnichnyy: RSFSR. *Coal* (Kizel' Group).*	†	59N	57E
Rudnichnyy: RSFSR. *Iron ore* (Ivdel' Group).*	†	60N	60E
Rudnitsa: Ukrainian SSR.	9	48N	29E
Rudnya: RSFSR.	8	55N	31E
Rudolstadt: E. Germany	6	51N	12E
Rügen, *i.*: E. Germany	6	55N	14E
Ruhland: E. Ger. *Lignite.*	6	52N	14E
Ruiena [*Lat.* Rujiena]: Latvian SSR.	5	58N	25E
Ruma: Yugoslavia	10	45N	20E
Rumburk: Czechoslovakia	6	51N	15E
Rumyantsevo: Ukrainian SSR. *Coal* (Donbass).*	†	48N	38E
Rupea: Romania	11	46N	25E
Ruse [Ruschuk]: Bulgaria. *Pop. 1946: 53,420. River port.*	11	44N	26E
Rushan: Tadzhik SSR.	17	38N	72E
Russian Soviet Federated Socialist Republic	EP	–	–
Rustavi: Georgian SSR. *Pop. 1948: 10,000; planned 50,000. Iron and steel.*	12	42N	45E
Rutchenkovo: Ukrainian SSR. *Coal* (Donbass).*	9	48N	38E
Ruzayevka: RSFSR.	8	54N	45E
Ruzhany: Byelorussian SSR.	7	53N	25E
Ružomberok: Czech. *Paper.*	6	49N	19E
Ryapina [Rapina]: Est. SSR.	8	58N	28E
Ryazan': RSFSR. *Pop. 1954: 125,000. Engineering.*	8	55N	40E
Ryazan' Oblast: RSFSR.	EP	54N	41E
Ryazhsk: RSFSR.	8	54N	40E
Rybach'ye: Kirgiz SSR. *Lake port.*	17	42N	76E
Rybinsk Reservoir: RSFSR.	8	59N	38E
Rybnik: Poland	6	50N	18E
Rybnitsa: Moldavian SSR.	10	48N	29E
Rybnoye: RSFSR.	8	55N	40E
Ryl'sk: RSFSR.	9	52N	34E
Rýmařov: Czechoslovakia	6	50N	17E
Rypin [*Ger.* Rippin]: Poland	6	53N	20E
Rzepin: Poland	6	52N	15E
Rzeszów: Poland. *Eng., manganese.*	7	50N	22E
Rzhava: RSFSR.	9	51N	37E
Rzhev: RSFSR. *Pop. 1954: 60,000. Engineering*.*	8	56N	34E
Saalfeld. E. Germany	6	51N	11E
Saatly: Azerbaydzhan SSR.	12	40N	48E
Šabac [Shabats]: Yugoslavia	10	45N	20E
Sabinov: Czechoslovakia.	7	49N	21E
Sabirabad [Petropavlovka]: Azerbaydzhan SSR.	12	40N	48E
Sablinskoye: RSFSR.	9	44N	43E
Sabunchi: Azerbaydzhan SSR. *Petroleum.*	12	40N	50E
Sachkere: Georgian SSR.	12	42N	44E
Săcueni: Romania .	11	47N	22E
Sadon: RSFSR. *Lead, zinc, lead smelting.*	12	43N	44E
Sadovoye: RSFSR.	9	47N	44E
Safakulevo: RSFSR.	13	57N	62E
Safonovo: RSFSR. *Lignite.*	8	55N	34E
Sagiz: Kazakh SSR.	16	48N	55E
Sahy: Czechoslovakia	6	48N	19E

	Page	Lat.	Long.
Sakhalin Bay: RSFSR.	19	54N	142E
Sakhalin Oblast: RSFSR.	EP	50N	142E
Sakhnovshchina: Ukr. SSR.	9	49N	36E
Saki: Ukrainian SSR.	9	45N	34E
Saksaul'skiy: Kazakh SSR.	16	47N	61E
Sala: Czechoslovakia	10	48N	18E
Salair: RSFSR. *Lead-zinc, barium.*	15	54N	85E
Salavat: RSFSR. *Petroleum.*	†	53N	56E
Saldus: Latvian SSR. *Textiles.*	‡5	56N	22E
Salekhard: RSFSR. *Port.*	18	67N	66E
Salgótarján: Hungary. *Coal*, lignite*, iron and steel.*	6	48N	20E
Salm I.: RSFSR.	18	80N	60E
Salonta: Romania	11	47N	22E
Sal'sk: RSFSR.	9	46N	42E
Sal'yany: Azerbaydzhan SSR.	12	40N	49E
Salzwedel: E. Ger. *Rock-salt*, chemicals.*	6	53N	11E
Sama: RSFSR. *Iron ore* (Ivdel' Group).*	13	60N	60E
Samac: Yugoslavia	10	45N	18E
Samarga: RSFSR.	19	48N	138E
Samarkand: Uzbek SSR. *Pop. 1954: 165,000. Textiles, eng..*	17	40N	67E
Samarkand Oblast: Uzbek SSR.	EP	41N	68E
Samarovo: RSFSR.	14	61N	69E
Samarskoye: Kazakh SSR. *Gold*.*	†	49N	84E
Sambor: Ukrainian SSR..	7	49N	23E
Samoded: RSFSR.	18	64N	40E
Samokov: Bulgaria.	11	42N	24E
Šamorín: Czechoslovakia	10	48N	17E
Samtredia: Georgian SSR.	12	42N	42E
Samus': RSFSR.	15	57N	85E
Sandanski: Bulgaria	11	42N	23E
Sandomierz: Poland	7	51N	22E
Sandovo: RSFSR.	8	58N	36E
Sangar: RSFSR. *Coal.*	19	64N	128E
Sangerhausen: E. Ger. *Copper.*	6	51N	11E
Sangyyakhtakh: RSFSR.	19	61N	124E
Sanok: Poland. *Manganese.*	7	49N	22E
Sapozhok: RSFSR.	8	54N	41E
Sarajevo: Yugoslavia. *Pop. 1953: 135,657. Eng. (machine tools).*	10	44N	18E
Saraktash: RSFSR.	13	52N	56E
Sarala: RSFSR.	15	55N	89E
Saran': Kazakh SSR. *Coal.*	17	50N	73E
Sarandë: Albania	10	40N	20E
Saransk: RSFSR. *Pop. 1954: 50,000. Engineering.*	8	54N	45E
Sarany: RSFSR. *Chromium, chemicals.*	13	59N	59E
Sarapul: RSFSR. *Pop. 1954: 50,000.*	13	56N	54E
Sarata: Ukrainian SSR.	11	46N	30E
Saratov: RSFSR. *Pop. 1954: 600,000. River port. Eng. (particularly ball-bearings), textiles, petroleum refining, elec. power, sawmilling..*	9	52N	46E
Saratov Oblast: RSFSR.	EP	52N	47E
Sárbogárd: Hungary	10	47N	19E
Sarema, *i.*: Est. SSR.	‡5	58N	22E
Sarich, C.: Ukrainian SSR.	9	44N	34E
Sarisu, R. RSFSR.	22	47N	68E
Sarny: Ukrainian SSR.	7	51N	27E
Sars: RSFSR.	13	57N	57E
Sartas: Turkmen SSR.	16	42N	53E
Sartyn'ya: RSFSR.	18	63N	63E
Sárvár: Hungary	10	47N	17E
Saryagach: Kazakh SSR.	17	41N	69E
Sásd: Hungary	10	46N	18E
Saskylakh: RSFSR.	19	72N	114E
Sasovo: RSFSR.	8	54N	42E
Sassnitz: E. Germany	6	55N	14E
Sasyk, *lake*: Ukrainian SSR.	9	46N	29E
Sasyk Kol', *lake*: Kazakh SSR..	15	47N	82E
Satka: RSFSR. *Magnesite, chemicals.*	13	55N	59E
Sátoraljaújhely: Hungary	7	48N	22E
Satu-Mare: Romania. *Pop. 1948: 46,519. Engineering.*	11	48N	23E
Satyga: RSFSR.	13	60N	65E
Sava, R.: RSFSR. *Petroleum.*	22	46N	16E
Săvârşin: Romania	11	46N	22E
Săveni: Romania	11	48N	27E
Savino: RSFSR.	8	56N	41E
Šavnik: Yugoslavia	10	43N	19E
Savran': Ukrainian SSR.	9	48N	30E
Sayat: Turkmen SSR.	16	39N	64E
Sayn Shanda: Mongolian People's Republic. *Coal.*	19	45N	110E
Sazan, *i.*: Albania	10	40N	19E
Scânteia: Romania	11	47N	28E
Schleiz: E. Germany	6	50N	12E
Schneeberg: E. Ger. *Uranium, nickel, cobalt.*	6	51N	12E
Schwedt: E. Germany	6	53N	14E

	Page	Lat.	Long.
Schwerin: E. Germany. *Pop. 1946: 88,164. Chemicals*.*	6	54N	12E
Sebeş: Romania	11	46N	24E
Sebezh: RSFSR.	8	56N	28E
Sebiş: Romania	11	46N	22E
Sečenovo	8	55N	46E
Sečovce: Czechoslovakia	7	49N	22E
Segezha [*Fin.* Sekehen]: Karelo-Finnish SSR. *Sawmilling.*	†	63N	34E
Seini: Romania	11	48N	23E
Selenga, R.: RSFSR.	23	49N	103E
Selety Tengiz, *lake*: Kazakh SSR.	15	53N	73E
Seligdar: RSFSR. *Gold*, elec. power.*	19	58N	125E
Seliger, *lake*: RSFSR.	8	57N	33E
Selishche: RSFSR. *Lignite* (Schzharovo Group).*	8	57N	33E
Selizharovo: RSFSR. *Lignite.*	8	57N	34E
Selyakino: RSFSR. *Columbium.*	†	55N	60E
Semenov: RSFSR.	8	57N	44E
Semenovka: Ukrainian SSR.	9	52N	33E
Semenovskoye: RSFSR.	18	63N	43E
Semiluki: RSFSR.	9	52N	39E
Semiozernoye: Kazakh SSR.	13	52N	64E
Semipalatinsk: Kazakh SSR. *Pop. 1954: 120,000. River port. Textiles, sawmilling, eng., elec. power.*	15	50N	80E
Semipalatinsk Oblast: Kazakh SSR.	EP	48N	80E
Semiyarskoye: Kazakh SSR.	15	51N	78E
Semiz-Bugu: Kazakh SSR.	15	50N	75E
Semki: Ukrainian SSR.	7	50N	28E
Sengiley: RSFSR.	14	54N	49E
Senj: Yugoslavia	10	45N	15E
Sennaya: RSFSR.	9	45N	37E
Senno: Byelorussian SSR.	8	55N	30E
Senta: Yugoslavia	11	46N	20E
Sępolno: Poland	6	53N	18E
Septemvri [Saranovo]: Bulg.	11	42N	24E
Serafimovich: RSFSR.	9	50N	43E
Sercaia: Romania	11	46N	25E
Serdobsk: RSFSR.	8	52N	44E
Serdtse-Kamen, C.: RSFSR. *Lead-silver.*	†	67N	172W
Sered: Czechoslovakia	6	48N	18E
Seregovo: RSFSR.	18	62N	50E
Sergach: RSFSR.	8	56N	45E
Serge Kirov Is.: RSFSR.	18	78N	90E
Sergeyevka: Ukrainian SSR.	9	49N	38E
Sergiyevsk: RSFSR.	13	54N	51E
Sergiyevskiy: RSFSR.	13	58N	53E
Sernyy-Zavod: Turkmen SSR. *Sulphur, chemicals.*	16	40N	59E
Serov [Nadezhdinsk, Kabakovsk]: RSFSR. *Pop. 1954: 110,000. Bauxite, iron & steel, eng.*	13	59N	61E
Serpukhov: RSFSR. *Pop. 1954: 125,000. Eng., textiles.*	8	55N	37E
Sestroretsk: RSFSR.	8	60N	30E
Sevan: Armenian SSR. *Site of Ozernaya underground hydro-elec. power station.*	12	41N	45E
Sevastopol': Ukr. SSR. *Pop. 1954: 200,000. Port (naval base). Engineering.*	9	45N	34E
Severnaya Zemlya: RSFSR.	19	79N	95E
Severny Kommunar: RSFSR.	13	58N	54E
Severokamsk *see* Krasnokamsk			
Severo-Kuril'sk: RSFSR. *Port..*	19	51N	156E
Severoural'sk [Petropavlovskiy]: RSFSR. *Pop. 1939: under 2,000; 1947: 30,000. Bauxite.*	13	60N	60E
Severo-Yeniseyskiy: RSFSR. *Gold.*	15	60N	93E
Severskiy: RSFSR. *Steel.*	†	56N	60E
Sevlievo: Bulgaria	11	43N	25E
Sevsk: RSFSR.	9	52N	34E
Seymchan: RSFSR. *Gold.*	19	63N	152E
Sfânta Ana: Romania	11	46N	22E
Sfânta Gheorghe: Romania	11	46N	26E
Shabla: Bulgaria	11	44N	28E
Shadrinsk: RSFSR. *Textiles.*	13	56N	64E
Shaim: RSFSR.	13	60N	64E
Shakhrisyabz: Uzbek SSR.	17	39N	67E
Shakhta: RSFSR. *Coal* (Kizel' Group).*	†	59N	58E
Shakhta: *The name of a number of coal mining villages in Stalino and Voroshilovgrad, Ukr. SSR. (Donbass Coalfield).*	†	48N	39E
Shakhty [Aleksandrovsk-Grushevskiy]: RSFSR. *Pop. 1954: 200,000. Coal, elec. power*, eng.*	9	48N	40E
Shakhun'ya: RSFSR.	14	58N	46E
Shakyay: Lithuanian SSR.	7	55N	23E
Shalya: RSFSR. *Gold.*	13	57N	59E
Shalym: RSFSR. *Iron ore.*	15	53N	88E
Shantar Is.: RSFSR.	19	55N	138E
Shapola: Ukrainian SSR.	9	49N	31E

* Located by symbol, but not named, on special topic map. ‡ See Inset map. EP Front Endpaper map.
† On relevant special topic map only. For list of topics see p. VIII. Place names in [square] brackets are former, or alternative, names.
↓ See appendix on p.134

127

SHARAPOVY—STALINABAD

	Page	*Lat.*	*Long.*
Sharapovy Koshki I.: RSFSR.	18	71N	67E
Shargorod: Ukrainian SSR.	7	51N	28E
Shar'ya: RSFSR.	8	58N	46E
Sharyn: Kazakh SSR.	17	44N	80E
Shatsk: RSFSR.	8	54N	42E
Shatura: RSFSR. *Pop. 1954: 50,000. Elec. power**.	8	56N	39E
Shchekino: RSFSR. *Lignite, elec. power**.	8	54N	38E
Shchelkovo: RSFSR. *Pop. 1954: 50,000.*	8	56N	38E
▲ Shcherbakov : RSFSR. *Pop. 1954: 160,000. Hydro-elec., eng., textiles.*	8	58N	39E
Shchigry: RSFSR. *Phosphates.*	9	52N	37E
Shchors [Snovsk]: Ukr. SSR.	9	52N	32E
Shchorsk: Ukrainian SSR.	9	48N	34E
Shchuchin: Byelorussian SSR.	7	54N	25E
Shchuchinsk: Kazakh SSR.	15	53N	70E
▲ Shchurovo: RSFSR.	8	55N	39E
Shebekino: RSFSR.	9	50N	37E
Shelekov Bay: RSFSR.	19	60N	158E
Shelyakino: RSFSR.	9	50N	39E
Shemakha: Azerbaydzhan SSR.	12	41N	49E
Shemonaikha: Kazakh SSR.	15	51N	82E
Shëngjin; Albania	10	42N	20E
Shenkursk: RSFSR.	18	62N	43E
Shepetovka: Ukrainian SSR.	7	50N	27E
Sherlovaya Gora: RSFSR. *Tin, beryllium.*	12	51N	116E
Shilka: RSFSR.	12	52N	116E
Shilka, R.: RSFSR.	23	53N	118E
Shilovo: RSFSR.	8	53N	38E
Shilute: Lithuanian SSR.	‡5	55N	21E
Shimanovsk: RSFSR.	19	52N	128E
Shirabad: Uzbek SSR.	17	38N	67E
Shiringushi: RSFSR.	8	54N	43E
Shirvan Steppe: Azerbaydzhan SSR.	12	40N	48E
Shkodër [Scutari]: Albania	10	42N	20E
Shklov: Byelorussian SSR.	7	54N	30E
Shmidt I.: RSFSR.	19	81N	90E
Shor-Su: Uzbek SSR. *Sulphur, chemicals.*	‡17	40N	71E
Shortandy: Kazakh SSR.	15	52N	71E
Shostka: Ukrainian SSR.	9	52N	33E
Shpola: Ukrainian SSR.	9	49N	32E
Shtergres: Ukr. SSR. *Elec. power.*	9	48N	39E
Shterovka: Ukrainian SSR.	9	48N	39E
Shubany: Azer. SSR. *Petroleum.*	†	40N	50E
Shubar-Kuduk: Kazakh SSR. *Petroleum.*	6	49N	56E
Shugurovo: RSFSR. *Petroleum.*	13	55N	52E
Shumerlya: RSFSR.	14	55N	46E
Shumikha: RSFSR.	13	55N	63E
Shumilino: Byelorussian SSR.	8	55N	30E
Shurab: Tadzhik SSR.	17	40N	71E
Shuroabad: Tadzhik SSR.	17	38N	70E
Shusha: Azer. SSR. *Pop. 1920: over 40,000; 1926: 5,104. (Population massacred by Moslems.)*	12	40N	47E
Shushtalep: RSFSR. *Coal.*	†	54N	87E
Shuya: RSFSR. *Pop. 1954: 75,000. Eng., textiles.*	8	57N	41E
Shvenchenis: Lithuanian SSR.	‡5	55N	26E
Shyaulyay [Šiauliai]: Lith. SSR. *Pop. 1954: 5,000. Eng. chemicals.*	‡5	56N	23E
Sibay: RSFSR. *Copper, zinc.*	13	53N	59E
Šibenik: Yugoslavia. *Lignite, hydro-elec., chemicals.*	10	44N	16E
Sibiriyakov I.: RSFSR.	18	73N	79E
Sibiu: Romania. *Pop. 1948: 60,602. Engineering.*	11	46N	24E
Sid: Yugoslavia	10	45N	19E
Sidorovsk: RSFSR.	18	67N	82E
Siedlce: Poland	7	52N	22E
Siemiatycze: Poland.	7	52N	23E
Sieradz: Poland	6	52N	19E
Sierpc: Poland	6	53N	20E
Sighet: Romania	11	48N	24E
Sighişoara: Romania	11	46N	25E
Sigulda: Latvian SSR. *Inset.*	5	57N	25E
Siklós: Hungary	10	46N	18E
Siktyakh: RSFSR.	19	70N	125E
Silistra: Bulgaria	11	44N	27E
Sim: RSFSR.	13	55N	58E
Simenz: Ukrainian SSR.	9	44N	34E
Simferopol': Ukr. SSR. *Pop. 1954: 175,000. Engineering.*	9	45N	34E
Şimleul Silvaniei: Romania	11	47N	23E
Simushir, *i.*: RSFSR.	19	47N	151E
Sinaia: Romania	11	45N	26E
Sinancha: RSFSR. *Tin.*	†	45N	136E
Sindel: Bulgaria	11	43N	28E
Sindi: Estonian SSR.	‡5	58N	24E
Sinegorsk [*Jap.* Kawakami-Tanzan]: RSFSR. *Coal.*	19	47N	143E
Sinegorskiy: RSFSR. *Coal* (Donbass).*	9	48N	41E
Sinel'nikovo: Ukrainian SSR.	9	48N	36E
Siniy Shikhan: RSFSR.	13	52N	60E
Siniyvir: Ukrainian SSR.	7	48N	24E
Sinj: Yugoslavia	10	44N	17E
Sinskoye: RSFSR.	19	62N	126E
Sinyavino: RSFSR.	8	60N	32E
Siófok: Hungary	10	47N	18E
Sipotele Sucevei: Romania	11	48N	25E
Siria: Romania	11	46N	22E
Sisak: Yugoslavia. *River port. Petroleum*, natural gas*, petroleum refining* (Sava River Field)*	10	46N	16E
Sivaki: RSFSR.	19	53N	127E
Siverskiy: RSFSR.	8	59N	30E
Sjenica: Yugoslavia	11	43N	20E
Skadovsk: Ukrainian SSR.	9	46N	33E
Skal'nyy: RSFSR. *Coal*.*	13	58N	58E
Skarżysko: Poland	7	51N	21E
Skawina: Poland. *Aluminium.*	†	50N	20E
Skidel': Byelorussian SSR.	7	54N	24E
Skierniewice: Poland	7	52N	20E
Skole: Ukrainian SSR.	7	49N	24E
Skopin: RSFSR. *Lignite, eng.*	8	54N	40E
Skopje [Skoplje]: Yugo. *Pop. 1953: 121,551. Asbestos, chemicals.*	11	42N	22E
Skovorodino: RSFSR. *Gold centre.*	19	54N	124E
Skradin: Yugoslavia	10	44N	16E
Skuodas: Lith. SSR.	‡5	56N	21E
Skvira: Ukrainian SSR.	9	50N	30E
Skwierzyna [*Ger.* Schwerin]: Poland	6	53N	16E
Slănic: Romania	11	45N	26E
Slantsy: RSFSR. *Oil shale.*	8	59N	28E
Slaný: Czechoslovakia	6	50N	14E
Slatina: Romania	11	44N	24E
Slăveni: Romania	11	44N	24E
Slavgorod: Byelorussian SSR	8	54N	31E
Slavgorod: RSFSR.	15	53N	79E
Slavgorod: Ukrainian SSR.	9	48N	36E
Slavkov: Czechoslovakia	6	49N	17E
Slavyansk: Ukr. SSR. *Pop. 1954: 80,000. Engineering.*	9	49N	38E
Slavyanskaya: RSFSR.	12	45N	38E
Slavuta: Ukrainian SSR.	7	50N	27E
Sławno [*Ger.* Schlawe]: Poland.	6	54N	17E
Slezská Ostrava: Czech. *Coal, eng.**	†	50N	18E
Sliac: Czechoslovakia	6	49N	19E
Sliven: Bulgaria. *Textiles.*	11	43N	26E
Slivnitsa: Bulgaria	11	43N	26E
Sloboda: RSFSR.	8	56N	32E
Slobodskoy: RSFSR.	14	59N	50E
Slobodzeya: Moldavian SSR.	9	47N	30E
Slobozia: Romania	11	44N	28E
Slonim: Byelorussian SSR.	7	53N	25E
Slovenska Bistrica: Yugoslavia. *Lignite*.*	10	45N	16E
Slunj: Yugoslavia	10	45N	16E
Słupsk [*Ger.* Stolp]: Poland. *Pop. 1939: 50,377; 1946: 33,948*	6	55N	17E
Slutsk: Byelorussian SSR.	7	53N	28E
Slyudyanka: RSFSR. *Mica, uranium.*	12	52N	104E
Smederevo: Yugoslavia. *Steel.*	11	45N	21E
Smela: Ukrainian SSR.	9	49N	32E
Smeloye: Ukrainian SSR.	9	51N	34E
Smidovich: RSFSR.	19	48N	134E
Smilovichi: Byelorus. SSR.	7	54N	28E
Smirnovo: Kazakh SSR.	14	55N	70E
Smolensk: RSFSR. *Pop. 1954: 175,000. Eng., textiles.*	8	55N	32E
Smolensk Oblast: RSFSR.	EP	55N	33E
Smolevichi: Byelorussian SSR.	7	54N	28E
Smolyan: Bulgaria	11	42N	25E
Smorgon' [*Pol.* Smorgonie]: Byelorussian SSR.	7	54N	26E
Snezhnoye: Ukr. SSR. *Coal* (Donbass).*	†	48N	39E
Snigirevka: Ukrainian SSR.	9	47N	33E
Snina: Czechoslovakia	7	49N	22E
Snyatyn: Ukrainian SSR.	7	48N	26E
Sobinka: RSFSR.	8	56N	40E
Sobolevo: RSFSR.	19	54N	156E
Sobrance: Czechoslovakia	7	49N	22E
Sochaczew: Poland	7	52N	20E
Sochi: RSFSR. *Pop. 1954: 50,000. Port. (Noted health resort.)*	12	44N	40E
SOFIA [*Bulg.* Sofiya]: Bulg. *Pop. 1946: 434,888. Eng., textiles, chemicals.*	11	43N	23E
Sofiysk: RSFSR. *Gold.*	19	53N	134E
Sokal': Ukrainian SSR.	7	51N	24E
Sokol: RSFSR. *Paper.*	8	60N	40E
Sokółka: Poland	7	53N	24E
Sokolov: Czechoslovakia. *Lignite, chemicals.*	6	50N	13E
Sokolov: Kazakh SSR. *Iron ore.*	†	53N	63E
Sokołow Podlaski: Poland	7	52N	22E
Sokol'skoye: RSFSR.	8	57N	43E
Solčava: Yugoslavia	10	46N	15E
Soligalich: RSFSR.	8	59N	42E
Solikamsk: RSFSR. *Potash, magnesium, salt, chemicals.*	13	60N	57E
Sol'Iletsk: RSFSR. *Potash.*	13	51N	55E
Solnechnogorsk: RSFSR.	8	56N	37E
Solntsevo: RSFSR.	9	51N	37E
Solntsdar: RSFSR.	9	44N	38E
Solotobe: Kazakh SSR.	17	45N	66E
Solotvin: Ukrainian SSR.	7	49N	25E
Solov'yevsk: RSFSR.	19	55N	125E
Solta, *i.*: Yugoslavia	10	16N	43E
Sol'tsy: RSFSR.	8	58N	30E
Sol'vychegodsk: RSFSR.	†	61N	47E
Sombor: Yugoslavia	10	46N	19E
Somcuţa-Mare: Romania	11	48N	23E
Scmovit: Bulgaria	11	44N	25E
Sondershausen: E. Germany	6	51N	11E
Sonkovo: RSFSR.	8	58N	37E
Sonneberg: E. Germany	6	50N	11E
Sonskiy: RSFSR.	15	54N	90E
Sopkaklyuchevskaya, *mtn.*: RSFSR.	19	56N	160E
Sopot [*Ger.* Zoppot): Poland	6	54N	19E
Sopron: Hungary	10	48N	16E
Sorochinsk: RSFSR.	13	52N	53E
Soroki [*Rom.* Soroca]: Mold. SSR.	7	48N	28E
Sorokino: RSFSR.	15	54N	85E
Sortavala [*Rus.* Serdobol']: Karelo-Finnish SSR. *Sawmilling.*	†	62N	31E
Sosnitsa: Ukrainian SSR.	9	51N	33E
Sosnovka: RSFSR.	8	53N	41E
Sosnovo: RSFSR.	8	61N	30E
Sosnovo-Ozerskoye: RSFSR.	12	52N	111E
Sosnowiec: Poland. *Pop. 1950: 95,147. Elec. power, eng.**	6	50N	19E
Sos'va: RSFSR.	13	59N	62E
Southern Fergana Canal.	‡17	40N	72E
▲ South Kazakhstan Oblast: Kazakh SSR.	EP	42N	67E
South Osetian AO.: Georgian SSR.	EP	42N	44E
Sovetsk [Kukarka]: RSFSR.	14	58N	48E
Sovetsk [*Ger.* Tilsit]: RSFSR. *Pop. 1954: 50,000. River port. Sawmilling.*	7	55N	22E
Sovetskiy [Johannes]: RSFSR.	8	61N	29E
Sovetskiy [Ichki-Grammatikovo]: Ukrainian SSR.	9	45N	35E
Soviet Harbour [Sovetskaya Gavan'): RSFSR. *Pop. 1954: 75,000. Port (naval base, civil harbour at Vanino). Sawmilling.*	19	49N	140E
Sozopol: Bulgaria	11	52N	13E
Spas-Demensk: RSFSR.	8	54N	34E
Spas-Klepiki: RSFSR.	8	55N	40E
Spassk-Dal'niy: RSFSR. *Engineering.*	†	44N	133E
Spassk-Ryazanskiy: RSFSR.	8	54N	40E
Speedwell, C.: RSFSR.	18	75N	55E
Spirovo: RSFSR.	8	58N	35E
Spišská Nová Ves: Czech. *Iron ore.*	7	49N	21E
Split: Yugoslavia. *Pop. 1953: 75,377. Port. Engineering.*	10	43N	16E
Spokoynyy: RSFSR. *Gold* (Aldan Field).*	19	58N	128E
Sporyy Navolok, C.: RSFSR.	18	77N	70E
Spremberg: E. Germany	6	52N	14E
Srbobran: Yugoslavia	10	46N	20E
Srebrnica: Yugoslavia	10	44N	19E
Sredets: Bulgaria	11	42N	27E
Sredne Kolymsk: RSFSR.	19	68N	153E
Sredne Ural'sk: RSFSR. *Elec. power.*	13	57N	60E
Sredniy Urgal: RSFSR. *Coal.*	19	51N	133E
Srednyaya Akhtuba: RSFSR.	9	49N	45E
Srednyaya Nyukzha [Blyukherovsk]: RSFSR.	19	55N	121E
Śrem: Poland	6	52N	17E
Sremska Mitrovica: Yugo.	10	45N	20E
Srnetica: Yugoslavia	10	44N	17E
Sretensk: RSFSR. *River port.*	12	52N	118E
Środa: Poland	6	52N	17E
Stalać: Yugoslavia	11	44N	22E
Stalin [Kuçovë]: Albania. *Petroleum.*	10	41N	20E
▲ Stalin : Bulgaria. *Pop. 1946: 77,792. Port. Eng., textiles.*	11	43N	28E
Stalin *see* Braşov Romania. *Pop. 1948: 82,984. Petroleum refining, eng.*, textiles.*	11	46N	26E
▲ **STALINABAD**: Tadzhik SSR. *Pop. 1954: 150,000. Eng. textiles.*	17	39N	69E

* Located by symbol, but not named, on special topic map.
† On relevant special topic map only. For list of topics see p. VIII
▲ See appendix on p.134
‡ See Inset map.
EP Front Endpaper map.
Place names in [square] brackets are former, or alternative, names.

	Page	Lat.	Long.
⬩Stalinabad Oblast: Tadzhik SSR.	EP	38N	68E
⬩Stalingrad [Tsaritsyn]: RSFSR. *Pop. 1954: 650,000. River port. Elec. power, eng., steel, chemicals, petroleum refining, sawmilling.*	9	49N	44E
⬩Stalingrad Oblast: RSFSR.	EP	50N	45E
⬩Staliniri: Georgian SSR.	12	42N	44E
⬩Stalino [Yuzovka]: Ukr. SSR. *Pop. 1954: 600,000. Coal, iron & steel, eng., chemicals.*	9	48N	38E
⬩Stalino: Uzbek SSR.	‡17	41N	42E
⬩Stalino Oblast: Ukr. SSR.	EP	48N	38E
⬩Stalinogorsk [Bobriki]: RSFSR. *Pop. 1954: 125,000. Chemicals, eng.*	8	54N	38E
⬩Stalinogród : Poland *Pop. 1950: 170,000. Coal, lead, zinc, lead & zinc foundries, chemicals.*	6	50N	19E
⬩Stalin Peak: USSR	17	39N	72E
⬩Stalinsk : RSFSR. *Pop. 1954: 316,000. River port. Town developed under first 5 yr. plan. Coal, iron & steel, eng., chemicals, aluminium.*	15	54N	87E
⬩Stalinstadt : E. Ger. *Ferroalloys, iron & steel (new plant).*	6	52N	15E
Stanchik: RSFSR.	19	71N	150E
⬩Stanislav [Pol. Stanisławow]: Ukrainian SSR. *Pop. 1954: 70,000. Petroleum refining*.	7	49N	25E
⬩Stanislav Oblast: Ukr. SSR.	EP	48N	25E
Star': RSFSR.	8	54N	34E
Stara Gradiška: Yugoslavia	10	45N	17E
Stara Pazova: Yugoslavia	11	45N	20E
Staraya Russa: RSFSR. *Engineering*.	8	58N	31E
Stara Zagora: Bulgaria	11	42N	26E
Stargard: Poland. *Pop. 1939: 39,760; 1946: 9,733.*	6	53N	15E
Staritsa: RSFSR.	8	56N	35E
Staritsa: RSFSR.	15	58N	80E
Starobel'sk: Ukrainian SSR.	9	49N	39E
Staro-Beshevo: Ukrainian SSR.	9	48N	38E
Starobin: Byelorussian SSR.	7	53N	27E
Starodub: RSFSR.	9	53N	33E
Starogard [Ger. Preussisch Stargard]: Poland	6	54N	18E
Staro Konstantinov: Ukr. SSR.	7	50N	27E
Staro-Minskaya: RSFSR.	9	46N	39E
Staro Oryakhovo: Bulgaria	11	43N	28E
Staro-Shcherbinovskaya: RSFSR.	9	47N	39E
Starotitarovskaya: RSFSR.	9	45N	37E
Staroutkinsk: RSFSR.	13	57N	59E
Staro-Yur'yevo: RSFSR.	8	53N	41E
Staryy Dorogi: Byelorus. SSR.	7	53N	28E
Staryy Krym: Ukrainian SSR.	9	45N	35E
Staryy Oskol: RSFSR. *Pop. 1954: 50,000.*	9	51N	38E
Staryy Salavan: RSFSR.	14	55N	50E
Stassfurt: E. Ger. *Potash*, magnesite & magnesium salts.*	6	52N	12E
Staszów: Poland	7	51N	21E
Stavishche: Ukrainian SSR.	9	49N	30E
⬩Stavropol': RSFSR. *Petroleum.*	14	54N	49E
Stavropol': RSFSR. *Pop. 1954: 100,000. Eng., natural gas.*	9	45N	42E
⬩Stavropol' Kray: RSFSR.	EP	45N	43E
Stavropol' Uplands: RSFSR.	12	45N	43E
Stebnik: Ukr. SSR. *Potash.*	7	49N	23E
Steierdorfanina: Romania. *Coal.*	11	45N	22E
Stendal: E. Germany. *Engineering.*	6	53N	12E
Stepan': Ukrainian SSR.	7	55N	26E
Stepanakert [Khankendy]: Azer. SSR. *Textiles.*	12	40N	47E
Stepanavan [Dzelaloglu]: Armenian SSR.	12	41N	44E
⬩Stepnoy: RSFSR.	9	46N	44E
Stepnyak: Kazakh SSR. *Gold.*	15	53N	71E
Sterlibashevo: RSFSR.	13	53N	55E
Sterlitamak: RSFSR. *Pop. 1954: 50,000. Petroleum.*	13	54N	56E
Šternberk: Czech. *Iron ore.*	6	50N	17E
Stettin see Szczecin			
Štip: Yugoslavia	11	42N	22E
Stolac: Yugoslavia	10	43N	18E
Stolbovaya: RSFSR.	8	55N	38E
Stolbtsy [Pol. Stolbce]: Byelorussian SSR.	7	53N	27E
Stolin: Byelorussian SSR.	7	52N	27E
Stolp see Słupsk			
Stony Tunguska, R.: RSFSR.	23	62N	95E
Storozhinets [Rom. Storojinet]: Ukrainian SSR.	7	48N	26E
Stoyba: RSFSR. *Gold.*	19	53N	131E
Stralsund: E. Germany. *Pop. 1946: 50,389. Port.*	6	54N	13E

	Page	Lat.	Long.
Strandzha: Bulgaria	11	42N	27E
Strazhitsa: Bulgaria	11	43N	26E
Strážnice: Czechoslovakia	6	49N	17E
Strehaia: Romania	11	45N	23E
Strelka: RSFSR.	19	62N	103E
Streshin: RSFSR.	8	53N	30E
Stříbro: Czech. *Lead, zinc.*	6	50N	13E
Strnišče: Yugo. *Aluminium refinery.*	†	46N	16E
Stropkov: Czechoslovakia	7	49N	22E
Struga: Yugoslavia	11	41N	21E
Strugi Krasnyye: RSFSR.	8	58N	29E
Strumica: Yugoslavia	11	41N	23E
Strunino: RSFSR.	8	57N	38E
Stryków: Poland	6	52N	20E
Stryy [Pol. Stryj]: Ukr. SSR.	7	49N	24E
Strzegom [Ger. Streigau]:Poland.	6	51N	16E
Strzelce Kraj: Poland	6	53N	16E
Strzelin [Ger. Strehlen]:Poland.	6	51N	17E
Strzelno: Poland	6	52N	18E
Strzyżów: Poland	7	50N	22E
Stupino [Elektrovoz]: RSFSR. *Pop. 1954: 50,000.*	8	55N	38E
Subotica: Yugo. *Pop. 1953: 115,402. Engineering.*	10	46N	20E
Suceava: Romania	11	48N	26E
Sucha: Poland	6	50N	20E
Suchan: RSFSR. *Pop. 1954: 50,000. Coal.*	19	43N	133E
Suda: RSFSR.	8	59N	38E
Sudak: Ukrainian SSR.	9	45N	35E
Suday: RSFSR.	8	59N	43E
Sudogda: RSFSR. *Textiles.*	8	56N	41E
Sudzha: RSFSR.	9	51N	35E
Suhl: E. Germany	6	51N	11E
Šujica: Yugoslavia	10	44N	17E
Sukhana: RSFSR.	19	68N	118E
Sukhe-Bator: Mongolian People's Republic	12	50N	106E
Sukhinichi: RSFSR.	8	54N	35E
Sukhoy Log: RSFSR. *Coal* (Artemovskiy Group).*	13	57N	62E
Sukhumi: Georgian SSR. *Pop. 1954: 50,000. Port. Sawmilling.*	12	43N	41E
Suksun [Suksunskiy Zavod]: RSFSR.	13	57N	57E
Sulechów [Ger. Zullichau]: Poland	6	52N	16E
Sulęcin[Ger. Zielenzig]: Poland.	6	52N	15E
Sulina: Romania. *Port (naval base). Engineering.*	†	45N	30E
Sultangulovo: RSFSR. *Natural gas.*	14	54N	52E
Sulyukta: Kirgiz SSR. *Lignite.*	17	40N	70E
Sumečani: Yugoslavia. *Petroleum*.*	10	46N	16E
Sümeg: Hungary	10	47N	17E
Sumgait: Azer. SSR. *Pop. 1954: 50,000. Elec. power*, steel, chemicals, synthetic rubber, aluminium.*	12	41N	50E
Šumperk: Czech. *Textiles.*	6	50N	17E
Sumy: Ukrainian SSR. *Pop. 1954: 70,000. Eng., chemicals, textiles.*	9	51N	35E
Sumy Oblast: Ukrainian SSR.	EP	51N	35E
Sunskiy: Karelo-Finnish SSR. *Hydro-elec.*	†	62N	34E
Suntar: RSFSR.	19	62N	117E
Suok: Mongolian People's Republic	15	49N	89E
Suordakh: RSFSR.	19	67N	140E
Šurany: Czechoslovakia	6	48N	18E
Surakhany: Azerbaydzhan SSR. *Petroleum.*	†	40N	50E
Surazh: Byelorussian SSR.	8	55N	31E
Surdulica: Yugoslavia	11	43N	22E
Surgut: RSFSR. *River port.*	15	61N	73E
Surkhan-Dar'ya Oblast: Uzbek SSR.	EP	38N	68E
Surovikino: RSFSR.	9	49N	43E
⬩Sur Sari, i.: RSFSR.	‡5	60N	27E
Sušac, i.: Yugoslavia	10	43N	17E
Sušak: Yugoslavia	10	45N	15E
Susanino: RSFSR.	8	58N	42E
Sushchevo: RSFSR.	8	57N	30E
Sušice: Czechoslovakia	6	49N	14E
Suvorovo: Ukrainian SSR.	9	46N	29E
Suwałki: Poland	7	54N	23E
Suzak: Kazakh SSR.	17	44N	68E
Suzdal': RSFSR.	8	56N	40E
Svalyava [Hung. Szolyva, Czech. Svalava]: Ukrainian SSR. *Chemicals.*	7	48N	23E
Svatovo: Ukrainian SSR.	9	50N	38E
Svätý Kříž. Czech. *Steel, aluminium works.*	6	49N	19E
Svboda: RSFSR.	9	52N	36E
Sverdlovsk: RSFSR. *Pop. 1954: 700,000. Steel, eng., chemicals, textiles.*	13	57N	61E
Sverdlovsk: Ukr. SSR. *Coal.*	9	48N	40E

	Page	Lat.	Long.
Sverdlovsk Oblast: RSFSR.	EP	58N	61E
Sverdrup I.: RSFSR.	18	75N	80E
Svessa: Ukrainian SSR.	9	52N	34E
Svet Peter: Yugoslavia	10	46N	14E
Svetlaya: RSFSR.	19	47N	138E
Svetlogorsk: RSFSR.	7	55N	20E
Svetlyy: RSFSR. *Gold* (Lena-Vitim Field).*	19	58N	116E
Svetogorsk: RSFSR. *Paper.*	†	61N	29E
Svetozarevo: Yugoslavia	11	44N	21E
Svilajnac: Yugoslavia	11	44N	21E
Svilengrad: Bulgaria. *Textiles.*	11	42N	26☐
Sviritsa: RSFSR.	8	60N	33E
Svir'stroy: RSFSR. *Hydroelec., eng.*	8	60N	33E
Svishtov: Bulgaria	11	44N	25E
Svitavy: Czechoslovakia	6	50N	16E
Svobodnyy [Alekseyevsk]: RSFSR. *Engineering.*	19	52N	128E
Swidnica [Ger. Schweidnitz]: Poland. *Eng., textiles.*	6	51N	16E
Swiebodzin [Ger. Schwiebus]: Poland. *Chemicals.*	6	52N	16E
Swinoujście [Ger. Swinemünde]: Poland. *Pop. 1939: 30,239; 1946: 5,771. Port.*	6	54N	14E
Syas'troy: RSFSR.	8	60N	33E
Syava: RSFSR.	8	58N	46E
Sychevka: RSFSR.	8	56N	34E
Syców: Poland	6	51N	18E
Syktyvkar [Ust'-Sysol'sk]: RSFSR. *Sawmilling, eng.*	14	62N	51E
Sylva: RSFSR.	13	58N	57E
Syntul: RSFSR.	8	55N	41E
Syr-Dar'inskiy: Uzbek SSR.	17	41N	69E
Syr Darya, R.: USSR.	23	44N	67E
Syrskiy: RSFSR.	8	53N	39E
Sysert: RSFSR.	13	57N	61E
Syzran': RSFSR. *Pop. 1954: 160,000. River port. Petroleum, petroleum refining, engineering.*	14	53N	48E
Szamosszeg: Hungary	7	48N	22E
Szamotuły [Ger. Samter]: Poland	6	53N	17E
Szarvas: Hungary	11	47N	20E
Szczebrzeszyn: Poland	7	51N	23E
Szczecin [Ger. Stettin]: Poland. *Pop. 1950: 159,122. Port. Elec. power, eng., iron, textiles, paper.*	6	53N	14E
Szczecinek [Ger. Neustettin]: Poland	6	54N	17E
Szczekociny:Poland. *Chemicals.*	6	51N	20E
Szczuczyn Białostocki: Poland	7	54N	22E
Szczytno [Ger. Ortelsburg]: Poland	7	54N	21E
Szécsény: Hungary	6	48N	19E
Szeged: Hungary. *Pop. 1941: 136,752.*	11	46N	20E
Szegvár: Hungary	11	47N	20E
Székesfehérvár: Hungary. *Pop. 1941: 47,968.*	10	47N	18E
Szekszárd: Hungary	10	46N	19E
Szentendre: Hungary	10	48N	19E
Szentes: Hungary	11	47N	20E
Szentetornya: Hungary	11	47N	21E
Szentgotthárd: Hungary	10	47N	16E
Szerencs: Hungary	11	48N	21E
Szigetszentmiklós: Hungary	10	47N	19E
Szigetvár: Hungary	10	46N	18E
Szikszó: Hungary	11	48N	21E
Szób: Hungary	10	48N	19E
Szolnok: Hungary. *River port.*	11	47N	20E
Szombathely: Hungary. *Textiles.*	10	47N	17E
Szopienice [Ger. Schoppinitz]: Poland. *Iron.* ꭓ	†	50N	19E
Szőreg: Hungary. *Petroleum refining.*	10	46N	20E
Szprotawa [Ger. Sprottau] Poland	6	52N	16E
Sztalinváros: Hungary. *Iron & steel (founded as new town in 1950).*	10	47N	19E
Szubin [Ger. Schubin]: Poland	6	53N	18E
Szydłowiec: Poland	7	51N	21E
Tábor: Czechoslovakia	6	49N	15E
Tabory: RSFSR.	13	58N	64E
Taboshar: Tadzhik SSR. *Lead-zinc, uranium, vanadium.*	17	41N	70E
Tadzhik: SSR.	EP	39N	70E
Tadzhikabad [Kalai-Lyabiob]: Tadzhik SSR.	17	39N	71E
Taganrog: RSFSR. *Pop. 1954: 205,000. Port. Steel, eng.*	9	47N	39E
Takeli: Tadzhik SSR.	17	40N	69E
Takhta: RSFSR.	19	53N	140E
Takhta-Kupyr: Uzbek SSR.	16	43N	60E
⬩Talas Oblast: Kirgiz SSR.	EP	43N	72E
Talass: Kirgiz SSR.	17	43N	72E

---

* Located by symbol, but not named, on special topic map.
† On relevant special topic map only. For list of topics see p. VIII.
⬩ See appendix on p.134

‡ See Inset map.

EP Front Endpaper map.
Place names in [square] brackets are former, or alternative, names.

	Page	Lat.	Long.
Taldy Kurgan: Kazakh SSR.	17	45N	78E
▲Taldy-Kurgan Oblast: Kazakh SSR.	EP	45N	80E
Talgi: RSFSR. Petroleum* (Izberbash Group).	12	43N	48E
TALLIN [Tallinn, Rus. Revel']: Est. SSR. Pop. 1954: 280,000. Port. Cotton text., eng., chemicals, paper. elec. power.	5	59N	25E
Tălmaciu: Romania	11	46N	24E
Tal'menka: RSFSR.	15	54N	83E
Tal'noye: Ukrainian SSR.	9	49N	31E
Talovaya: RSFSR.	9	51N	41E
Talsy: Latvian SSR.	‡5	57N	23E
Tamak: RSFSR.	19	60N	90E
Tamak: RSFSR.	19	61N	172E
Taman': RSFSR.	9	45N	37E
Tamasi: Hungary	10	47N	18E
Tambey: RSFSR.	18	72N	72E
Tambov: RSFSR. Pop. 1954: 140,000. Eng., chemicals, synthetic rubber.	9	53N	42E
Tambov Oblast: RSFSR.	EP	53N	41E
Tamdy-Bulak: Uzbek SSR.	16	42N	65E
Tamsak Bulak: Mongolian People's Republic	19	47N	118E
Tândărei: Romania	11	45N	28E
Tangermünde: E. Germany	6	52N	12E
Tanguy: RSFSR.	12	56N	101E
Tapa: Estonian SSR.	‡5	59N	26E
Tapolca: Hungary	10	47N	17E
Tara: RSFSR.	15	57N	75E
Tarak: RSFSR. Thorium.	15	60N	92E
Tarakliya: Moldavian SSR.	11	46N	29E
Tarashcha: Ukrainian SSR.	9	50N	30E
Târgovişte: Romania	11	45N	25E
Târgu Bereşti: Romania	11	46N	28E
Târgu-Frumos: Romania	11	47N	27E
Târgu Jiu: Romania	11	45N	23E
Târgu Mureş: Romania. Pop. 1948: 47,043. Petroleum refining.	11	46N	25E
Târgu Ocna: Romania	11	46N	26E
Târgu-Săcuesc: Romania	11	46N	26E
Targyn: Kazakh SSR. Gold, tin.	17	50N	82E
Tarkhankut, C.: Ukr. SSR.	9	45N	32E
Tarkosale: RSFSR.	18	65N	78E
Târnăveni [Diciosanmartin]: Romania	11	46N	24E
Tarnobrzeg: Poland	7	50N	22E
Tarnów: Poland	7	50N	21E
Tartu [Dorpat]: Estonian SSR. Pop. 1954: 70,000. Engineering.	8	58N	27E
Tarusa: RSFSR.	8	55N	37E
Tarutino: Ukrainian SSR.	11	46N	29E
Taseyevo: RSFSR.	19	57N	95E
Tashauz: Turkmen SSR.	16	42N	60E
▲Tashauz Oblast: Turkmen SSR.	EP	41N	58E
TASHKENT: Uzbek SSR. Pop. 1954: 750,000. Textiles, eng.	17	41N	69E
Tashkent Oblast: Uzbek SSR.	EP	42N	70E
Tashkumyr: Kirgiz SSR. Coal.	‡17	41N	72E
Tashtagol: RSFSR. Iron ore.	15	53N	88E
Tashtyp: RSFSR.	15	53N	90E
Taskan: RSFSR.	19	63N	150E
Taşnad: Romania	11	47N	23E
Tata: Hungary	10	48N	18E
Tatabánya: Hungary. Lignite.	10	48N	18E
Tatar ASSR.: RSFSR.	EP	55N	50E
Tatarka: RSFSR. Bauxite.	15	58N	94E
Tatar Pass: Ukrainian SSR.	7	48N	24E
Tatarsk: RSFSR.	15	55N	76E
Tatsinskaya: RSFSR.	9	48N	41E
Tauchik: Kazakh SSR. Lignite.	16	44N	52E
Taurage: Lithuanian SSR.	‡5	55N	22E
Tauz: Azerbaydzhan SSR.	12	41N	46E
Tavda: RSFSR. River port.	13	58N	65E
Tavolzhan: Kazakh SSR. Chemicals, salt.	15	52N	77E
Taymyr N.O.: Krasnoyarsk Kray, RSFSR.	EP	73N	95E
Taymyr Penin: RSFSR.	19	75N	105E
▲Tayncha: Kazakh SSR.	15	54N	70E
Tayshet: RSFSR. Sawmilling.	19	56N	98E
Taz, R.: RSFSR.	23	65N	83E
Taza: RSFSR.	12	55N	111E
TBILISI [Tiflis]: Georgian SSR. Pop. 1954: 580,000. Petroleum refining, eng., textiles, elec. power.	12	42N	45E
Tczew [Ger. Dirschau]: Poland.	6	54N	19E
Teaca: Romania	11	47N	25E
Tecuci: Romania	11	46N	27E
Tedzhen: Turkmen SSR.	16	37N	60E
Tegul'det: RSFSR.	15	57N	88E
Teiuş: Romania	11	46N	24E
Tekeli: Kazakh SSR. Lead-zinc, lead smelting.	17	45N	79E
Telavi: Georgian SSR.	12	42N	45E
Tel'bes: RSFSR.	15	53N	87E
Telč: Czechoslovakia	6	49N	15E
Telekhany: Byelorussian SSR.	7	52N	26E

	Page	Lat.	Long.
Tel'manovo [Ostheim]: Ukr. SSR.	9	47N	38E
Tel'shyay [Lith. Telsiai]: Lith. SSR.	‡5	56N	22E
Tembenchi: RSFSR.	19	65N	100E
Temerin: Yugoslavia	10	45N	20E
Temir: Kazakh SSR.	16	49N	57E
Temir-Tau [Samarkand]: Kazakh SSR. Pop. 1954: 60,000. Hydro-elec., steel, chemicals, synthetic rubber.	15	50N	73E
Temir-Tau [Timer-Tau]: RSFSR. Iron ore.	15	53N	87E
Temkino: RSFSR.	8	55N	35E
Temnikov: RSFSR.	8	55N	43E
Temryuk: RSFSR.	9	45N	37E
Tengiz, lake: Kazakh SSR.	14	50N	69E
Tepelenë [Tepelena]: Albania	11	40N	20E
Teplaya Gora: RSFSR. Iron.	13	59N	59E
Teplice [Ger. Teplitz Schönau]: Czech. Lignite, paper.	6	51N	14E
Teplyy Stan: RSFSR.	8	54N	42E
Terebovlya: Ukrainian SSR.	7	49N	26E
Teregova: Romania	11	45N	22E
Terekhovka: Byelorus. SSR.	9	52N	31E
Teriberka: RSFSR.	18	69N	35E
Termez: Uzbek SSR. Textiles*.	17	38N	67E
Ternopol' [Pol. Tarnopol]: Ukr. SSR. Engineering.	7	50N	26E
Ternopol' Oblast: Ukr. SSR.	EP	49N	27E
Teschen see Cieszyn.			
Tesha: RSFSR.	8	56N	43E
Tetevene: Bulgaria	11	43N	24E
Tetovo: Yugoslavia	11	42N	21E
Tetyukhe: RSFSR. Lead, zinc. Linked by rail to smelting centre & port of Tetyukhe-Pristan'.	19	44N	136E
Tetyushi: RSFSR.	14	55N	49E
Tevriz: RSFSR.	15	58N	73E
Teykovo: RSFSR.	8	57N	40E
Tianeti: Georgian SSR.	12	42N	45E
Tiflis see Tbilisi			
Tigăneşti: Romania	11	45N	26E
Tigil': RSFSR.	19	58N	159E
Tihany: Hungary	10	47N	18E
Tikhoretsk: RSFSR.	9	46N	40E
Tikhvin: RSFSR. Pop. 1954: 50,000.	8	60N	34E
Tiksi: RSFSR. Port.	19	72N	128E
Tileagd: Romania	11	47N	22E
Tilichiki: RSFSR.	19	61N	166E
Tilsit see Sovetsk			
Timashevo: RSFSR.	14	54N	51E
Timashevskaya: RSFSR.	12	46N	39E
Timişoara: Romania. Pop. 1948: 111,987. Iron ore, textiles, chemicals.	11	46N	21E
Tinca: Romania	11	47N	22E
TIRANË [Tirana]: Albania. Pop. 1945: 59,897. Elec. power, textiles.	10	41N	20E
Tiraspol': Mold. SSR. Pop. 1954: 60,000. Engineering*.	11	47N	30E
Tirlyanskiy: RSFSR.	13	54N	58E
Tisovec: Czech. Iron.	6	48N	19E
Tisul': RSFSR.	15	56N	88E
Tiszalök: Hungary. River port.	†	48N	21E
Titel: Yugoslavia	11	45N	20E
Titograd: Yugoslavia	10	42N	19E
Titovo Užice: Yugo. Hydro-elec., manganese, eng.	10	44N	20E
Titov Veles: Yugoslavia	11	42N	22E
Tkhab, mtn.: RSFSR.	9	44N	38E
Tkibuli: Georgian SSR. Coal.	12	42N	43E
Tkvarcheli: Georgian SSR. Coal, elec. power.	12	43N	42E
Tlyarata: RSFSR.	12	42N	46E
Tobol', R.: RSFSR.	22	54N	63E
Tobol'sk: RSFSR. River port.	14	58N	68E
Toguchin: RSFSR. Coal.	†	55N	84E
Togyz: Kazakh SSR.	16	47N	60E
Tokaj: Hungary. (Wine.)	11	48N	21E
Tokarevka: Kazakh SSR.	15	50N	73E
Tokmak: Kirgiz SSR. Textiles.	17	43N	76E
Tolba, R.: RSFSR. Petroleum.	†	60N	123E
Tolbukhin: Bulgaria	11	44N	28E
Tolmachevo: RSFSR.	8	59N	30E
Tolmin: Yugoslavia	10	46N	14E
Tolna: Hungary	10	46N	19E
Tolstoy, C.: RSFSR.	19	59N	155E
Tomari [Jap. Tomarioru]: RSFSR. Chemicals.	19	48N	142E
Tomarovka: RSFSR.	9	51N	36E
Tomaszów: Poland	7	50N	23E
Tomaszów Mazowiecki: Poland	7	52N	20E
Tommot: RSFSR. Gold* (Aldan Field).	19	59N	126E
Tompo: RSFSR.	19	64N	137E
Tomsk: RSFSR. Pop. 1954: 225,000. River port. Eng., chemicals, elec. power.	15	56N	85E

	Page	Lat.	Long.
Tomsk Oblast: RSFSR.	EP	58N	82E
Topalu: Romania	11	45N	28E
Topki: RSFSR.	15	55N	85E
Topliţa: Romania	11	47N	25E
Topolčane: Yugoslavia	11	41N	21E
Topol'čany: Czechoslovakia	6	48N	18E
Topolovgrad: Bulgaria	11	42N	26E
Torchin: Ukrainian SSR.	7	51N	25E
Torgau: E. Ger. Chemicals*.	6	51N	13E
Torkovichi: RSFSR.	8	59N	30E
Tormosin: RSFSR.	9	48N	43E
Törökszentmiklós: Hungary	11	47N	20E
Toropets: RSFSR.	8	56N	32E
Toruń [Ger. Thorn]: Pop. 1950: 80,000. Engineering.	6	53N	19E
Torzhok: RSFSR. Pop. 1926: 14,449; 1939: 31,800.	8	57N	35E
Tosno: RSFSR.	8	60N	31E
Tovil'-Dora: Tadzhik SSR.	17	39N	70E
Trans-Carpathian Oblast: Ukr. SSR.	EP	48N	23E
Transylvanian Alps: Bauxite.	11	45N	25E
Travnik: Yugoslavia	10	44N	18E
Třebíč: Czechoslovakia	6	49N	16E
Trebinje: Yugoslavia	10	43N	18E
Trebišov: Czechoslovakia	7	49N	22E
Trebnje: Yugoslavia	10	46N	15E
Třeboň: Czechoslovakia	6	49N	15E
Trenčin: Czechoslovakia	6	49N	18E
Trepča: Yugoslavia. Lead, zinc, silver, pyrites.	11	43N	21E
Trikhaty: Ukrainian SSR.	9	47N	32E
Trn: Bulgaria	11	43N	23E
Trnava: Czechoslovakia	6	48N	18E
Trnovo: Yugoslavia	10	44N	18E
Trogir: Yugoslavia	10	44N	16E
Troitsk: RSFSR. Textiles.	19	54N	62E
Troitskiy: RSFSR.	13	57N	64E
Troitsko-Pechorsk: RSFSR.	18	63N	56E
Troitskoye: RSFSR.	15	53N	85E
Troitskoye: RSFSR.	19	50N	136E
Trostyanets: Ukr. SSR.	7	49N	25E
Trostyanets: Ukrainian SSR.	9	48N	29E
Trostyanets: Ukrainian SSR.	9	51N	35E
Troyan: Bulgaria	11	43N	25E
Trstenik: Yugoslavia	11	44N	21E
Trubchevsk: RSFSR.	9	52N	34E
Trŭn: Bulgaria	11	43N	23E
Truskavets: Ukrainian SSR.	7	49N	24E
Trutnov: Czechoslovakia	6	51N	16E
Tryavna: Bulgaria	11	43N	26E
Trzcianka: Poland	6	53N	16E
Trzebiatów [Ger. Treptow]: Poland	6	54N	15E
Trzebinia: Poland. Petroleum refining.	6	50N	20E
Trzebnica: Poland	6	51N	17E
Tsalka: Georgian SSR.	12	42N	44E
Tsarichanka: Ukrainian SSR.	9	49N	34E
Tsatsa: RSFSR.	9	48N	45E
Tsementnyy: RSFSR.	8	54N	34E
Tsementnyy: RSFSR.	13	57N	60E
Tsentral'nyy: RSFSR. Gold* (Kiya Valley).	15	55N	88E
Tsesis [Lat. Cēsis]: Latvian SSR.	‡5	57N	25E
Tsetserlig: Mongolian People's Republic	19	47N	102E
Tsimlyansk Reservoir: RSFSR.	9	48N	43E
Tsimlyanskiy: RSFSR. Tsimlyansk hydro-elec. station.	9	48N	42E
Tsinandali: Georgian SSR.	12	42N	46E
Tsipikan: RSFSR. Gold* (Lena-Vitim Field).	12	55N	114E
Tskhakaya: Georgian SSR.	12	42N	42E
Tskhaltubo: Georgian SSR.	12	42N	42E
Tsnori: Georgian SSR.	12	42N	46E
Tsulukidze [Khoni]: Georgian SSR.	12	42N	42E
Tsyr: Ukrainian SSR.	7	52N	25E
Tsyurupinsk [Aleshki]: Ukr. SSR.	9	46N	33E
Tuapse: RSFSR. Pop. 1954: 50,000. Port. Petroleum refining, eng.	12	44N	39E
Tubinikha: Ukrainian SSR.	9	49N	35E
Tubinskiy: RSFSR. Gold.	13	53N	58E
Tuchola: Poland	6	54N	18E
Tuchów: Poland. Iron ore.	7	50N	21E
Tugur: RSFSR.	19	53N	137E
Tukan: RSFSR.	13	54N	58E
Tukum [Lat. Tukums]: Latvian SSR.	‡5	57N	23E
Tula: RSFSR. Pop. 1954: 325,000. Eng., iron ore nearby.	8	54N	38E
Tula Oblast: RSFSR.	EP	54N	38E
Tulcea: Romania. Pop. 1948: 21,642. Chemicals, copper in hills around.	11	45N	29E
Tul'chin: Ukrainian SSR.	7	49N	29E
Tulovo: Bulgaria	11	42N	26E
Tulun: RSFSR. Textiles.	12	55N	101E
Tumanyan: Armenian SSR.	12	41N	45E

* Located by symbol, but not named, on special topic map. ‡ See Inset map. EP Front Endpaper map.
† On relevant special topic map only. For list of topics see p. VIII. Place names in [square] brackets are former, or alternative, names.
▲ See appendix on p.134

	Page	Lat.	Long.
Tuora: RSFSR.	19	59N	125E
Tupik: RSFSR.	19	55N	120E
Tura: RSFSR.	19	64N	100E
Turan: RSFSR.	19	52N	94E
Turanian Plain: USSR.	16	42N	60E
Turčiansky Svätý Martin: Czech. *Sawmilling, paper.*	6	49N	19E
Turda: Romania. *Chemicals.*	11	47N	24E
Turek: Poland	6	52N	18E
Turgay: Kazakh SSR. *Antimony.*	15	52N	73E
Turgay: Kazakh SSR.	16	50N	63E
Turgay Uplands: Kazakh SSR.	13	51N	62E
Türgovishte: Bulgaria	11	43N	26E
Turinsk: RSFSR.	13	58N	64E
Turka: Ukrainian SSR.	7	49N	23E
Turkestan: Kazakh SSR. *Pop. 1939: 54,000 (including Borisovka).*	17	43N	68E
Turkmen SSR.	EP	40N	60E
↓Türnovo: Bulgaria.	11	43N	26E
Turnu Măgurele: Romania	11	44N	25E
Turnu Roşu: Romania	11	46N	24E
Turnu Severin: Romania. *River port. Engineering.*	11	45N	23E
Turochak: RSFSR. *Manganese.*	15	52N	87E
Turtkul' [Petro-Aleksandrovsk]: Uzbek SSR.	16	41N	61E
Turtu: Mongolian People's Republic	12	52N	101E
Turukhansk: RSFSR. *River port. Petroleum.*	18	66N	88E
Turzovka: Czechoslovakia	6	49N	19E
Tushino: RSFSR. *Pop. 1954: 100,000.*	8	56N	37E
Tutayev: RSFSR.	8	58N	40E
Tutin: Yugoslavia	11	43N	20E
Tutrakan: Bulgaria	11	44N	27E
↑ Tuva A.O.: RSFSR.	EP	52N	95E
Tuymazy: RSFSR.	13	55N	54E
Tuzla: Yugoslavia. *Petroleum, natural gas.*	10	45N	19E
Tvardeyskoye: Ukrainian SSR.	9	45N	34E
Tvŭrditsa: Bulgaria	11	42N	25E
Tyan'-Shan' Oblast: Kirgiz SSR.	EP	42N	75E
Tyazhin: RSFSR. *Gold* (Kiya Valley).*	15	56N	89E
Tychany: RSFSR.	19	62N	97E
Tychy: Poland	6	50N	19E
Tygda: RSFSR.	19	53N	126E
Týn: Czechoslovakia	6	49N	14E
Tyndinskiy: RSFSR. *Gold.*	19	55N	125E
Tyrny-Auz: RSFSR. *Molybdenum, tungsten.*	12	43N	43E
Tyrva: Estonian SSR.	‡5	58N	26E
Tyukalinsk: RSFSR.	15	56N	72E
Tyumen': RSFSR. *Pop. 1954: 110,000. River port. Sawmilling, eng., textiles.*	14	57N	66E
Tyumen' Oblast: RSFSR.	EP	57N	68E
Tyup: Tadzhik SSR.	17	43N	78E
Tyuri [Est. Türi]: Estonian SSR. *Paper, textiles*.*	5	59N	25E
Tyuya-Muyun: Kirgiz SSR. *Uranium, vanadium.*	17	40N	73E
Ubaredmet: Kazakh SSR. *Tin, tungsten.*	15	50N	82E
Ubinskoye: RSFSR.	15	55N	80E
Ubsu Nor, *lake:* Mongolian People's Republic	19	50N	93E
Uchimchak: Kirgiz SSR. *Arsenic.*	†	42N	72E
Uch-Kurgan: Uzbek SSR.	‡17	41N	72E
Udmurt ASSR.: RSFSR.	EP	57N	53E
Udskoye: RSFSR.	19	55N	135E
Udzhary: Azerbaydzhan SSR.	12	41N	48E
Uelen: RSFSR.	19	67N	170W
Uel'kal': RSFSR.	19	66N	180
Ufa: RSFSR. *Pop. 1954: 250,000. River Port. Eng., chemicals, textiles (see Chernikovsk).*	13	55N	56E
Ufa Plateau: RSFSR.	13	56N	58E
Uglegorsk [*Jap.* Esutoru]: RSFSR. *Port· Coal.*	19	49N	142E
Ugleural'sk: RSFSR. *Coal.*	†	59N	57E
Uglich: RSFSR. *Elec.power.*	8	58N	38E
Uglovka: RSFSR.	8	58N	34E
Ugnev: Ukrainian SSR.	7	50N	24E
Ugol'nyy: RSFSR. *Coal.*	19	63N	178E
Ugol'nyy [*Jap.* Taiei]: RSFSR. *Coal.*	†	48N	142E
Uil: Kazakh SSR. *Petroleum.*	16	49N	55E
Újpest: Hungary. *Pop. 1941: 76,001. Chemicals*.*	10	48N	19E
Uka: RSFSR.	19	58N	162E
↓Ukhta [*Fin.* Uhtua]: Karelo-Finnish SSR.	18	65N	30E
Ukhta [Chib'-Yu]: RSFSR. *Petroleum, petroleum refining.*	14	64N	54E
Ukmerge: Lith. SSR.	‡5	55N	24E

	Page	Lat.	Long.
Ukrainian SSR.	EP	50N	30E
ULAN BATOR [Urga]: Mongolian People's Republic. *Pop. 1951: 70,000. Industrial combine:—elec. power, eng., textiles.*	19	48N	107E
Ulangom: Mongolian People's Republic	19	50N	92E
Ulan-Ude: RSFSR. *Pop. 1954: 175,000. Eng., textiles, sodium sulphate, sawmilling, elec. power.*	12	52N	107E
Ulbanskiy Bay: RSFSR.	19	54N	138E
Ulcinj: Yugoslavia.	10	42N	19E
Uldza: Mongolian People's Republic	19	49N	113E
Ulla: Byelorussian SSR.	8	55N	29E
Ulutau: Kazakh SSR.	17	49N	67E
Ul'yanovka: Ukrainian SSR.	9	48N	30E
Ul'yanovsk[Simbrisk]: RSFSR. *Pop. 1954· 175,000. River port. Elec. power*, eng., textiles.*	14	54N	48E
Ul'yanovsk Oblast: RSFSR.	EP	54N	49E
Umal'tinsky [Polovinka] RSFSR.	19	52N	133E
Uman': Ukrainian SSR. *Pop. 1954: 50,000.*	9	49N	30E
Undory: RSFSR. *Oil shale.*	14	55N	48E
Unecha: RSFSR.	8	53N	33E
Ungeny [*Rom.* Ungheni] Moldavian SSR	11	47N	28E
Uniejów: Poland	6	52N	19E
Unterwellenborn: E. Ger. *Iron & steel.*	†	50N	11E
Ural Mountains: RSFSR.	13	60N	59E
Ural, R.: RSFSR.	22	49N	52E
Uralets: RSFSR. *Gold*, platinum.*	13	58N	60E
Ural'sk: Kazakh SSR. *Pop. 1954: 80,000. River port. Engineering.*	13	51N	52E
Ura Tyube: Tadzhik SSR.	17	40N	69E
Urazovo: RSFSR.	9	50N	38E
Urda: Kazakh SSR.	16	49N	47E
Urech'ye: Byelorussian SSR.	7	53N	28E
Urgench: Uzbek SSR.	16	42N	60E
Urgut: Uzbek SSR.	17	39N	67E
Uritsk: RSFSR.	8	60N	30E
Urkut: Hungary. *Manganese.*	10	47N	18E
Uroševac: Yugoslavia	11	42N	21E
Urshel'skiy: RSFSR.	8	56N	40E
Ursk: RSFSR. *Gold*.*	15	55N	85E
Urup, *i.*: RSFSR.	19	46N	150E
Urussu: RSFSR.	13	55N	53E
Uryupinsk: RSFSR. *Iron ore.*	9	51N	42E
Urzhum: RSFSR.	14	57N	50E
Urzicani: Romania	11	45N	27E
Usa: RSFSR. *Manganese.*	15	54N	89E
Usa, R.: RSFSR.	22	66N	61E
Usa-Su: Turkmen SSR. *Sodium sulphate.*	†	39N	55E
Ušče: Yugoslavia	11	43N	21E
Ushachi: Byelorussian SSR.	8	55N	28E
Ushanov I.: RSFSR.	18	61N	80E
Ush-Tobe: Kazakh SSR.	17	45N	78E
Usman': RSFSR.	9	52N	40E
Usol'ye: RSFSR.	13	59N	57E
Usol'ye-Sibirskoye: RSFSR. *Pop. 1954: 50,000. Petroleum, chemicals.*	12	53N	104E
Usovo: Ukrainian SSR.	7	51N	28E
Uspenka: Ukrainian SSR.	9	48N	39E
Uspenskiy: Kazakh SSR. *Copper.*	17	49N	73E
Ussuri, R.: RSFSR.	23	48N	134E
Ust'-Belaya: RSFSR.	19	66N	173E
Ust'-Bol'sheretsk: RSFSR.	19	53N	156E
Ust'-Bukhtarma: Kazakh SSR. *Hydro-elec.*	17	50N	83E
Ust-Buzulukskaya: RSFSR.	9	50N	42E
Ústí: Czechoslovakia	6	50N	16E
Ústí-nad-Labem: Czech. *Pop. 1947: 56.328.*	6	51N	14E
Ustka [*Ger.* Stolpmünde]: Poland	6	55N	17E
Ust'-Kamchatsk: RSFSR. *Pop. 1954: 100,000. Port. Sawmilling.*	19	56N	162E
Ust'-Kamenogorsk [Zashchita]: Kazakh SSR. *Pop. 1954: 100,000. River port. Hydroelec., eng., zinc.*	15	50N	83E
Ust'-Karsk: RSFSR. *Gold.*	12	53N	118E
Ust'-Katav: RSFSR. *Eng.*, steel.*	13	55N	58E
Ust'-Kut: RSFSR. *River port. Engineering.*	19	57N	105E
Ust'-Labinskaya: RSFSR.	9	45N	40E
Ust'-Luga: RSFSR.	8	60N	28E
Ust'-Maya: RSFSR.	19	60N	135E
Ust'-Niman: RSFSR. *Coal* (Bureya Field).*	19	52N	133E

	Page	Lat.	Long.
Ust'-Orda Buryat Mongol NO.: RSFSR.	EP	54N	105E
Ust'-Ordynskiy: RSFSR.	12	53N	105E
Ust'-Port: RSFSR.	18	70N	85E
Ust'-Sugoy: RSFSR.	19	65N	155E
Ust'-Tsil'ma: RSFSR.	18	65N	52E
Ust'-Uda: RSFSR.	12	54N	104E
Ust'-Ulagan: RSFSR.	15	51N	88E
Ust'-Urt Plateau: USSR.	16	42N	57E
Ust'-Usa: RSFSR.	18	66N	57E
Ust'-Uyskoye: RSFSR.	13	54N	64E
Ust'-Voya: RSFSR.	18	64N	57E
Ust'-Voyampolka: RSFSR. *Petroleum.*	19	58N	160E
Ustyuzhna: RSFSR.	8	59N	36E
Utena: Lith. SSR.	‡5	55N	25E
Utuncha: RSFSR.	19	71N	144E
Uvarovo: RSFSR.	9	52N	42E
Uyar: RSFSR. (*Meat packing.*)	19	56N	94E
Uyedineniye I.: RSFSR.	18	77N	82E
Uygur-Say: Uzbek SSR. *Uranium.*	‡17	41N	71E
Uyskoye: RSFSR.	13	54N	60E
Uzbek SSR.	EP	43N	62E
Uzda: Byelorussian SSR.	7	53N	27E
Uzgen: Kirgiz SSR. *Coal, light eng., cotton.*	‡17	41N	73E
Uzhgorod [*Czech.* Užhorod; *Hung.* Ungvar]: Ukr. SSR. *Engineering*.*	7	49N	22E
Uzhur: RSFSR.	15	55N	90E
Uzlovaya: RSFSR. *Pop. 1954: 50,000. Lignite.*	8	54N	38E
Uzman': RSFSR.	9	52N	40E
Vác: Hungary. *Eng. (surveying instruments), chemicals.*	10	48N	19E
Vacha: RSFSR.	8	56N	42E
Vagay: RSFSR.	14	58N	69E
Vakhshstroy: Tadzhik SSR. *Hydro-elec.*	17	38N	69E
Vakhtan: RSFSR.	14	58N	47E
Valamaz: RSFSR.	13	58N	52E
Valday: RSFSR.	8	58N	33E
Valday Hills: RSFSR.	8	57N	33E
Vale: Georgian SSR.	12	42N	43E
Văleni de Munte: Romania	11	45N	26E
Valerianovsk: RSFSR. *Gold*, platinum, titanium, iron ore*.*	13	59N	59E
Valga [*Lat.* Valka]: Est. SSR./ Latvian SSR. *Frontier town (pop. Valga 10,842; pop. Valka 3,268).*	‡5	58N	26E
Valjevo: Yugoslavia	10	44N	20E
Valki: Ukrainian SSR.	9	50N	36E
Valmiera: Latvian SSR.	‡5	57N	25E
Valuyki: RSFSR.	9	50N	38E
Vanavara: RSFSR. *Lignite (exploitation uncertain).*	19	60N	102E
Vanch: Tadzhik SSR.	17	38N	72E
Vankarem: RSFSR.	19	68N	178W
↓Vannovskiy: Uzbek SSR. *Petroleum refining*.*	‡17	40N	72E
Vapnyarka: Ukrainiar. SSR.	7	48N	28E
Varaždin: Yugoslavia	10	46N	16E
Varena: Lithuanian SSR.	7	54N	24E
Vareš: Yugoslavia. *Iron ore, manganese, iron.*	10	44N	18E
Varna: RSFSR.	13	53N	61E
Várpalota: Hungary	10	47N	18E
Varvarin: Yugoslavia	11	44N	21E
Varzob [Obi-Dzhuk]: Tadzhik SSR. *Tungsten.*	17	39N	69E
Vásárosna:mény: Hungary	7	48N	22E
Văscău: Romania	11	46N	22E
Vasilevichi: Byelorussian SSR..	9	52N	30E
Vasil'kov: Ukrainian SSR.	9	50N	30E
Vasil'yevskiy Mokh: RSFSR.	8	57N	36E
Vaslui: Romania	11	47N	28E
Vasvár: Hungary	10	47N	17E
Vasyugan: RSFSR.	15	59N	78E
Vasyugan'ye, *swamp:* RSFSR.	15	57N	80E
Vatra-Dornei: Romania	11	47N	25E
Vaygach I.: RSFSR. *Lead (exploitation uncertain).*	18	70N	60E
Važec: Czechoslovakia	6	49N	20E
Vejprty: Czechoslovakia	6	51N	13E
Vel'giya: RSFSR.	8	58N	34E
Veliki Popović: Yugoslavia	11	44N	21E
Velikiy Burluk: Ukr. SSR.	9	50N	37E
Velikiye Luki: RSFSR.	8	56N	30E
↓Velikiye Luki Oblast: RSFSR.	EP	57N	30E
Velikodvorskiy: RSFSR.	8	55N	41E
Velingrad: Bulgaria. *Sawmilling.*	†	42N	24E
Velizh: RSFSR.	8	56N	31E
Velké Meziříčí: Czech.	6	49N	16E
Vel'ki-Bereznyy: Ukr. SSR.	7	49N	22E
Vel'sk: RSFSR.	18	61N	42E
Venev: RSFSR.	8	54N	38E
Vengerovo: RSFSR.	15	56N	77E

---

* Located by symbol, but not named, on special topic map. ‡ See Inset map. EP Front Endpaper map.
† On relevant special topic map only. For list of topics see p. VIII. Place names in [square] brackets are former, or alternative, names.
↓ See appendix on p.134

Page Lat. Long.

Ventspils [Ger. Windau]: Latvian SSR. Port. Textiles*, sawmilling. ‡5 57N 21E
Venyukovskiy: RSFSR. 8 55N 38E
Vereshchagino: RSFSR. Engineering. 13 58N 55E
Veretski Pass: USSR. 7 49N 23E
Vereya: RSFSR. 8 55N 36E
Verkhne-Chusovskiye Gorodki: RSFSR. Petroleum, petroleum refining. 13 58N 57E
Verkhne-Dneprovsk: Ukr. SSR. 9 49N 34E
▲Verkhne Stalinsk: RSFSR. Gold* (Aldan Field). 19 59N 125E
Verkhne-Ural'sk: RSFSR. 13 54N 59E
Verkhne-Usinskoye: RSFSR. 19 52N 93E
Verkhne-Vilyuysk: RSFSR. Lignite. 19 63N 120E
Verkhneye: Ukrainian SSR. Chemicals. 9 49N 38E
Verkhniy Avzyan: RSFSR. 13 54N 58E
Verkhniy Baskunchak: RSFSR. Salt. 9 48N 46E
Verkhniy Mamon: RSFSR. 9 50N 40E
Verkhniy Rogachik: Ukr. SSR. 9 47N 34E
Verkhniy Ufaley: RSFSR. Nickel, cobalt, nickel refining, iron, chemicals. 13 56N 60E
Verkhnyaya Khortitsa: Ukr. SSR. 9 48N 35E
Verkhnyaya Pyshma: RSFSR. Copper, cobalt, copper smelting & refining. 13 57N 61E
Verkhnyaya Salda: RSFSR. Asbestos* (Asbest group), eng. 13 57N 61E
Verkhnyaya Sinyachikha: RSFSR. Bauxite, coal. 13 58N 62E
Verkhnyaya Tura: RSFSR. 13 58N 60E
Verkhovazh'ye: RSFSR. 8 61N 42E
Verkhov'ye: RSFSR. 9 53N 37E
Verkhoyansk: RSFSR. Usually considered coldest place in the world. Lowest recorded temp. —92° F. 19 68N 133E
Verkhoyansk Range: RSFSR. 19 65N 130E
Verkhozim: RSFSR. 14 53N 46E
Vernadovka: RSFSR. 8 53N 42E
Vershino-Shakhtaminskiy: RSFSR. Molybdenum. 12 51N 117E
Veselí: Czechoslovakia 6 49N 15E
Veseloye: Ukrainian SSR. 9 47N 35E
Veshenskaya: RSFSR. 9 50N 42E
Ves'yegonsk: RSFSR. 8 59N 38E
Veszprém: Hungary 10 47N 18E
Vetluga: RSFSR. 8 58N 46E
Vetluzhskiy: RSFSR. 8 57N 45E
Vetrino: Byelorussian SSR. 8 55N 28E
Veymarn: RSFSR. 8 59N 29E
Vichuga: RSFSR. Pop. 1954: 50,000. Engineering. 8 57N 42E
Vida: Romania 11 44N 25E
Vidin: Bulgaria 11 44N 23E
Vikulov, C.: RSFSR. 18 75N 60E
Vikulovo: RSFSR. River port. 15 57N 71E
Vileyka [Pol. Wilejka]: Byelorussian SSR. 7 54N 27E
Viliga: RSFSR. 19 61N 156E
Vilkavishkis [Rus. Volkovyshki]: Lithuanian SSR. 7 55N 23E
Vilkovo: Ukrainian SSR. 9 45N 30E
VIL'NYUS [Vilna: Pol. Wilno]: Lith. SSR. Pop. 1954: 210,000. Eng.*, chemicals, textiles, elec. power. 5 54N 25E
Vilyaka: Latvian SSR. ‡5 57N 28E
Vilyandi [Viljandi]: Est. SSR. ‡5 58N 25E
Vilyane: Latvian SSR. ‡5 56N 27E
Vilyuy Mts.: RSFSR. 19 67N 110E
Vilyuy, R. RSFSR. Platinum, gold. 23 64N 123E
Vilyuysk: RSFSR. 19 63N 121E
Vimperk: Czechoslovakia 6 49N 14E
Vinga: Romania 11 46N 21E
Vinkovci: Yugoslavia 10 45N 19E
Vinnitsa: Ukr. SSR. Pop. 1954: 110,000. Eng., chemicals, textiles. 7 49N 28E
Vinnitsa Oblast: Ukr. SSR. EP 49N 30E
Vinogradnoye: RSFSR. 12 44N 44E
Virbalis: Lith. SSR. ‡5 54N 22E
Virovitica: Yugoslavia 10 46N 17E
Virpazar: Yugoslavia 10 42N 19E
Virtsu: Estonian SSR. ‡5 58N 23E
Vis, i.: Yugoslavia 10 43N 16E
Višegrad: Yugoslavia 10 44N 19E
Vishniy-Volochek: RSFSR. Textiles, eng. † 57N 35E
Visim: RSFSR. Platinum. 13 58N 59E
Visoko: Yugoslavia 10 44N 18E
Vistula, R.: Poland 22 53N 20E
Vitebsk: Byelorus. SSR. Pop. 1954: 100,000. Textiles, eng. 8 55N 30E

▲Vitebsk Oblast: Byelorus. SSR. EP 55N 28E
Vitim: RSFSR. 19 60N 113E
Vitim Plateau: RSFSR. 12 53N 112E
Vitim, R.: RSFSR. 23 58N 114E
Vivi: RSFSR. 19 63N 97E
Vivikonna: Estonian SSR. 5 59N 27E
Viziru: Romania 11 45N 28E
Vizovice: Czechoslovakia 6 49N 18E
Vladimir: RSFSR. Pop. 1954: 90,000. Eng., chemicals. 8 56N 40E
Vladimir Oblast: RSFSR. EP 56N 41E
Vladimir-Volynskiy [Pol. Włodzimierz]: Ukr. SSR. 7 51N 24E
Vladislavovka: Ukrainian SSR. 9 45N 35E
Vladivostok: RSFSR. Pop. 1954: 350,000. Port. Eng., petroleum refining, sawmilling. 19 43N 132E
Vlonë [It. Valona]: Albania 10 40N 19E
Voislova: Romania 11 46N 22E
Vokhma: RSFSR. 8 59N 47E
Volchanka: RSFSR. Coal. 13 60N 60E
Volchansk: Ukrainian SSR. 9 50N 37E
Volchikha: RSFSR. 15 52N 80E
Volga: RSFSR. 8 58N 38E
Volga, R.: USSR. 22 52N 47E
Volga Uplands: RSFSR. 8 53N 45E
Volkhov: RSFSR. Pop. 1954: 50,000. Aluminium, hydro-elec. 8 60N 32E
Volkovysk [Pol. Wołkowysk]: Byelorussian SSR. 7 53N 24E
Volnovakha: Ukrainian SSR. 9 48N 38E
Volodarsk: Ukrainian SSR. 7 51N 28E
Volodarskoye: Kazakh SSR. 14 53N 68E
Volodary: RSFSR. 8 56N 43E
Vologda: RSFSR. Pop. 1954: 150,000. Eng., textiles, sawmilling. 8 59N 40E
Vologda Oblast: RSFSR. EP 60N 40E
Volokolamsk: RSFSR. 8 56N 36E
Volokonovka: RSFSR. 9 51N 38E
Voloshino: RSFSR. 9 49N 40E
Vološinovo: Yugoslavia 11 46N 20E
Volosovo: RSFSR. 8 59N 29E
Volovo: RSFSR. 8 54N 38E
Volozhin [Pol. Wołozyn]: Byelorussian SSR. Chemicals. 7 54N 27E
Vol'sk: Pop. 1954: 75,000. Chemicals. 14 52N 47E
Volyn' Oblast: Ukr. SSR. EP 53N 25E
Volzhsk [Lopatino]: RSFSR. 14 56N 48E
Vorkuta: RSFSR. Coal. 18 68N 64E
Vormsi, i.: Estonian SSR. 5 59N 23E
Vorokhta: Ukrainian SSR. 7 48N 24E
Voronezh: RSFSR. Pop. 1954: 350,000. Eng., synthetic rubber, elec. power. 9 52N 39E
Voronezh: Ukrainian SSR. 9 52N 33E
Voronezh Oblast: RSFSR. EP 52N 40E
Voronovitsa: Ukr. SSR. 7 51N 29E
Vorontsovka: RSFSR. Copper. 13 60N 60E
▲Voroshilov: RSFSR. Pop. 1954: 140,000. 19 44N 132E
▲Voroshilovgrad: Ukrainian SSR. Pop. 1954: 250,000. Eng., textiles. 9 49N 39E
▲Voroshilovgrad Oblast: Ukr. SSR. EP 48N 39E
▲Voroshilovsk [Alchevsk]: Ukr. SSR. Pop. 1954: 100,000. Iron & steel, chemicals. 9 48N 39E
Vorozhba: Ukrainian SSR. 9 51N 34E
Vorrë: Albania 10 41N 20E
Vorsma: RSFSR. 8 56N 43E
Voskresensk: RSFSR. Pop. 1954: 50,000. Chemicals. 8 55N 39E
Votice: Czechoslovakia 6 49N 15E
Votkinsk: RSFSR. Pop. 1954: 50,000. Eng.*, hydro-elec. 13 57N 54E
Voy-Vozh: RSFSR. Petroleum. † 62N 54E
Vozhayel': RSFSR. Sawmilling. † 63N 51E
Vozhe, lake: RSFSR. 8 61N 39E
Vozhega: RSFSR. 8 60N 40E
Voznesensk: Ukrainian SSR. 9 48N 31E
Voznesen'ye: RSFSR. 8 61N 35E
Vranje: Yugoslavia 11 43N 22E
Vratsa: Bulgaria. Textiles. 11 43N 24E
Vrbas: Yugoslavia 10 46N 19E
Vrbovec: Yugoslavia 10 46N 16E
Vršac: Yugoslavia 11 45N 21E
Vrútky: Czechoslovakia 6 49N 19E
Vsetín: Czechoslovakia 6 49N 18E
Vsevolodo-Vil'va: RSFSR. 13 59N 57E
Vucitrn: Yugoslavia 11 43N 21E
Vukovar: Yugoslavia 16 45N 19E
Vulkaneshty: Moldavian SSR. 11 46N 28E
Vŭrbitsa: Bulgaria 11 43N 27E
Vurnary: RSFSR. Phosphates chemicals. † 55N 47E
Vyartsilya [Fin. Värtsilä]: Karelo Finnish SSR. Steel. 18 62N 30E
Vyati: RSFSR. 13 56N 51E
Vyatskiye Polyany: RSFSR. 14 56N 51E

Vyazemskiy: RSFSR. 19 48N 135E
Vyaz'ma: RSFSR. 8 55N 34E
Vyazniki: RSFSR. Textiles. 8 56N 42E
Vyborg [Fin. Viipuri]: RSFSR. Pop. 1954: 50,000. Textiles. 18 61N 29E
Vychegda, R.: RSFSR. 22 62N 53E
Vyksa: RSFSR. Pop. 1954: 50,000. Steel. 8 55N 42E
Vyritsa: RSFSR. 8 59N 30E
Vyru [Est. Võru]: Estonian SSR. Textiles*. ‡5 57N 27E
Vysha: RSFSR. 8 54N 42E
Vyshniy-Volochek: RSFSR. Pop. 1954: 70,000. 8 58N 34E
Vyškov: Czechoslovakia 6 49N 17E
Vysokovsk: RSFSR. 8 56N 36E
Vysotsk: Ukrainian SSR. 7 52N 27E
Vytegra: RSFSR. 8 61N 36E

Wąbrzeźno: Poland 6 53N 19E
Wągrowiec: Poland 6 53N 17E
Wałbrzych [Ger. Waldenburg]: Poland. Pop. 1950: 81,260. Coal, chemicals. 6 51N 17E
Wałcz [Ger. Deutsch Krone]: Poland 6 53N 16E
Waldheim: E. Ger. Chemicals. 6 51N 13E
Waren: E. Germany 6 54N 13E
Warin: E. Germany 6 54N 12E
Warka: Poland 7 52N 21E
Warnemünde: E. Ger. Port. 6 54N 12E
WARSAW [Pol. Warszawa]: Poland. Pop. 1955: 965,000. River port. Eng., textiles, chemicals*, steel. 7 52N 21E
Warta: Poland 6 52N 18E
Węgliniec [Ger. Kohlfurt]: Poland 6 51N 15E
Węgrów: Poland 7 52N 22E
Weimar: E. Ger. Pop. 1946: 66,659. Paper. 6 51N 12E
Weissenfels: E. Ger. Pop. 1946: 50,995. Lignite. 6 51N 12E
Wejherowo [Ger. Neustadt]: Poland 6 55N 18E
Wernigerode: E. Germany. Chemicals, paper. 6 51N 11E
▲West Kazakhstan Oblast: Kazakh SSR. EP 50N 50E
West Siberian Plain: RSFSR. 15 60N 72E
White Sea: RSFSR. 18 66N 37E
Więcbork: Poland 6 53N 18E
Wielbark [Ger. Willenberg]: Poland 7 53N 21E
Wieleń: Poland 6 53N 16E
Wieluń: Poland 6 51N 18E
Wieruszów: Poland 6 51N 18E
Wierzbnik: Poland 7 51N 21E
Wiese I.: RSFSR. 18 59N 76E
Wilczek Land: RSFSR. 18 81N 62E
Wislinskiy Zaliw: Poland 7 54N 20E
Wismar: E. Ger. Engineering. 6 54N 12E
Wittenberg: E. Germany. Pop. 1946: 53,400. River port. 6 52N 13E
Wittenberge: E. Ger. River port. Textiles. 6 53N 12E
Wittstock: E. Germany 6 53N 12E
Włocławek: Poland. Pop. 1950: 54,650. Sulphur, sawmilling. 6 53N 19E
Włodawa: Poland 7 52N 24E
Włoszczowa: Poland 6 51N 20E
Wolin [Ger. Wollin]: Poland 6 54N 15E
Wołów: Poland 16 51N 17E
Wrangel I.: RSFSR. 19 71N 180
Wriezen: E. Germany 6 53N 14E
Wrocław [Ger. Breslau]: Poland. Pop. 1950: 341,500. River port. Eng., chemicals, textiles. 6 51N 17E
Wronki: Poland 6 53N 16E
Września: Poland 6 52N 18E
Wschowa: Poland 6 52N 16E
Wurzen: E. Germany 6 51N 13E
Wybraniec: Poland 6 52N 19E
Wyrzysk: Poland 6 53N 17E
Wysokie Mazowieckie: Poland 7 53N 23E
Wyszków: Poland 7 53N 22E

Yablonovyy Range: RSFSR. 12 52N 115E
Yadrin: RSFSR. 8 56N 46E
Yagarakhu: Est. SSR. ‡5 58N 22E
Yagman: Turkmen SSR. Lignite. 16 40N 54E
Yagotin: Ukrainian SSR. 9 50N 32E
Yagodnyy: RSFSR. Gold*. † 63N 149E

---

* Located by symbol, but not named, on special topic map.
† On relevant special topic map only. For list of topics see p. VIII.
▲ See appendix on p.134
‡ See Inset map.   EP Front Endpaper map.
Place names in [square] brackets are former, or alternative, names.

	Page	Lat.	Long.
Yakhroma: RSFSR.	8	56N	38E
Yakoruda: Bulgaria	11	42N	24E
Yaksha: RSFSR.	14	61N	57E
Yakshanga: RSFSR.	8	58N	46E
Yakut ASSR.: RSFSR.	EP	65N	130E
Yakutsk: RSFSR. *Pop. 1954: 60,000. Sawmilling, eng.*	19	62N	130E
Yalta: Ukrainian SSR.	9	44N	34E
Yalutorovsk: RSFSR.	14	57N	66E
Yama: Ukr. SSR. *Magnesite & magnesium salts.*	9	49N	38E
Yamal-Nenets NO.: RSFSR.	EP	65N	75E
Yamal Penin.: RSFSR.	8	72N	70E
Yamarovka: RSFSR.	12	51N	110E
Yambol: Bulgaria	11	42N	26E
Yamm: RSFSR.	8	58N	28E
Yampol': Ukrainian SSR.	7	48N	28E
Yampol': Ukrainian SSR.	7	50N	26E
Yamsk: RSFSR.	19	60N	154E
Yamaul: RSFSR.	13	56N	55E
Yangi-Yul' [Kaunchi]: Uzbek SSR.	17	41N	69E
Yanskiy: RSFSR.	19	68N	134E
Yantarnyy [Ger. Palmnicken]: RSFSR.	7	55N	20E
Yar: RSFSR.	13	58N	52E
Yaransk: RSFSR.	14	57N	48E
Yarega: RSFSR. *Petroleum.*	†	63N	54E
Yaremcha: Ukrainian SSR.	7	48N	24E
Yarmolintsy: Ukrainian SSR.	7	49N	27E
Yaroslavl': RSFSR. *Pop. 1954: 325,000. Elec. power, eng., synthetic rubber, textiles, chemicals, sawmilling.*	8	58N	40E
Yaroslavl' Oblast: RSFSR.	EP	58N	40E
Yartsevo: RSFSR. *Textiles.*	8	55N	33E
Yashkino: RSFSR.	13	53N	54E
Yashkino: RSFSR. *Elec. power.*	†	55N	85E
Yasinovataya: Ukr. SSR.	9	48N	38E
Yasinya: Ukrainian SSR.	7	48N	24E
Yasnoye: RSFSR.	5	55N	21E
Yasnyy: RSFSR.	19	53N	127E
Yaunpiyebalga: Latvian SSR.	‡5	57N	26E
Yavorov: Ukrainian SSR.	7	50N	23E
Yaya: RSFSR.	15	56N	87E
Yazykovo: RSFSR.	14	54N	47E
Yedintsy [Rom. Edineţi]: Moldavian SSR.	7	48N	27E
Yefremov: RSFSR. *Synthetic rubber.*	8	53N	38E
Yegendybulak: Kazakh SSR.	17	50N	77E
Yegorlykskaya: RSFSR.	9	47N	41E
Yegor'yevsk: RSFSR. *Pop. 1954: 100,000. Textiles, eng.**	8	55N	39E
Yekabpils: Latvian SSR.	‡5	56N	26E
Yekaterinoslavka: RSFSR.	19	50N	129E
Yelabuga: RSFSR.	13	56N	52E
Yelan': RSFSR.	9	51N	44E
Yelenskiy: RSFSR.	8	54N	36E
Yelets: RSFSR. *Pop. 1954: 67,000.*	9	53N	38E
Yelgava: Latvian SSR.	‡5	56N	24E
Yelizaveta, C.: RSFSR.	19	54N	143E
Yel'nya: RSFSR.	8	55N	33E
Yel'shanka: RSFSR. *Natural gas (pipe line to Moscow).*	16	52N	46E
Yel'sk: Byelorussian SSR.	7	52N	29E
Yemanzhelinsk: RSFSR. *Lignite.*	13	55N	61E
Yena: RSFSR. *Nickel, iron ore.*	18	68N	31E
Yenakiyevo: Ukr. SSR. *Pop. 1954: 125,000. Coal* (Donbass), iron & steel.*	9	48N	38E
Yenisey, G. of: RSFSR.	18	72N	82E
Yenisey, R.: RSFSR.	23	63N	88E
Yeniseysk: RSFSR. *River port. Sawmilling.*	15	58N	92E
Yenotayevka: RSFSR.	16	47N	47E
Yerbent: Turkmen SSR.	16	39N	58E
YEREVAN [Erivan]: Armenian SSR. *Pop. 1954: 340,000. Eng., chemicals, synthetic rubber, textiles, aluminium.*	12	40N	44E
Yergeni Hills: RSFSR.	9	48N	44E
Yermish: RSFSR.	8	55N	42E
Yermolayevo: RSFSR. *Coal.*	13	53N	56E
Yerofey Pavlovich: RSFSR. *Gold**.	19	54N	122E
Yeropol: RSFSR.	19	65N	169E
Yershov: RSFSR.	14	51N	48E
Yessentuki: RSFSR. *Pop. 1954: 50,000.*	12	44N	43E
Yessey: RSFSR.	19	68N	102E
Yevlakh: Azerbaydzhan SSR.	12	41N	47E
Yevpatoriya: Ukrainian SSR.	9	45N	33E
Yeysk: RSFSR. *Pop. 1954: 50,000. Port.*	9	47N	38E
Ynykchanskiy: RSFSR. *Gold** (Allakh-Yun Field).	19	60N	137E
Yoshkar-Ola [Tsarevokokshaysk, Krasnokokshaysk]: RSFSR. *Eng. (cinema projectors).*	14	57N	48E

	Page	Lat.	Long.
Yugodzyr: Mongolian People's Republic	19	46N	115E
Yugo-Kamskiy: RSFSR.	13	57N	56E
Yukagir Plateau: RSFSR.	19	67N	157E
Yukhnov: RSFSR.	8	55N	35E
Yunakovka: Ukrainian SSR.	9	51N	35E
Yur: RSFSR. *Gold** (Allakh-Yun Field).	19	60N	138E
Yurbarkas: Lith. SSR.	‡5	55N	23E
Yurga: RSFSR.	15	56N	85E
Yurkovka: Ukr. SSR. *Lignite.*	9	49N	31E
Yurlovka: RSFSR.	9	53N	41E
Yur'yevets: RSFSR. *Textiles.*	8	57N	43E
Yur'yev Pol'skiy: RSFSR.	8	56N	40E
Yuryuzan': RSFSR.	13	55N	58E
Yuzha: RSFSR.	8	56N	42E
Yuzhkuzbassgres: RSFSR. *Elec. power.*	†	54N	87E
Yuzhno-Sakhalinsk' [Jap. Toychara]: RSFSR. *Pop. 1954: 60,000. Paper, eng.*	19	47N	143E
Yuzhno Yeniseyskiy: RSFSR. *Gold** (Yenisey Fields).	19	59N	95E
Yuzhnyy Alamyslik: Uzbek SSR. *Petroleum**.	17	41N	73E
Yykhvi [Est. Jonvi]: Est. SSR.	5	59N	27E
Zabolotov [Pol. Zabłotów]: Ukrainian SSR.	7	48N	25E
Zabrze [Ger. Hindenburg]: Poland. *Pop. 1939: 126,220; 1946: 104,184; 1950: 132,900. Coal, steel, eng., chemicals.*	6	50N	19E
Zadar: Yugoslavia. *Port.*	10	44N	14E
Zadonsk: RSFSR. *River port.*	†	52N	38E
Zadon'ye: RSFSR. *Lignite** (Skopin Group).	8	54N	38E
Żagań [Ger. Sagan]: Poland. *Lignite.*	6	52N	15E
Zaglik: Azerbaydzhan SSR. *Alunite (Bauxite).*	12	41N	46E
Zagorsk: RSFSR. *Pop. 1954: 50,000.*	8	56N	38E
Zagreb [Ger. Agram]: Yugo. *Pop. 1953: 350,452. Eng., textiles, chemicals, paper, asbestos.*	10	46N	16E
Žagubica: Yugoslavia	11	44N	22E
Zagvozd: Yugoslavia	10	43N	17E
Zaječar: Yugoslavia. *Antimony, antimony smelting.*	11	44N	22E
Zakataly: Azerbaydzhan SSR.	12	42N	47E
Zakharovo: RSFSR.	8	54N	39E
Zakhmatabad: Tadzhik SSR.	17	39N	68E
Zakopane: Poland	6	49N	20E
Zakroczym: Poland	7	53N	21E
Zalaegerszeg: Hungary. *Oil refining.*	10	47N	17E
Zalău: Romania	11	47N	23E
Zalegoshch': RSFSR.	8	53N	37E
Zaleshchiki: Ukrainian SSR.	7	49N	26E
Zambrów: Poland	7	53N	22E
Zamość: Poland	7	51N	23E
Zangibasar [Ulukhanlu]: Armenian SSR.	12	40N	44E
Zapadnaya Dvina: RSFSR.	8	56N	32E
Zapokrovskiy: RSFSR. *Arsenic.*	12	51N	119E
Zaporozh'ye: Ukrainian SSR. *Pop. 1954: 375,000. Steel, ferro-alloys, aluminium, eng., chemicals.*	9	48N	35E
Zaporozh'ye Oblast: Ukr. SSR.	EP	47N	36E
Zarasay: Lith. SSR.	‡5	55N	26E
Zaraysk: RSFSR. *Textiles.*	8	55N	39E
Zărneşti: Romania. *Sawmilling.*	11	45N	25E
Zarubino: RSFSR.	8	59N	34E
Żary [Ger. Sorau]: Poland. *Lignite.*	6	52N	15E
Zaslavl': Byelorussian SSR.	7	54N	27E
Žatec: Czechoslovakia	6	50N	14E
Zatoka: Ukrainian SSR.	9	46N	30E
Zavetnoye: RSFSR.	9	47N	44E
Zavitaya: RSFSR.	19	50N	129E
Zawiercie: Poland	6	50N	19E
Zayarsk: RSFSR.	19	56N	103E
Zaysan: Kazakh SSR.	15	48N	85E
Zaysan, lake: Kazakh SSR.	17	48N	83E
Zbarazh [Pol. Zbaraż]: Ukr. SSR.	7	50N	26E
Zbąszyń: Poland	6	52N	16E
Zdice: Czechoslovakia. *Iron ore.*	6	50N	14E
Zdolbunov: Ukrainian SSR.	7	50N	26E
Zduńska Wola: Poland	6	52N	19E
Zdvinsk: RSFSR.	15	55N	79E
Zeitz: E. Germany. *Lignite.*	6	51N	12E
Żelechów: Poland	7	52N	22E
Zelenchukskaya: RSFSR.	12	44N	42E

	Page	Lat.	Long.
Zelenodol'sk: RSFSR.	14	55N	48E
Zelenogorsk: RSFSR.	8	60N	29E
Zelenogradsk: RSFSR.	7	55N	20E
Zemetchino: RSFSR.	8	53N	43E
Zemun: Yugoslavia	11	45N	20E
Ženica [Zenitsa]: Yugoslavia. *Iron & steel.*	10	44N	18E
Zen'kov: Ukrainian SSR.	9	50N	34E
Zepče: Yugoslavia	10	44N	18E
Zeravshan Range: USSR.	17	39N	68E
Zerbst: East Germany	6	52N	12E
Zernovoy: RSFSR.	9	47N	40E
Zestafoni [Kvirily]: Georgian SSR. *Ferro-alloys.*	12	42N	43E
Zeya: RSFSR. *Gold, zirconium in Zeya Valley.*	19	53N	127E
Zgierz: Poland	6	52N	20E
Zgorzelec *see* Görlitz			
Zgurovka: Ukrainian SSR.	9	50N	32E
Zhangiz-Tobe: Kazakh SSR. *Gold**.	17	49N	81E
Zharkamys: Kazakh SSR.	16	48N	57E
Zharma: Kazakh SSR. *Gold.*	17	49N	81E
Zhashkov: Ukrainian SSR.	9	49N	30E
Zhdanov [Mariupol']: Ukr. SSR. *Pop. 1954: 225,000. Port. Zirconium, iron & steel, chemicals.*	9	47N	38E
Zhelaniye, C.: RSFSR.	18	77N	70E
Zheleznodorozhnyy [Ger. Gerdauen]: RSFSR.	7	54N	21E
Zheleznodorozhnyy: RSFSR.	18	63N	50E
Zheleznovodsk: RSFSR.	12	44N	43E
Zheltaya Reka: Ukrainian SSR. *Iron ore.*	9	48N	34E
Zherdevka: RSFSR.	9	52N	42E
Zhety-Kol', lake: Kazakh SSR.	13	51N	60E
Zhidachov [Pol. Żydaczów]: Ukrainian SSR.	7	49N	24E
Zhigalovo: RSFSR.	12	55N	105E
Zhigansk: RSFSR. *River port.*	19	67N	123E
Zhigulevsk [Otvazhnyy]: RSFSR. *Petroleum.*	†	54N	50E
Zhilaya Kosa: Kazakh SSR.	16	47N	53E
Zhirkova: RSFSR.	19	68N	155E
Zhirnovskiy: RSFSR. *Petroleum.*	†	50N	54E
Zhitkovichi: Byelorus. SSR.	7	52N	28E
Zhitkur: RSFSR.	16	49N	46E
Zhitomir: Ukrainian SSR. *Pop. 1954: 110,000. Engineering.*	7	50N	28E
Zhitomir Oblast: Ukr. SSR.	EP	51N	30E
Zhlobin: Byelorussian SSR.	8	53N	30E
Zhmerinka: Ukrainian SSR.	7	49N	28E
Zhukovka: RSFSR.	8	54N	34E
Zhuravleka: Kazakh SSR. *Bauxite.*	15	52N	70E
Ziddy: Tadzhik SSR. *Lignite.*	17	39N	68E
Ziębice [Ger. Münsterberg]: Poland.	6	51N	17E
Zielóna Góra [Ger. Grünberg]: Poland. *Elec. power, lignite, eng., textiles.*	6	52N	16E
Zigazinskiy: RSFSR. *Iron ore.*	13	54N	58E
Zilair: RSFSR.	13	52N	58E
Žilina: Czech. *Textiles**, paper.	6	49N	19E
Zima: RSFSR. *Sawmilling.*	12	53N	102E
Zimnicea: Romania	11	44N	25E
Zimovniki: RSFSR.	9	47N	42E
Zingst: E. Germany	6	54N	13E
Zittau: E. Germany	6	51N	15E
Zlatograd: Bulgaria. *Lead, copper & zinc.*	11	41N	25E
Zlatoust: RSFSR. *Pop. 1954: 150,000. Steel, chemicals, sawmilling.*	13	55N	60E
Zlatoustovsk: RSFSR. *Gold**.	19	53N	134E
Zletovo: Yugoslavia. *Lead, zinc, copper.*	11	42N	22E
Zlin *see* Gottwaldov			
Złoczew: Poland	6	51N	18E
Złotów: Poland	6	53N	17E
Zlynka: RSFSR.	9	52N	32E
Zlynka: Ukrainian SSR.	9	48N	31E
Zmeinogorsk: RSFSR. *Lead, zinc, barium.*	15	51N	82E
Żmigród [Ger. Trachenberg]: Poland	6	51N	17E
Zmiyev: Ukrainian SSR.	9	50N	36E
Znamenka: Ukrainian SSR.	9	49N	33E
Znamensk [Ger. Wehlau]: RSFSR.	7	55N	21E
Znamenskoye: RSFSR.	8	53N	36E
Znin: Poland	6	53N	18E
Znojmo: Czechoslovakia	6	49N	16E
Zol'noye: RSFSR. *Petroleum.*	†	53N	50E
Zolochev [Pol. Złoczów]: Ukr. SSR. *Lignite.*	7	49N	24E
Zolochev: Ukrainian SSR.	9	50N	36E
Zolotarevka: RSFSR.	8	53N	45E
Zolotonosha: Ukrainian SSR.	9	50N	32E
Zolotoye: RSFSR.	16	51N	46E

* Located by symbol, but not named, on special topic map.   ‡ See Inset map.   EP Front Endpaper map.
† On relevant special topic map only. For list of topics see p. VIII.   Place names in [square] brackets are former, or alternative, names.

133

	Page	Lat.	Long.
Zolotoye: Ukr. SSR. *Coal**.	†	49N	38E
Zolotoy Potok [*Pol. Potok Zloty*] Ukrainian SSR.	7	49N	25E
Zrenjanin [Petrograd]: Yugo.	11	45N	20E
Zubtsov: RSFSR.	8	56N	34E
Zugdidi: Georgian SSR.	12	42N	42E
Zugres: Ukrainian SSR. *Elec. power.*	9	48N	38E
Zurash: RSFSR.	8	53N	32E
Zuyevka: RSFSR.	14	58N	50E

	Page	Lat.	Long.
Zvenigorod: RSFSR.	8	56N	37E
Zvenigorodka: Ukr. SSR. *Lignite.*	9	49N	31E
Zverevo: RSFSR.	9	48N	40E
Zvolen: Czechoslovakia.	10	49N	19E
Zvornik: Yugoslavia.	10	44N	19E
Zwickau: E. Germany. *Pop. 1946: 122,862. Coal, textiles, eng., chemicals.*	6	51N	12E
Zwoleń: Poland.	7	51N	22E
Zychlin: Poland.	6	52N	20E

	Page	Lat.	Long.
Zykh: Azerbaydzhan SSR. *Petroleum.*	†	40N	50E
Żyrardów: Poland. *Lignite, eng.*	7	52N	20E
Zyryanka: RSFSR. *Lignite.*	19	66N	150E
Zyryanovsk: Kazakh SSR. *Pop. 1954: 80,000. Lead, zinc, silver.*	17	50N	84E
Zyryanovskiy: RSFSR.	13	58N	62E
Zyryanskoye: RSFSR.	15	57N	87E
Zywiec: Poland. *Hydro-elec.*	6	50N	19E

* Located by symbol, but not named, on special topic map. On relevant special topic map only. For list of topics see p. VIII. ‡ See Inset map. EP Front Endpaper map. Place names in [square] brackets are former, or alternative, names.

APPENDIX

Akmolinsk: Kazakh. SSR. *now Tselinograd.*
Akmolinsk Oblast: Kazakh. SSR. *now Tselinograd Oblast.*
Arzamas Oblast: RSFSR. *incorporated in Gorkiy Oblast.*
Ashkabad Oblast: Turkmen SSR. *abolished.*
Astrakhan-Bazar: Azer. SSR. *now Dzhalilabad.*

Balashov Oblast: RSFSR. *incorporated in Tambov, Saratov, Volgograd and Voronezh Oblasts.*
Berdyansk: Ukrainian SSR. *formerly Osipenko.*
Bereznik: RSFSR. *formerly Semenovskoye.*
Bessarabka: Moldavian SSR. *formerly Romanovka.*
Braşov: Romania, *formerly Orasul Stalin.*
Bukhara Oblast: Uzbek SSR. *incorporated in Syr-Dar'ya Oblast.*
Bulgaria: *internal administrative divisions have been reorganised (1964).*

Chardzhou Oblast: Turkmen SSR. *abolished.*
Chemnitz: E. Germany. *now Karl-Marx-Stadt.*
Chesnokovka: RSFSR. *now Novoaltaysk.*
Chimkent Oblast: Kazakh. SSR. *formerly South Kazakhstan Oblast.*
Chistyakovo: Ukrainian SSR. *now Thorez.*
Chkalov: RSFSR. *now Orenburg.*
Chkalov Oblast: RSFSR. *now Orenburg Oblast.*

Donetsk: Ukrainian SSR. *formerly Stalino.*
Donetsk Oblast: Ukrainian SSR. *formerly Stalino Oblast.*
Drissa: Byelorussian SSR. *now Verkhnedvinsk.*
Drogobych Oblast: Ukrainian SSR. *incorporated into Lvov-Oblast.*
DYUSHAMBE: Tadzhik SSR. *formerly Stalinabad.*
Dyushambe Oblast: Tadzhik SSR. *formerly Stalinabad Oblast.*
Dzhala-Abad Oblast: Kirghiz SSR. *incorporated in Osh Oblast.*
Dzhalilabad: Azer. SSR. *formerly Astrakhan-Bazar.*

Eisenhuttenstadt: E. Germany. *formerly Furstenberg and Stalinstadt.*
Elista: RSFSR. *formerly Stepnoy.*

Fizuli: Azer. SSR. *formerly Karyagino.*
Furstenberg and Stalinstadt: E. Germany. *now Eisenhuttenstadt.*

Garm Oblast: Tadzhik SSR. *incorporated in Dyushambe Oblast.*
Gogland I. RSFSR. *formerly Sur Sari I.*

Issyk-Kul Oblast: Kirghiz SSR. *incorporated in Frunze Oblast.*
Ivano-Frankovsk: Ukrainian SSR. *formerly Stanislav.*
Ivano-Frankovsk Oblast: Ukrainian SSR. *formerly Stanislav Oblast.*

Kaganovich: Moscow Oblast: RSFSR. *now Novokashirsk.*
Kaganovich: Tula Oblast: RSFSR. *now Tovarkovsky.*
Kaganovichabad: Tadzhik SSR. *now Kolkhozabad.*
Kaganovichevsk: Turkmen SSR. *now Komsomolsk.*
Kaganovichi: Ukrainian SSR. *now Polesskoye.*
Kalevala: Karelo-Finnish ASSR. *formerly Ukhta.*
Kalmyk Oblast: RSFSR. *created from parts of Astrakhan and Rostov Oblasts and Stavropol' Kray.*
Kamensk Oblast: RSFSR. *incorporated in Rostov Oblast.*

Kantagi: Kazakh. SSR. *now Kentau.*
Kapsukas: Lithuanian SSR. *formerly Mariyampole.*
Karelo-Finnish SSR. *now Karelo-Finnish ASSR.*
Karl-Marx-Stadt: E. Germany. *formerly Chemnitz.*
Karyagino: Azer. SSR. *now Fizuli.*
Kashka-Dar'ya Oblast: Uzbek SSR. *now Surkhan-Dar'ya Oblast.*
Katowice: Poland. *formerly Stalinogrod.*
Kentau: Kazakh. SSR. *formerly Kantagi.*
Kermine: Uzbek SSR. *now Novoi.*
Khamza Khakimzada: Uzbek SSR. *formerly Vannovskiy.*
Kolarovgrad: Bulgaria. *now Sumen.*
Kolkhozabad: Tadzhik SSR. *formerly Kaganovichabad.*
Kommunarsk: Ukrainian SSR. *formerly Voroshilovsk.*
Komsomolsk: Turkmen SSR. *formerly Kaganovichevsk.*
Krasnoarmeysk: Kazakh. SSR. *formerly Tayncha.*
Krasnogvardeyskoye: RSFSR. *formerly Molotovskoye.*
Krasnovodsk Oblast: Turkmen SSR. *abolished.*
Kruglyakov: RSFSR. *now Oktyabr'skiy.*

Leninskiy: RSFSR. *formerly Verkhne Stalinsk.*
Lensk: RSFSR. *formerly Mukhtuya.*
Lower Amur: RSFSR. *fully incorporated into Khabarovsk Kray.*
Lugansk: Ukrainian SSR. *formerly Voroshilovgrad.*
Lugansk Oblast: Ukrainian SSR. *formerly Voroshilovgrad Oblast.*

Marek: Bulgaria. *now Stanke Dimitrov.*
Mariyampole: Lithuanian SSR. *now Kapsukas.*
Mary Oblast: Turkmen SSR. *abolished.*
Molodechno Oblast: Byelorussian SSR. *incorporated in Vitebsk Oblast.*
Molotov: RSFSR. *now Perm.*
Molotovabad: Kirghiz SSR. *now Uch Korgon.*
Molotov Oblast: RSFSR. *now Perm Oblast.*
Molotovo: Georgian SSR. *now Trialeti.*
Molotovo: RSFSR. *now Oktyabrskoye.*
Molotovsk: Arkhangel'sk Oblast: RSFSR. *now Severodvinsk.*
Molotovsk: Kirov Oblast: RSFSR. *now Nolinsk.*
Molotovskoye: RSFSR. *now Krasnogvardeyskoye.*
Moskovskiy: Uzbek SSR. *formerly Stalino.*
Mukhtuya: RSFSR. *now Lensk.*
Muztor: Kirghiz SSR. *now Toktogul.*

Namangan Oblast: Uzbek SSR. *incorporated in Andizhan Oblast.*
Nolinsk: RSFSR. *formerly Molotovsk (Kirov Oblast).*
Novoaltaysk: RSFSR. *formerly Chesnokovka.*
Novoi: Uzbek SSR. *formerly Kermine.*
Novokashirsk: RSFSR. *formerly Kaganovich (Moscow Oblast).*
Novokuznetsk: RSFSR. *formerly Stalinsk.*
Novomoskovsk: RSFSR. *formerly Stalinogorsk.*

Oktyabr'skiy: RSFSR. *formerly Kruglyakov.*
Oktyabrskoye: RSFSR. *formerly Molotovo.*
Orasul Stalin: Romania. *now Braşov.*
Orenburg: RSFSR. *formerly Chkalov.*
Orenburg Oblast: RSFSR. *formerly Chkalov Oblast.*
Osipenko: Ukrainian SSR. *now Berdyansk.*

Perm: RSFSR. *formerly Molotov.*
Perm Oblast: RSFSR. *formerly Molotov Oblast.*
Pik Kommunizma: USSR. *formerly Stalin Peak.*
Polesskoye: Ukrainian SSR. *formerly Kaganovichi.*
Romania: *internal administrative divisions have been reorganised. (17th Feb.1968).*
Romanovka: Moldavian SSR. *now Bessarabka.*
Rybinsk: RSFSR. *formerly Shcherbakov.*

Semenovskoye: RSFSR. *now Bereznik.*
Severodvinsk: RSFSR. *formerly Molotovsk (Arkhangel'sk Oblast).*
Shcherbakov: RSFSR. *now Rybinsk.*
Shchurovo: RSFSR. *incorporated in Kolomna.*
South Kazakhstan Oblast: Kazakh. SSR. *now Chimkent Oblast.*
Stalin: Bulgaria. *now Varna.*
STALINABAD: Tadzhik SSR. *now Dyushambe.*
Stalinabad Oblast: Tadzhik SSR. *now Dyushambe Oblast.*
Stalingrad: RSFSR. *now Volgograd.*
Stalingrad Oblast: RSFSR. *now Volgograd Oblast.*
Staliniri: Georgian SSR. *now Tskhinvali.*
Stalino: Ukrainian SSR. *now Donetsk.*
Stalino Oblast: Ukrainian SSR. *now Donetsk Oblast.*
Stalino: Uzbek SSR. *now Moskovskiy.*
Stalinogorsk: RSFSR. *now Novomoskovsk.*
Stalinogrod: Poland. *now Katowice.*
Stalin Peak. USSR. *now Pik Kommunizma.*
Stalinsk: RSFSR. *now Novokuznetsk.*
Stalinstadt and Furstenberg. *now Eisenhuttenstadt.*
Stanislav: Ukrainian SSR. *now Ivano-Frankovsk.*
Stanislav Oblast: Ukrainian SSR. *now Ivano-Frankovsk Oblast.*
Stanke Dimitrov: Bulgaria. *formerly Marek.*
Stavropol': RSFSR. *new Togliatti.*
Stepnoy: RSFSR. *now Elista.*
Sumen: Bulgaria. *formerly Kolarovgrad.*
Surkhan-Dar'ya Oblast: Uzbek SSR. *formerly Kashka-Dar'ya Oblast.*
Sur Sari: RSFSR. *now Gogland.*
Syr-Dar'ya Oblast: Uzbek SSR. *new Oblast.*

Talas Oblast: Kirghiz SSR. *incorporated in Frunze Oblast.*
Taldy-Kurgan Oblast: Kazakh. SSR. *incorporated in Alma-Ata Oblast.*
Tashauz Oblast: Turkmen SSR. *abolished.*
Tayncha: Kazakh. SSR. *now Krasnoarmeysk.*
Thorez: Ukrainian SSR. *formerly Chistyakovo.*
Togliatti: RSFSR. *formerly Stavropol'.*
Toktogul: Kirghiz SSR. *formerly Muztor.*
Tovarkovsky: RSFSR. *formerly Kaganovich (Tula Oblast).*
Trialeti: Georgian SSR. *formerly Molotovo.*
Tselinograd: Kazakh. SSR. *formerly Akmolinsk.*
Tselinograd Oblast: Kazakh. SSR. *formerly Akmolinsk Oblast.*
Tskhinvali: Georgian SSR. *formerly Staliniri.*
Turnovo: Bulgaria. *now Veliko Turnovo.*
Tuva A.O.: RSFSR. *now Tuva ASSR.*

Uch Korgon: Kirghiz SSR. *formerly Molotovabad.*
Ukhta: Karelo-Finnish SSR. *now Kalevala.*
Ural'sk Oblast: Kazakh. SSR. *formerly West Kazakhstan Oblast.*
Ussuriysk: RSFSR. *formerly Voroshilov.*

Vannovskiy: Uzbek SSR. *now Khamza Khakimzada.*
Varna: Bulgaria. *formerly Stalin.*
Velikiye Luki: RSFSR. *incorporated in Pskov and Kalinin Oblasts.*
Velike Turnovo: Bulgaria. *formerly Turnovo.*
Verkhnedvinsk: Byelorussian SSR. *formerly Drissa.*

Verkhne Stalinsk: RSFSR. *now Leninskiy.*
Volgograd: RSFSR. *formerly Stalingrad.*
Volgograd Oblast: RSFSR. *formerly Stalingrad Oblast.*
Voroshilov: RSFSR. *now Ussuriysk.*
Voroshilovgrad: Ukrainian SSR. *now Lugansk.*
Voroshilovgrad Oblast: Ukrainian SSR. *now Lugansk Oblast.*
Voroshilovsk: Ukrainian SSR. *now Kommunarsk.*

West Kazakhstan Oblast: Kazakh. SSR. *now Ural'sk Oblast.*

POPULATION OF TOWNS OVER 50,000 FROM

Abakan: RSFSR	63,000	Bugul'ma: RSFSR	64,000	Gus'-Khrustal'nyy: RSFSR	59,000	Komsomol'sk: RSFSR	192,000
Aktyubinsk: Kazakh SSR	110,000	Bukhara: Uzbek SSR	84,000	Irkutsk: RSFSR	385,000	Konotop: Ukr. SSR	56,000
Alma Ata: Kazakh SSR	534,000	Buzuluk: RSFSR	57,000	Ivanovo: RSFSR	360,000	Konstantinovka: Ukr. SSR	94,000
Andizhan: Uzbek SSR	145,000	Chapayevsk: RSFSR	85,000	Izhevsk: RSFSR	322,000	Kopeysk: RSFSR	168,000
Angarsk: RSFSR	160,000	Chardzhou: Turkmen SSR.	75,000	Kadiyevka: Ukr. SSR	192,000	Korkino: RSFSR	87,000
Angren: Uzbek SSR	65,000	Cheboksary: RSFSR	134,000	Kalinin: RSFSR	286,000	Kostroma: RSFSR	189,000
Anzhero-Sudzhensk:		Chelyabinsk: RSFSR	751,000	Kaliningrad: RSFSR	232,000	Kovrov: RSFSR	105,000
RSFSR	120,000	Cheremkhovo: RSFSR	119,000	Kaluga: RSFSR	151,000	Kramatorsk: Ukr. SSR	126,000
Arkhangel'sk: RSFSR	276,000	Cherepovets: RSFSR	124,000	Kamensk-Shakhtinskiy:		Krasnodar: RSFSR	354,000
Armavir: RSFSR	123,000	Cherkassy: Ukr. SSR	99,000	RSFSR	62,000	Krasnokamsk: RSFSR	56,000
Artem: RSFSR	61,000	Chernigov: Ukr. SSR	107,000	Kamensk-Ural'skiy:		Krasnotur'insk: RSFSR	64,000
Artemovsk: Ukr. SSR	66,000	Chernogorsk: RSFSR	54,000	RSFSR	152,000	Krasnoyarsk: RSFSR	465,000
Asbest: RSFSR	65,000	Chernovtsy: Ukr. SSR	150,000	Kamyshin: RSFSR	65,000	Krasnyy Luch: Ukr. SSR	98,000
Ashkhabad: Turkmen SSR.	197,000	Chimkent: Kazakh SSR	173,000	Kansk: RSFSR	90,000	Kremenchug: Ukr. SSR	100,000
Astrakhan': RSFSR .	320,000	Chirchik: Uzbek SSR	80,000	Karaganda: Kazakh SSR	459,000	Krivoy Rog: Ukr. SSR	448,000
Baku (with suburbs):		Chistopol': RSFSR	57,000	Kaunas: Lith. SSR	247,000	Kropotkin, RSFSR	59,000
Azer: SSR	1,067,000	Chistyakovo: Ukr. SSR	92,000	Kazan': RSFSR	711,000	Kungur: RSFSR	67,000
Balashov: RSFSR	68,000	Chita: RSFSR	185,000	Kemerovo: RSFSR	305,000	Kurgan: RSFSR	173,000
Balkhash: Kazakh SSR	64,000	Chusovoy: RSFSR	63,000	Kerch': Ukr. SSR	107,000	Kursk: RSFSR	228,000
Baranovichi: Byelo, SSR	64,000	Daugavpils: Latvian SSR	74,000	Khabarovsk: RSFSR	363,000	Kustanay: Kazakh SSR	102,000
Barnaul: RSFSR	347,000	Donetsk: Ukr. SSR .	760,000	Khar'kov: Ukr. SSR	990,000	Kutaisi: Georgian SSR	141,000
Bataysk: RSFSR	72,000	Dneprodzerzhinsk:		Kherson: Ukr. SSR	183,000	Kuybyshev: RSFSR	881 000
Batumi: Georgian SSR	89,000	Ukr. SSR	207,000	Khmel'nitskiy: Ukr. SSR	69,000	Kuznetsk: RSFSR	64,000
Belaya Tserkov': Ukr. SSR.	77,000	Dnepropetrovsk: Ukr. SSR	722,000	Kiev: Ukr. SSR	1,208,000	Kzyl-Orda: Kazakh SSR	74,000
Belgorod: RSFSR	87,000	Dyushambe: Tadzhik SSR.	260,000	Kineshma: RSFSR	91,000	Leninabad: Tadzhik SSR	86,000
Beloretsk: RSFSR	62,000	Dzerzhinsk: RSFSR	180,000	Kirov: RSFSR	277,000	Leninakan: Armenian SSR	117,000
Belovo: RSFSR	118,000	Dzhambul: Kazakh SSR	136,000	Kirovabad: Azer. SSR	126,000	Leningrad (with suburbs):	
Bel'tsy: Mold. SSR	73,000	Elektrostal': RSFSR .	105,000	Kirovograd: Ukr. SSR	138,000	RSFSR	3,498,000
Berdichev: Ukr. SSR	57,000	Engel's: RSFSR	106,000	Kiselevsk: RSFSR	142,000	Leninogorsk: Kazakh SSR	69,000
Berdyansk: Ukr. SSR	73,000	Fergana: Uzbek SSR	81,000	Kishinev: Mold. SSR	244,000	Leninsk-Kuznetskiy:	
Berezniki: RSFSR	120,000	Frunze: Kirgiz SSR .	312,000	Kislovodsk: RSFSR	83,000	RSFSR	140,000
Biysk: RSFSR	165,000	Glazov: RSFSR	62,000	Kizel: RSFSR .	60,000	Leipaya (Liepāja):	
Blagoveshchensk: RSFSR	101,000	Gomel': Byero, SSR	193,000	Klaypeda (Klaipeda):		Latvian SSR.	77,000
Bobruysk: Byelo, SSR	108,000	Gor'kiy: RSFSR	1,025,000	Lith. SSR	105,000	Lipetsk: RSFSR	194,000
Borisoglebsk: RSFSR	57,000	Gorlovka: Ukr. SSR	309,000	Klin: RSFSR	60,000	Lugansk: Ukr. SSR	306,000
Borisov: Byelo SSR	65,000	Grodno: Byelo SSR .	85,000	Kokand: Uzbek, SSR	117,000	L'vov: Ukr. SSR	447,000
Bratsk: RSFSR	82,000	Groznyy: RSFSR	280,000	Kokhtla-Yarve: Est. SSR	60,000	Lys'va: RSFSR	76,000
Brest: Byelo, SSR	80,000	Gukovo: RSFSR	59,000	Kolomna: RSFSR	125,000	Lyubertsy: RSFSR	100,000
Bryansk: RSFSR	241,000	Gur'yev: Kazakh SSR	89,000	Kommunarsk: Ukr. SSR	110,000	Magadan: RSFSR	68,000

EASTERN

POPULATION OF TOWNS

Arad: Romania	122,331	Buzău: Romania .	75,569	Elbląg: Poland 1959	75,300	Iaşi: Romania	123,000
Belgrade: Yugo. 1961	587,899	Bydgoszcz: Poland	231,500	Erfurt: E. Ger.	186,448	Jena: E. Ger.	81,190
Berlin: Germany:		Bytom: Poland	182,500	Frankfurt: E. Ger.	56,638	Kalisz: Poland 1959	68,300
East	1,071,775	České Budějovice,		Galaţi: Romania .	105,000	Karl-Marx-Stadt	286,329
West	2,202,200	Czech. 1959	65,447	Gdańsk: Poland	286,500	(Chemnitz): E. Ger.	286,329
Białystok: Poland .	120,800	Chorzów: Poland .	146,700	Gdynia: Poland	147,800	Katowice (Stallinogród):	
Bielsko Biala:		Cluj: Romania	162,000	Gera: E. Ger.	101,373	Poland	268,900
Poland 1959	72,400	Constanţa: Romania	130,000	Gliwice: Poland	134,900	Kielce: Poland	85,000
Brăila: Romania	109,394	Cottbus: E. Ger. .	66,813	Görlitz: E. Ger.	89,909	Kladno: Czech. 1959	51,205
Brandenburg: E. Ger.	86,722	Craiova: Romania.	70,000	Gotha: E. Ger.	56,218	Košice: Czech. 1959	82,862
Braşov: Romania .	129,000	Częstochowa:		Gottwaldov: Czech. 1959	59,447	Kraków: Poland	479,000
Bratislava: Czech.. 1961	255,000	Poland	163,800	Grudziądz: Poland. 1959	64,000	Legnica: Poland 1959	63,300
Brno: Czech. 1961	320,000	Debrecen: Hung. .	129,000	Halle: E. Ger.	277,855	Leipzig: E. Ger.	589,632
Bucharest: Romania	1,349,000	Dessau: E. Ger.	93,459	Hradec Králové:		Liberec: Czech. 1959	68,513
Budapest: Hung. .	1,807,000	Dresden: E. Ger. .	493,603	Czech. 1959	57,273	Ljubljana: Yugo. 1961	133,386

U.S.S.R. 1962 OFFICIAL ESTIMATES

City	Population	City	Population
Magnitogorsk: RSFSR	333,000	Osh: Kirgiz SSR	85,000
Makeyevka: Ukr. SSR	381,000	Osinniki: RSFSR	71,000
Makhachkala: RSFSR	135,000	Pavlodar: Kazakh SSR	115,000
Margelan: Uzbek SSR	80,000	Pavlovo: RSFSR	55,000
Maykop: RSFSR	92,000	Pavlovskiy-Posad: RSFSR	58,000
Melitopol': Ukr. SSR	104,000	Penza: RSFSR	286,000
Michurinsk: RSFSR	85,000	Perm: RSFSR	701,000
Minsk: Byelo. SSR	599,000	Pervoural'sk: RSFSR	104,000
Mogilev: Byelo. SSR	139,000	Petropavlovsk: Kazakh SSR	146,000
Moscow (with suburbs): RSFSR	6,296,000	Petropavlovsk-Kamchatskiy: RSFSR	100,000
Murmansk: RSFSR	245,000	Petrozavodsk: Karelo-Finnish ASSR	142,000
Murom: RSFSR	83,000	Podol'sk: RSFSR	144,000
Mytishchi: RSFSR	107,000	Poltava: Ukr. SSR	154,000
Nakhodka: RSFSR	73,000	Prokop'yevsk: RSFSR	292,000
Nal'chik: RSFSR	102,000	Pskov: RSFSR	98,000
Namangan: Uzbek SSR	138,000	Pushkin: RSFSR	55,000
Nikolayev: Ukr. SSR	258,000	Pyatigorsk: RSFSR	74,000
Nikopol': Ukr. SSR	95,000	Revda: RSFSR	57,000
Nizhniy Tagil: RSFSR	359,000	Riga: Latvian SSR	620,000
Noginsk: RSFSR	98,000	Rostov-na-Donu: RSFSR	661,000
Noril'sk: RSFSR	117,000	Rovno: Ukr. SSR	72,000
Novgorod: RSFSR	72,000	Rubtsovsk: RSFSR	127,000
Novocherkassk: RSFSR	104,000	Rustavi: Georgian SSR	72,000
Novokuznetsk: RSFSR	410,000	Ryazan': RSFSR	252,000
Novomoskovsk: RSFSR	114,000	Rybinsk: RSFSR	195,000
Novorossiysk: RSFSR	104,000	Rzhev: RSFSR	55,000
Novoshakhtinsk: RSFSR	108,000	Samarkand: Uzbek SSR	215,000
Novosibirsk: RSFSR	985,000	Saransk: RSFSR	118,000
Novo-Troitsk: RSFSR	69,000	Sarapul: RSFSR	78,000
Odessa: Ukr. SSR	704,000	Saratov: RSFSR	631,000
Oktyabr'skiy: RSFSR	70,000	Semipalatinsk: Kazakh SSR	182,000
Omsk: RSFSR	650,000	Serov: RSFSR	102,000
Ordzhonikidze: RSFSR	183,000	Serpukhov: RSFSR	113,000
Orekhovo-Zuyevo: RSFSR	113,000	Sevastopol': Ukr. SSR	169,000
Orel': RSFSR	174,000	Severodvinsk: RSFSR	97,000
Orenburg: RSFSR	288,000	Shadrinsk: RSFSR	62,000
Orsha: Byelo. SSR	74,000	Shakhty: RSFSR	201,000
Orsk: RSFSR	199,000		

City	Population	City	Population
Shuya: RSFSR	67,000	Ust-Kamenogorsk: Kazakh SSR	181,000
Shyaulyay (Siauliai): Lith. SSR	68,000	Uzhgorod: Ukr. SSR	53,000
Simferopol': Ukr. SSR	202,000	Uzlovaya: RSFSR	54,000
Slavyansk: Ukr. SSR	86,000	Velikiye Luki: RSFSR	68,000
Smolensk: RSFSR	164,000	Vichuga: RSFSR	53,000
Sochi: RSFSR	174,000	Vil'nyus (Vilnius): Lith. SSR	264,000
Stanislav: Ukr. SSR	75,000	Vinnitsa: Ukr. SSR	136,000
Stavropol': RSFSR	154,000	Vitebsk: Byelo. SSR	169,000
Sterlitamak: RSFSR	131,000	Vladimir: RSFSR	174,000
Sukhumi: Georgian SSR	80,000	Vladivostok: RSFSR	325,000
Sumgait: Azer. SSR	68,000	Volgograd: RSFSR	649,000
Sumy: Ukr. SSR	113,000	Vologda: RSFSR	149,000
Sverdlovsk: RSFSR	853,000	Vol'sk: RSFSR	67,000
Sverdlovsk: Ukr. SSR	66,000	Vorkuta: RSFSR	60,000
Svobodnyy: RSFSR	58,000	Voronezh: RSFSR	516,000
Syktyvkar: RSFSR	79,000	Votkinsk: RSFSR	68,000
Syzran': RSFSR	159,000	Vyborg: RSFSR	57,000
Taganrog: RSFSR	220,000	Vyshniy-Volochek: RSFSR	71,000
Tallin (Tallinn): Est. SSR	305,000	Yakutsk: RSFSR	79,0000
Tamboy: RSFSR	189,000	Yaroslavl': RSFSR	443,000
Tartu: Est. SSR	77,000	Yegor'yevsk: RSFSR	61,000
Tashkent: Uzbek SSR	1,002,000	Yelets: RSFSR	85,000
Tbilisi: Georgian SSR	743,000	Yenakiyevo: Ukr. SSR	92,000
Ternopol': Ukr. SSR	59,000	Yerevan: Armenian SSR	583,000
Tikhoretsk: RSFSR	53,000	Yevpatoriya: Ukr. SSR	61,000
Tiraspol': Mold. SSR	75,000	Yeysk: RSFSR	61,000
Tomsk: RSFSR	275,000	Yoshkar-Ola: RSFSR	110,000
Troitsk: RSFSR	80,000	Yurga: RSFSR	52,000
Tselinograd: Kazakh SSR	127,000	Yuzhno-Sakhalinsk': RSFSR	86,000
Tula: RSFSR	342,000	Zagorsk: RSFSR	78,000
Tyumen': RSFSR	174,000	Zaporozh'ye: Ukr. SSR	490,000
Ufa (incl. Chernikovsk): RSFSR	610,000	Zelenodol'sk: RSFSR	66,000
Ulan-Ude: RSFSR	196,000	Zhdanov: Ukr. SSR	321,000
Ul'yanovsk: RSFSR	239,000	Zhitomir: Ukr. SSR	117,000
Ural'sk: Kazakh SSR	109,000	Zlatoust: RSFSR	167,000
Usol'ye-Sibirskoye: RSFSR	60,000		
Ussuriysk: RSFSR	113,000		

EUROPE

OVER 50,000, 1960

City	Year	Population	City	Year	Population
Łódź: Poland		708,400	Pécs: Hung.		115,000
Lublin: Poland		180,700	Piotrków: Poland	1959	51,800
Magdeburg: E. Ger.		261,594	Plauen: E. Ger.		79,056
Maribor: Yugo.	1961	82,388	Ploești: Romania		108,000
Miskolc: Hung.		144,000	Plovdiv: Bulg.	1959	171,391
Moravská Ostrava: Czech.	1961	250,000	Plzeň: Czech.		136,854
Novi Sad: Yugo.	1961	102,385	Potsdam: E. Ger.		115,004
Olomouc: Czech.	1959	75,965	Poznań: Poland		407,800
Olsztyn: Poland	1959	65,900	Prague: Czech.	1961	1,000,000
Opole: Poland	1959	57,700	Radom: Poland		130,100
Oradea: Romania		106,000	Rijeka: Yugo.	1961	100,339
Osijek: Yugo.	1961	71,843	Ruda Śląska: Poland		131,300
Pabianice: Poland.	1959	55,200	Rostock: E. Ger.		158,630
Pardubice: Czech..	1959	56,946	Sarajevo: Yugo.	1961	142,423
			Satu-Mare: Romania		58,169

City	Year	Population	City	Year	Population
Schwerin: E. Ger.		92,508	Tiranë: Albania		130,000
Sibiu: Romania		97,211	Toruń: Poland		104,800
Skopje: Yugo.	1959	161,983	Usti-nad-Labem: Czech.	1959	66,682
Słupsk: Poland	1959	54,200	Varna (Stalin): Bulg.	1959	123,830
Sofia: Bulg.	1959	671,192	Wałbrzych: Poland		117,100
Sosnowiec: Poland		131,600	Warsaw: Poland		1,136,000
Split: Yugo.	1961	99,462	Weimar: E. Ger.		63,943
Stralsund: E. Ger.		65,758	Wismar: E. Ger.		55,400
Subotica: Yugo.		74,832	Włocławek: Poland	1959	68,400
Suceava: Rom.		125,000	Wrocław: Poland		429,200
Szczecin: Poland		268,900	Zabrze: Poland		188,800
Szeged: Hung.		99,000	Zagreb: Yugo.		427,319
Tarnów: Poland		68,100	Zwickau: E. Ger.		129,138
Timișoara: Romania		147,000			